BASIC ELECTRICITY

A TEXT-LAB MANUAL

Seventh Edition

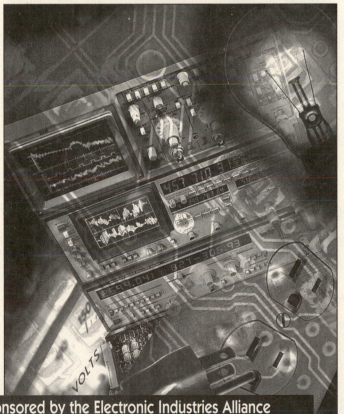

Paul B. Zbar

Gordon Rockmaker

David J. Bates

EIA Sponsored by the Electronic Industries Alliance

Glencoe
McGraw-Hill

New York, New York   Columbus, Ohio   Woodland Hills, California   Peoria, Illinois

Cover photos: (*tl*) Lester Lefkowitz/The Stock Market; (*tr*) Telegraph Colour Library/FPG; (*bl*) Antonio Rosario/The Image Bank; (*bc*) PhotoDisk; (*br*) Doug Martin; (*background*) Court Mast/FPG.

Other Glencoe Books of Interest

Basic Electronics: A Text-Lab Manual, Sixth Edition by Paul B. Zbar, Albert P. Malvino, and Michael A. Miller
Electricity-Electronics Fundamentals: A Text-Lab Manual, Fourth Edition by Paul B. Zbar and Joseph G. Sloop
Industrial Electronics: A Text-Lab Manual, Fourth Edition by Paul B. Zbar and Richard Koelker
Basic Television: Theory and Servicing: A Text-Lab Manual, Third Edition by Paul B. Zbar and Peter W. Orne
Industrial Electricity: Principles and Practices, Third Edition by James Adams and Gordon Rockmaker

Library of Congress Cataloging-in-Publication Data
Zbar, Paul B.
    Basic electricity: a text-lab manual / Paul B. Zbar, Gordon
  Rockmaker, David J. Bates—7th ed.
            p.   cm.
    ISBN 0-07-821275-8
    1. Electronics—Laboratory manuals. I. Rockmaker, Gordon.
II Title.
TK7818.Z18 1991
621.31'078—dc20   99-11784
                                                                        CIP

Basic Electricity: A Text-Lab Manual, Seventh Edition   ISBN 0-07-821275-8
Instructor's Manual for Basic Electricity, Seventh Edition   ISBN 0-07-821276-6

Glencoe/McGraw-Hill

A Division of The McGraw·Hill Companies

Basic Electricity: A Text-Lab Manual, Seventh Edition

Send all inquiries to:
Glencoe/McGraw-Hill
936 Eastwind Drive
Westerville, OH 43081

ISBN 0-07-821275-8

Printed in the United States of America.

1 2 3 4 5 6 7 8 9 10 021 05 04 03 02 01 00 99

CONTENTS

NOTE ON EXPERIMENT CONTENT

Each of the experiments is set up like this:

OBJECTIVES   The objectives are enumerated and clearly stated.

BASIC INFORMATION   The theory and basic principles involved in the experiment are clearly stated.

SUMMARY   A summary of the salient points is given.

SELF TEST   A self test, based on the material included in Basic Information, helps students evaluate their understanding of the principles covered, prior to the experiment proper. The self test should be taken before the experiment is undertaken. Answers to the self-test questions are given at the end of the Procedures section.

MATERIALS REQUIRED   All the materials required to do the experiment—including test equipment and components—are listed.

PROCEDURE   A detailed step-by-step procedure is given for performing the experiment.

ANSWERS TO SELF-TEST PERFORMANCE SHEETS   Special tear-out pages with fill-in tables and a series of questions related to the students' experimental results.

SERIES PREFACE

Electronics is at the core of a wide variety of specialized technologies that have been developing over several decades. Challenged by rapidly expanding technology and the need for increasing numbers of technicians, the Consumer Electronics Group Technical Education and Services Committee of the Consumer Electronics Manufacturers Association (CEMA) along with the Electronic Industries Alliance (EIA) and various publishers have been active in creating and developing educational materials to meet these challenges.

In recent years, a great many consumer electronic products have been introduced and the traditional radio and television receivers have become more complex. As a result, the pressing need for training programs to permit students of various backgrounds and abilities to enter this growing industry has induced EIA to sponsor the preparation of an expanding range of materials. Three branches of study have been developed in two specific formats. The tables list the books in each category; the paragraphs following them explain these materials and suggest how best to use them to achieve the desired results.

THE BASIC ELECTRICITY-ELECTRONICS SERIES

Title	Author
Electricity-Electronics Fundamentals	Zbar/Sloop
Basic Electricity	Zbar/Rockmaker/Bates
Basic Electronics	Zbar/Malvino/Miller

The laboratory text-manuals in the Basic Electricity-Electronics Series provide in-depth, detailed, completely up-to-date technical material by combining a comprehensive discussion of the objectives, theory, and underlying principles with a closely coordinated program of experiments. *Electricity-Electronics Fundamentals* provides material for an introductory course especially suitable for preparing service technicians; it can also be used for other broad-based courses. *Basic Electricity* and *Basic Electronics* are planned for 270–hour courses, one to follow the other, providing a more thorough background for all levels of technician training. A related instructor's guide is available for each course.

THE TELEVISION-AUDIO SERVICING SERIES

Title	Author
Audio Servicing— Theory and Practice	Wells

Audio Servicing—	Wells
Text-Lab Manual	
Basic Television:	Zbar/Orne
Theory and Servicing	
Cable Television	Deschler
Technology	

The Television-Audio Servicing Series includes materials in two categories: those designed to prepare apprentice technicians to perform in-home servicing and other apprenticeship functions, and those designed to prepare technicians to perform more sophisticated and complicated servicing such as bench-type servicing in the shop.

Audio Servicing (theory and practice, text-lab manual, and instructor's guide) covers each component of a modern home stereo with an easy-to-follow block diagram and a diagnosis approach consistent with the latest industry techniques.

Basic Television: Theory and Servicing provides a series of experiments, with preparatory theory, designed to provide the in-depth, detailed training necessary to produce skilled television service technicians for both home and bench servicing of all types of television. A related instructor's guide is also available.

Cable Television Technology (text, instructor's guide) covers all aspects of cable television operation, from the traditional "Lineman"-oriented topics to the high technology subjects that come into play with satellite antennas and fiber-optics links.

Basic laboratory courses in industrial control and computer circuits and laboratory standard measuring equipment are covered by the Industrial Electronics Series and their related instructor's guides. *Industrial Electronics* is concerned with the fundamental building blocks in industrial electronics technology, giving the student an understanding of the basic circuits and their applications. *Electronic Instruments and Measurements* fills the need for basic training material in the complex field of industrial instrumentation. Prerequisites for both courses are *Basic Electricity* and *Basic Electronics*.

The foreword to the first edition of the EIA-cosponsored basic series states: "The aim of this basic instructional series is to supply schools with a well-integrated, standardized training program, fashioned to produce a technician tailored to industry's needs." This is still the objective of the varied training program that has been developed through joint industry-educator-publisher cooperation.

Gary Shapiro, President
Electronic Industries Alliance

THE INDUSTRIAL ELECTRONICS SERIES

Title	Author
Industrial Electronics	Zbar/Koelker
Electronic Instruments	Zbar
and Measurements	

PREFACE

Basic Electricity: A Text-Lab Manual, Seventh Edition is an introductory textbook in electrical technology for students of electricity and electronics. It provides a comprehensive laboratory program in basic electrical theory, electric circuits, and passive devices in both direct and alternating current. As in previous editions, the focus is on practical and analytical techniques essential to the modern technician. By emphasizing hands-on activities, the text helps students develop their troubleshooting and circuit-design skills in a systematic fashion.

For this edition, we have expanded the coverage of oscilloscope measurement techniques. Use of the function generator has been increased in many of the ac experiments in this seventh edition. Also we recognize the emergence of circuit simulation software by incorporating optional activities using this equipment into many of the experiments.

The fifty-eight experiments are presented so that each new concept builds on the previous one. Topics range from an introduction to experimental methods, basic components, instruments and measurements, simple series, parallel, and series-parallel circuits to the more advanced circuit theorems, troubleshooting, and circuit design.

Comments and suggestions from students and instructors who have used previous editions of *Basic Electricity: A Text-Lab Manual* have contributed to fine-tuning this seventh edition. The format and content of this edition carefully address the needs of today's students as well as those of the industries that will employ them. Topics no longer pertinent to a modern electronics curriculum have been discarded and new topics have been added in their place. However, the basic strengths of the previous editions remain.

The organization of each experiment consists of the following:

Objectives. Each experiment opens with a concise statement of the goals of the experiment. Thus, from the very beginning of the experiment, the direction and purpose of the laboratory procedures are made clear to the student as well as the instructor.

Basic Information. Before performing an experiment, the student must understand the underlying principles as well as the practical aspects of the concepts being investigated. This section focuses on the theory and concepts essential to successfully completing the experimental assignments.

Summary. The purpose of the summary is to highlight the key elements of the basic information for quick reference.

Self-Test. Following the summary is a self-test designed to measure the student's understanding of the basic theory and practices involved in the experiment and to reinforce the student's knowledge of certain concepts. Students are expected to understand the correct answers to all the questions before proceeding to the experiment. Answers to the self-tests are given at the end of the procedure section.

Materials Required. As the initial step in the experiment, the student is provided with a list of the power supplies, instruments, and components necessary to conduct the experiment.

Procedure. Step-by-step instructions are given for each part of the experiment. During the experiment, the student is expected to wire circuits according to schematic diagrams, make electrical measurements using meters and instruments similar to those used in industry, tabulate data, and use formulas to calculate unknown quantities. By following the sequence of steps given in the procedure, the student will develop practical, hands-on experience, learn safe laboratory practices, and apply analytical skills to the solution of practical problems. In selected experiments, this section ends with a new optional activity, which utilizes circuit simulation software. Circuit simulation is gaining popularity in classrooms as well as in industry.

Performance sheets are provided at the end of each experiment with tables to be completed and space for calculations. The performance sheets also contain questions and problems directly related to the student's experimental results. On these sheets, the student will report on the theoretical and practical aspects underlying the experiment in addition to the tabulated results. In some cases students are required to plot graphs based on their experimental data. Through this, students learn to demonstrate technical communication and analytical skills directly on the performance sheets. The sheets are perforated so that they can be readily removed from the manual and submitted to the instructor. The rest of the manual will remain intact for review and possible re-working of the experiment.

As in previous editions, the text is fully illustrated with circuit diagrams, tables, and graphs designed to supplement and support the basic information. Detailed circuit diagrams are provided as required for performing the experiments. Sample problems and their step-by-step solutions are included in the basic information sections wherever appropriate.

The authors would like to acknowledge the cooperation and support this text has received from the electronics industry. The preceding Series Preface details the long-standing close working relationship with CEMA and the EIA that this series of text-lab manuals enjoys. Special thanks go to the members of the CEMA education subcommittee, who reviewed this manuscript: Mike Begala (Thompson Consumer Electronics), Marcel Rialland (Toshiba America Consumer Products), and Brian Ott (CEMA). Thanks to their guidance, this text combines the latest industry practices with sound educational and training principles.

Last but not least, the authors would like to thank their wives, May Zbar, Ellen Rockmaker, and Jackie Bates, for their patience, encouragement, and inspiration.

Paul B. Zbar
Gordon Rockmaker
David J. Bates

SAFETY

Electronics technicians work with electricity, electronic devices, motors, and other rotating machinery. They are often required to use hand and power tools in constructing prototypes of new devices or in setting up experiments. They use test instruments to measure the electrical characteristics of components, devices, and electronic systems. They are involved in any of a dozen different tasks.

These tasks are interesting and challenging, but they may also involve certain hazards if technicians are careless in their work habits. It is therefore essential that the student technicians learn the principles of safety at the very start of their career and that they practice these principles.

Safe work requires a careful and deliberate approach to each task. Before undertaking a job, technicians must understand what to do and how to do it. They must plan the job, setting out on the workbench in a neat and orderly fashion, tools, equipment, and instruments. Extraneous items should be removed, and cables should be securely fastened.

When working on or near rotating machinery, loose clothing should be anchored, ties firmly tucked away.

Line (power) voltages should be isolated from ground by means of an isolation transformer. Powerline voltages can kill, so these should *not* come in contact with the hands or body. Line cords should be checked before use. If the insulation on line cords is brittle or cracked, these cords must *not* be used. TO THE STUDENT: Avoid direct contact with any voltage source. Measure voltages with one hand in your pocket. Wear rubbersoled shoes or stand on a rubber mat when working at your experiment bench. Be certain that your hands are dry and that you are not standing on a wet floor when making tests and measurements in a live circuit. Shut off power before connecting test instruments in a live circuit.

Be certain that line cords of power tools and nonisolated equipment use safety plugs (polarized 3-post plugs). Do not defeat the safety feature of these plugs by using ungrounded adapters. Do not defeat the purpose of any safety device, such as a fuse or circuit breaker, by shorting across it or by using a higher amperage fuse than that specified by the manufacturer. Safety devices are intended to protect you and your equipment.

Handle tools properly and with care. Don't indulge in horseplay or play practical jokes in the laboratory. When using power tools, secure your work in a vise or jig. Wear gloves and goggles when required.

Exercise good judgment and common sense and your life in the laboratory will be safe, interesting, and rewarding.

First Aid

If an accident should occur, shut off the power immediately. Report the accident at once to your instructor. It may be necessary for you to give emergency care before a physician can come, so you should know the principles of first aid. You can learn the basics by taking a Red Cross first-aid course.

Some first-aid suggestions are set forth here as a simple guide.

Keep the injured person lying down until medical help arrives. Keep the person warm to prevent shock. Do not attempt to give water or other liquids to an unconscious person. Be sure nothing is done to cause further injury. Keep the injured one comfortable until medical help arrives.

Artificial Respiration

Severe electric shock may cause someone to stop breathing. Be prepared to start artificial respiration at once if breathing has stopped. The two recommended techniques are:

1. Mouth-to-mouth breathing, considered more effective
2. Schaeffer method

These techniques are described in first-aid books. You should master one or the other so that if the need arises you will be able to save a life by applying artificial respiration.

These safety instructions should not frighten you but should make you aware that there are hazards in the work of an electronics technician—as there are hazards in every job. Therefore you must exercise common sense and good judgment, and maintain safe work habits in this, as in every other job.

INTRODUCTION TO EXPERIMENTS

BASIC INFORMATION

The electricity/electronics laboratory is a place of learning, whether it be an industrial, commercial, or school laboratory. The experiments in this book support the overall learning process by helping you do the following:

1. Develop practical skills in the use of power supplies, instruments, and components in a hands-on test environment
2. Develop the techniques of gathering accurate experimental information and communicating that information to others through written reports and tabulated data

Although specific objectives will be described in the course of the experiments that follow, the preceding two items are the keys to all of them. To build the circuits, perform the experiments, gather and interpret data, and report your conclusions you will need to be familiar with the following basic equipment, tools, and techniques of the electricity/electronics laboratory:

1. Schematic diagrams and symbols
2. Power sources and supplies
3. Instruments
4. Tables and reports

Schematic Diagrams and Symbols

Schematic diagrams are a form of language that helps us communicate information about the electrical relationships in a circuit. Schematic diagrams do not indicate physical positions or sizes of components, nor do they show actual points of interconnection (though certain drawing conventions do try to indicate all these in somewhat simplified form). The symbols used in schematic diagrams represent the components and conductors of the circuit, but again, the symbols do not try to show the physical shape or size of the actual component. They do in some respects convey the idea of a characteristic feature of the component.

Figure 1–1 shows some of the graphic symbols used in this book. When specialized or unique symbols are used in schematic diagrams, they are labeled and identified.

OBJECTIVES

1. To become familiar with basic laboratory equipment and components

2. To read and interpret electrical and electronic drawings

3. To gather and tabulate data in the laboratory

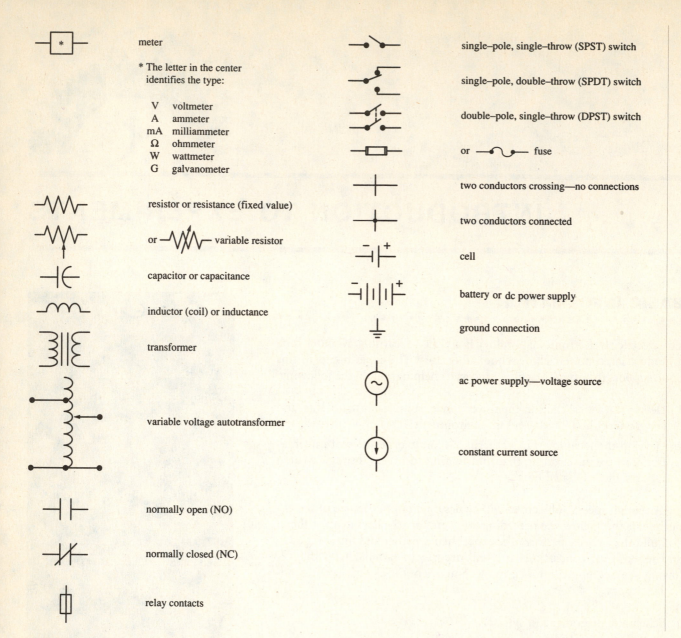

✱ meter	single–pole, single–throw (SPST) switch
* The letter in the center identifies the type:	single–pole, double–throw (SPDT) switch
V voltmeter	double–pole, single–throw (DPST) switch
A ammeter	
mA milliammeter	or ⏤⌇⏤ fuse
Ω ohmmeter	
W wattmeter	two conductors crossing—no connections
G galvanometer	
	two conductors connected
resistor or resistance (fixed value)	
	cell
or variable resistor	
	battery or dc power supply
capacitor or capacitance	
	ground connection
inductor (coil) or inductance	
	ac power supply—voltage source
transformer	
	constant current source
variable voltage autotransformer	
normally open (NO)	
normally closed (NC)	
relay contacts	

Figure 1–1. Common schematic symbols of components, devices, and conductors used in this book.

Power Sources and Supplies · · · · · · · · · ·

Direct Current. The two basic power sources used in the experiments in this book are direct current and alternating current. Direct current can be supplied by electric cells such as the familiar D, C, and AA cells used in many portable electronic devices. Batteries consist of combinations of cells designed to produce voltages that cannot be supplied by a single cell. Automobile batteries and the small 9-V batteries used in portable radios and wireless alarm systems actually contain combinations of cells. Although cells and batteries are convenient and widely used in the laboratory, especially to power portable instruments, they do not produce a constant voltage and they are not easy to use over a wide range of voltages. For such purposes a variable voltage-regulated supply is most often used.

A variable voltage-regulated power supply is one that can be manually adjusted to deliver any required voltage within its range of operation. The term *voltage-regulated* means that the voltage delivered by the supply remains constant despite changes in load current, within specified limits. Some power supplies are designed to provide two or more independent dc voltages, in which case the instrument has separate controls and separate output terminals.

The polarity of the dc terminals on the supply is usually marked either −, +, or GND and V+. By convention a red jack is used for the positive terminal, and a black jack is used for the negative terminal of the supply.

Alternating Current. Alternating current sources include the wall outlet supplying a nominal 120 V as well as variable voltage-regulated supplies similar to those for dc. Single-voltage ac sources are also common for particular power requirements, such as telephone answering machines and video games. Variable-voltage autotransformers are also widely used. Although the output of these supplies is usually unregulated, it is constant enough for most laboratory applications.

Using a Power Supply.
The following general precautions should be observed in using a power supply:

1. Read the operating manual carefully and be certain that you understand the instructions before turning the power supply on for the first time.

2. Never create a short circuit across the output terminals, or you may damage the supply. Keep the leads connected to the output terminals from making contact with each other.

3. If any component on the experimental circuit appears to be overheating after power is applied, turn the supply off and determine the cause before you proceed with the experiment.

4. To prevent damage to the power supply, do not switch the supply on and off excessively. If an experiment requires power to be interrupted frequently, use an external switch on the breadboard to apply and remove power from the circuit.

5. Do not operate the supply beyond its *rated* current capacity. If an ammeter indicates that you are exceeding the current capability of the power supply, turn it off and check the experimental circuit to determine why it is drawing excessive current.

6. Always be alert to shock hazards. Even equipment that has what may appear to be relatively safe voltages can produce dangerous if not lethal currents under certain conditions.

Instruments

Electrical and electronic instruments are used in the laboratory to measure quantities, to provide external signal stimuli to circuits, and to view graphically the behavior of circuits.

Measuring Instruments.
The two types of measuring instruments in common use are analog meters and digital meters. Analog meters can be further classified as electronic meters or electromechanical meters. In the analog meter (both electronic and electromechanical) a pointer indicates the measured value on a calibrated scale. This usually requires making an accurate estimate of the exact position of the pointer on the scale. The value measured by a digital meter appears as a number consisting of three or four digits on a lighted display. Anyone who can read numbers can therefore read the quantity being measured by a digital meter. Voltage measurements are made primarily with an instrument called a *voltmeter*. Other instruments may also be used to measure voltage— for example, the oscilloscope—but voltage measurement is not the primary purpose of these devices.

Current is measured with an *ammeter*. If the ammeter is designed to measure very small quantities, it is known specifically as a *milliammeter* or *microammeter*.

Resistance is measured with an *ohmmeter*. Whereas the voltmeter and ammeter are always used in circuits to which power has been applied, ohmmeters are always used in unpowered circuits. The ohmmeters carry their own power supply, usually in the form of one or two 1.5-V cells. Another common laboratory meter, the *wattmeter*, measures power.

Meters may be single-function devices, such as the voltmeter, which can measure only voltage. Or they may be multipurpose devices, such as the multimeter, which is used to measure a variety of electrical quantities, such as voltage, current, and resistance. The DMM (digital multimeter) and the VOM (volt-ohm-milliammeter) in Figure 1–2 are examples of multiple-purpose meters.

Signal Generators.
Many test procedures and experiments require application of voltages with frequencies other than the 60 Hz supplied by the power companies. In addition, certain applications call for alternating current with waveforms other than a sine wave. Signal generators and function generators are designed to produce variable-frequency outputs as well as sine, sawtooth, and square waves. The voltage output of these generators is not intended to be a continuous power source. Rather, the outputs provide signal stimuli for circuits so that the behavior of the circuit may be observed under these signal conditions. For example, a signal generator may simulate the modulated radio wave that enters the antenna of a radio receiver in order to troubleshoot various circuits of the radio. The signal generator can produce precise, stable signals, but it does not measure frequencies or voltages.

Oscilloscope.
Modern oscilloscopes are used to observe circuit behavior graphically by displaying voltage waveforms on the screen of a cathode-ray tube. In addition to making it possible to view the shape of the wave, scopes can also be used to measure the voltage and frequency of the wave. If more than one waveform is present, the oscilloscope also permits determining the phase displacement between the waves. Dual-trace oscilloscopes make it possible to view two waveforms simultaneously on the same screen.

The Internet

An electronics technician requires many sources of information, along with technical and communications skills, to properly perform on the job. Information covering new product information, service bulletins, component specifications, training and workshops, and professional organizations is vital to keep up with current technology. The Internet, with its massive amounts of online information, can be used to provide the necessary technical resources.

Information dealing with professional organizations and electronics careers can be informative and helpful as

(a)

(b)

Figure 1–2. Multiple-purpose meters—(a) DMM; (b) Analog VOM.

you begin your study of electronics. Some helpful Internet World Wide Web addresses are:

> http://www.iee.org
> http://www.eia.org
> http://www.eia.org/cema
> http://www.iscet.org
> http://stats.bls.gov/ocohome.htm

Tables and Reports · · · · · · · · · · · · · · · ·

The purpose of most experiments is to observe reactions, measure quantities, record data, analyze the data, and, if possible, draw some conclusions. Attention to the following suggestions will result in more effective laboratory work:

1. Because gathering data is an important element of the experimental process, it must be done carefully and systematically.

2. The data should be tabulated in a logical fashion. The tables of data should be legible, the units of each quantity identified, and the level of accuracy consistent for like quantities.

3. Accuracy in reading meters and adjusting voltages of power supplies is essential if the data are to have any credibility. If special conditions arise or particular results are distorted, the facts and circumstances should be noted along with the tabulated data.

4. Many tables require calculations based on circuit analysis and measured data. In tabulating calculated values it is important to be consistent with the number of significant figures in your answers.

5. Some circuits should be analyzed and their performance calculated before performing an experiment. The measured quantities can then be compared with the calculated values. Discrepancies should be investigated and discussed.

6. If graphs are required, appropriate scales and type of graph paper should be used so that the plotted graph clearly shows trends and the particular relationships being demonstrated in the experiment.

7. The tabulated data, graphs, calculations, and the like should lead to some conclusions about the principles or concepts that are central to the experiment.

8. The conclusion should be written neatly, stated clearly, and refer to the supporting experimental data and observations.

9. The technical report should be free of spelling and grammatical errors and use the technical terminology appropriate to the subject of the experiment.

SUMMARY

1. The basic equipment, tools, and techniques of the electricity/electronics laboratory are schematic diagrams and symbols, power sources and supplies, instruments, and tables and reports.

2. Schematic diagrams and symbols are the shorthand language of electrical and electronic circuits.

3. Schematic diagrams show the electrical relationships of components and conductors in a circuit.

4. Schematic diagrams do not show the actual physical locations or connections of a circuit. Symbols do not represent the actual physical shape or size of the components.

5. The two main power systems are direct current and alternating current.

6. Electric cells such as the familiar D, C, and AA cells provide a source of direct current for portable devices.

7. Most laboratory dc requirements are provided by variable voltage-regulated power supplies.

8. The most common ac source is the wall outlet that provides a nominal 120 V.

9. Alternating current sources include variable voltage-regulated power supplies and single-voltage supplies.

10. A popular ac source in the laboratory is the variable-voltage autotransformer. This type of power source is usually unregulated.

11. The term voltage-regulated means that the voltage will remain constant at a particular setting despite changes in load current, within specified limits.

12. Instruments in the electricity/electronics laboratory are used to measure quantities, provide external signal sources to circuits, and to view the behavior of circuits graphically.

13. Voltmeters, ammeters, and ohmmeters can be either analog or digital meters.

14. Analog meters can be either electronic or electromechanical.

15. Meters can be single-function instruments, such as voltmeters or ammeters, or they can be multipurpose devices in which the voltmeter, ammeter, and ohmmeter functions are in the same housing and use the same display.

16. Signal generators produce variable frequencies from 1 Hz to as many as hundreds of megahertz, although no single signal generator produces all these frequencies.

17. A signal generator generally produces a sine wave, but some manufacturers produce models that also provide a square-wave output.

18. Function generators are capable of providing variable frequency and amplitude sinewave, sawtooth, and square waveforms.

19. The oscilloscope is used to observe the shape, frequency, and amplitude of waveforms in a circuit. Dual-trace oscilloscopes are capable of displaying two sets of waves on the same screen simultaneously. Oscilloscopes are used to measure voltage, frequency, and phase displacement of ac waveforms.

20. An essential part of experimental work in the laboratory is the gathering and tabulating of data.

21. The tabulated data, graphs, calculations, and observations made during an experiment should lead to some conclusion about the principles and concepts involved in the experiment.

22. Written reports are used to communicate the data, observations, and conclusions of the experimenter to others.

23. The Internet is a source of abundant information useful to the electronics technician.

SELF TEST

Check your understanding by answering the following questions:

1. The _____ diagram is a shorthand method of representing an electric circuit.

2. (True/False) Symbols used in electrical drawings usually are drawn to resemble the size and shape of the components they represent. _____

3. The two power systems used in the laboratory are _____ current and _____ current.

4. If a power supply voltage remains constant at a preset value despite changes in load current, the supply is said to be voltage-_____.

5. By convention the output terminals of a dc power supply are colored _____ for the positive terminal, and _____ for the negative, or ground, terminal.

6. The two types of measuring instruments are the _____ meters, which use a pointer and a _____ scale, and the _____ meters, which use a numeric readout display.

7. Never use a(n) _____ on a circuit to which power is applied.

8. The signal generator provides variable _____ signals at relatively low voltage.

9. The typical function generator can supply sine, square, and _____ waveforms.

10. The _____ oscilloscope is used to display two sets of waveforms on the same screen simultaneously.

11. (True/False) When tabulating calculated values it is not necessary to use the same number of significant figures. _____

MATERIALS REQUIRED

Power Supplies:
- Variable ac
- Variable 0–15 V dc, regulated
- Variable-voltage autotransformer (Variac or equivalent)

PROCEDURE

Instruments:
A varied selection of the following should be available:

- DMM
- Analog voltmeter (range not critical)
- VOM
- Analog ammeter (range not critical)

Miscellaneous:

- A group of components provided by the instructor (components should be mounted and labeled)
- Operating manuals for all instruments provided

A. Schematic Diagrams and Symbols

A1. Your instructor will provide you with a group of components, only a few of which are pictured in the schematic diagram of Figure 1–3. In Table 1–1 (p. 7) identify those of your components that are represented in the figure. Next to the symbol number in the table, record the name of the component and sketch what the component actually looks like.

B. Power Sources and Supplies

B1. Your instructor will give you a power supply of the type used in the laboratory. Obtain a copy of the operating manual for the supply.

B2. Read the manual to become thoroughly familiar with the controls and terminals on the front panel of the power supply. At this point it is not necessary to become familiar with the actual operation of the supply.

B3. Complete Table 1–2 (p. 7), Power Supply Controls and Features.

C. Instruments

C1. Your instructor will give you two instruments with which you are to become familiar. Obtain a copy of the operating manuals for these two instruments.

C2. Read the manuals carefully to become familiar with the various switches and jacks on the front panel.

C3. Complete Table 1–3(*a*)(p. 8), Meter 1 Functions and Features, and Table 1–3(*b*)(p. 8), Meter 2 Functions and Features.

D. Tables and Reports

D1. Select a piece of electronic entertainment equipment that you own (radio, electronic game, portable CD player, TV set, or the like). On a separate sheet of 8½ × 11 paper write a set of operating instructions for

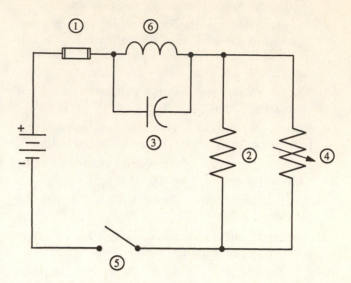

Fig. 1–3. Circuit for procedure step A1.

the basic features of the device. If certain types of instructions lend themselves to it, tabulate specific detailed operations. Assume that the instructions will be used by someone with the equipment present but with no previous experience operating that type of device. Further, assume that the user is unfamiliar with any of the technical terminology associated with the equipment.

Optional Activity · · · · · · · · · · · · · · · ·

Use the Internet to explore career opportunities for an electronics technician.

ANSWERS TO SELF TEST

1. schematic
2. false
3. direct; alternating
4. regulated
5. red; black
6. analog; calibrated; digital
7. ohmmeter
8. frequency
9. sawtooth
10. dual-trace
11. false

TABLE 1–1. Schematic Diagram Symbols

Symbol No. (Figure 1–3)	Component Name	Sketch of Component	Symbol No. (Figure 1–3)	Component Name	Sketch of Component
1			4		
2			5		
3			6		

TABLE 1–2. Power Supply Controls and Features

Manufacturer _____

Model no. _____

Type of Power Supply

_____ DC   _____ Regulated

_____ AC   _____ Unregulated

_____ Single voltage (_____ V)

_____ Multiple voltage (voltages: _____)

_____ Variable voltage (range: _____)

No. of output terminals _____

Controls and Switches	Function
_____	_____
_____	_____
_____	_____
_____	_____
_____	_____

TABLE 1–3(a). Meter 1 Functions and Features

Manufacturer _____

Model no. _____

Type of meter _____

_____ DC	_____ Electronic	_____ Battery operated
_____ AC	_____ Analog	_____ Line operated
_____ DC/AC	_____ Digital	_____ Battery/line operation
_____ Single meter		
_____ Multimeter		

Switches, Ranges, Terminals, Special Features	*Function*
_____	_____
_____	_____
_____	_____
_____	_____
_____	_____
_____	_____

TABLE 1–3(b). Meter 2 Functions and Features

Manufacturer _____

Model no. _____

Type of meter _____

_____ DC	_____ Electronic	_____ Battery operated
_____ AC	_____ Analog	_____ Line operated
_____ DC/AC	_____ Digital	_____ Battery/line operation
_____ Single meter		
_____ Multimeter		

Switches, Ranges, Terminals, Special Features	*Function*
_____	_____
_____	_____
_____	_____
_____	_____
_____	_____
_____	_____

Name _____ Date _____

QUESTIONS

1. Is is possible to use the same calibrated scale on the 3-V range as on the 300-V range of a voltmeter? Explain.

_____

_____

_____

_____

2. Is the scale on your analog ohmmeter linear or nonlinear? Give examples to support your answer.

_____

_____

_____

3. An unknown voltage (known only to be less than 600 V dc) is to be measured using a dc voltmeter having the following ranges: 600; 300; 60; 15; 3. Explain how you would measure this voltage with your voltmeter.

_____

_____

_____

_____

4. What is meant by zeroing an ohmmeter? How is it done?

_____

_____

_____

_____

5. What is the difference between the readout of an analog electronic voltmeter and a digital voltmeter?

_____

_____

_____

_____

6. List two advantages of a variable regulated dc power supply over cells and batteries.

_____

_____

_____

_____

7. Select two features of the power supply investigated in Table 1–2 and describe why they are particularly useful for laboratory experiments.

_____

_____

_____

_____

8. Select two features of the meters investigated in Tables 1–3(*a*) and (*b*) and describe why these features are particularly useful for laboratory experiments.

_____

_____

_____

_____

RESISTOR COLOR CODE AND MEASUREMENT OF RESISTANCE

BASIC INFORMATION

Color Code

The *ohm* is the unit of resistance, and it is represented by the symbol Ω (Greek letter omega). Resistance values are indicated by a standard color code that manufacturers have adopted. This code uses color bands on the body of the resistor. The colors and their numerical values are given in the resistor color chart, Table 2–1 (p. 13). This code is used for ⅛-W, ¼-W, ½-W, 1-W, and 2-W resistors.

The basic resistor is shown in Figure 2–1. The standard color-code marking consists of four bands around the body of the resistor. The color of the first band indicates the first significant figure of the resistance value. The second band indicates the second significant figure. The color of the third band indicates the number of zeros that follow the first two significant figures. If the third band is gold or silver, the resistance value is less than 10 Ω. For resistors less than 10 Ω, the third band indicates a fractional value of the first two significant figures:

- A *gold* band means the resistance is ⅒ the value of the first two significant figures.
- A *silver* band means the resistance is ⅟₁₀₀ the value of the first two significant figures.

The fourth band indicates the *percent tolerance* of the resistance. Percent tolerance is the amount the resistance may vary from the value indicated by the color code. Because resistors are mass produced, variations in materials will affect their actual resistance. Many circuits can still operate as designed even if the resistors in the circuit do not have the precise value specified. Tolerances are usually given as plus or minus the nominal, or color-code, value.

High-precision resistors have five bands. The first three bands indicate the first three significant figures of the resistance; the fourth band

OBJECTIVES

1 To determine the value of resistors from their color code

2 To measure resistors of different values

3 To measure a resistor using the various resistance ranges of an ohm-meter

4 To measure the resistance across each combination of two of the three terminals of a potentiometer and to observe the resistance change as a shaft of the potentiometer is rotated

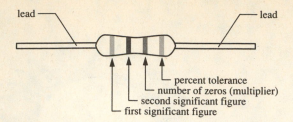

Figure 2–1. Resistor color code.

indicates the number of zeros; the fifth band is the percent tolerance. Percent tolerances for these resistors are:

Brown ± 1%

Red ± 2%

Green ± 0.5%

Blue ± 0.25%

Violet ± 0.1%

Resistors manufactured to military specification (MILSTD) also contain a fifth band. The fifth band in this case is used to indicate reliability. The figure given is the percentage of defective parts per 1000 hours of operation.

Examples of color-coded resistors are given in Table 2–2.

Wirewound, high-wattage resistors usually are not color coded but have the resistance value and wattage rating printed on the body of the resistor.

To avoid having to write all the zeros for high-value resistors the metric abbreviations of k (for 1000) and M (for 1,000,000) are used. For example,

- 33,000 Ω can be written as 33 kΩ (pronounced 33 kay, or 33 kilohms).

- 1,200,000 Ω can be written as 1.2 MΩ (pronounced 1.2 meg, or 1.2 megohms)

On many of today's electronic circuit boards, chip resistors are used. A typical chip resistor is shown in Figure 2–2. These resistors are surface-mount components as compared to thru-hole axial lead types. The resistance value of a chip resistor is determined from a three- or four-digit number printed on its body.

Variable Resistors

In addition to fixed-value resistors, variable resistors are used extensively in electronics. The two types of variable resistors are the *rheostat* and *potentiometer*. Volume con-

trols used in radio, and the balance and tone control of stereo receivers, are typical examples of potentiometers.

A rheostat is essentially a two-terminal device. Its circuit symbol is shown in Figure 2–3. Points A and B connect into the circuit. A rheostat has a maximum resistance value, specified by the manufacturer, and a minimum value, usually 0 Ω. The arrowhead in Figure 2-3 indicates a mechanical means of adjusting the rheostat so that the resistance, measured between points A and B, can be adjusted to any intermediate value within the range of variation.

The circuit symbol for a potentiometer [Figure 2–4(a), p. 14] shows that this is a three-terminal device. The resistance between points A and B is fixed. Point C is the variable arm of the potentiometer. The arm is a metal contactor that slides along the uninsulated surface of the resistance element. The amount of resistance material between the point of contact and one of the end terminals determines the resistance between those two points. Thus, the longer the surface between points A and C, the greater is the resistance between these two points. In other words, the resistance between points A and C varies as the length of element included between points A and C. The same is true for points B and C.

The resistance R_{AC} from A to C plus the resistance R_{CB} from C to B make up the fixed resistance R_{AB} of the potentiometer. The action of the arm, then, is to increase the resistance between C and one of the end terminals and at the same time to decrease the resistance between C and the other terminal while the sum of the two resistances R_{AC} and R_{CB} remains constant.

A potentiometer may be used as a rheostat if the center arm and one of the end terminals are connected into the circuit and the other end terminal is left disconnected. Another method of converting a potentiometer into a rheostat is to connect a piece of hookup wire between the arm and one of the end terminals; for example, C can be connected to A. The points B and C now serve as the terminals of a rheostat. (When two points in a circuit are connected by hookup wire, these points are said to be *shorted* together.)

Measuring Resistance

A multimeter (often called a *volt-ohm-milliammeter*, or *VOM*) is capable of measuring a number of different electrical quantities. When the quantity being measured is resistance, the instrument is called an *ohmmeter*. Although

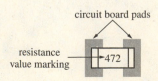

Figure 2–2. Chip resistor with number coding.

Figure 2–3. A rheostat is a variable resistor with two terminals.

TABLE 2–1.   Resistor Color Codes

Color	Significant Figure* (First and Second Bands)	No. of Zeros (Multiplier) (Third Band)	% Tolerance (Fourth Band)	% Reliability* (Fifth Band)
Black	0	0	—	—
Brown	1	1 (10^1)	—	1
Red	2	2 (10^2)	—	0.1
Orange	3	3 (10^3)	—	0.01
Yellow	4	4 (10^4)	—	0.001
Green	5	5 (10^5)	—	—
Blue	6	6 (10^6)	—	—
Violet	7	7 (10^7)	—	—
Gray	8	8 (10^8)	—	—
White	9	9 (10^9)	—	—
Gold	—	(0.1 or 10^{-1})	5	—
Silver	—	(0.01 or 10^{-2})	10	—
No color	—	—	20	—

*MILSTD five-band code

TABLE 2–2.   Examples of Color-Coded Resistors

First Band	Second Band	Third Band	Fourth Band	Resistor Value Ω	% Tolerance	Resistance Range, Ω
Orange	Orange	Brown	No color	330	20	264–396
Gray	Red	Gold	Silver	8.2	10	7.4–9.0
Yellow	Violet	Green	Gold	4.7 M	5	4.465 M–4.935 M
Orange	White	Orange	Gold	39 k	5	37.1 k–41 k
Green	Blue	Brown	No color	560	20	448–672
Red	Red	Yellow	Silver	220 k	10	198 k–242 k
Brown	Green	Gold	Gold	1.5	5	1.43–1.58
Blue	Gray	Green	No color	6.8 M	20	5.44 M–8.16 M
Green	Black	Silver	Gold	0.5	5	0.475–0.525

almost all ohmmeters have a number of common functions and operating features, you should refer to the manufacturer's operating manual before using an instrument with which you are not completely familiar.

To measure resistance, set the function switch of the multimeter to Ohms. Before taking any measurements on an electronic meter, adjust the Ohms and Zero controls of the meter according to the manufacturer's instructions. You are then ready to make resistance and continuity checks. To measure the resistance between two points, say A and B, connect one of the ohmmeter leads to point A and the other to point B. The meter pointer then indicates, on the ohms scale, the value of resistance between A and B. If the meter reading is 0 Ω, points A and B are *short-circuited,* or simply "shorted." If, however, the meter pointer does not move (that is, if the indicator points to infinity on the ohms scale), points A and B are *open-circuited;* that is, there is an infinite resistance between them.

Electronic analog meters contain a basic ohms scale from which readings are made directly on the $R \times 1$ range of the meter. Figure 2–5 (p. 14) shows that the ohms scale is nonlinear; that is, the subdivisions of the scale are not equally spaced. Thus, the space between 0 and 1 is much greater than the space between 9 and 10, though each space represents, in this case, a change of 1 Ω. The technician is required mentally to supply numbers for the unnumbered calibrations. If the pointer is on the second graduation to the right of 3, between 3 and 4, as in Figure 2–5, the corresponding ohms value on the $R \times 1$ range is 3.4 Ω.

Note that the ohms scale becomes fairly crowded to the right of the 100-Ω division. If a resistance greater than

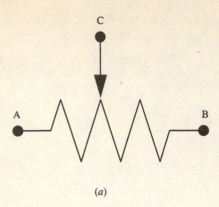

(a)

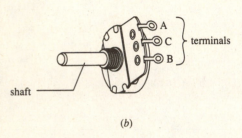

(b)

Figure 2–4. A potentiometer is a three-terminal variable resistor. (*a*) Schematic symbol. (*b*) Pictorial view.

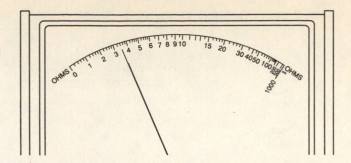

Figure 2–5. Ohmmeter scale of an analog meter. On the $R \times 1$ range setting, the pointer shows 3.4 Ω.

100 Ω is to be measured with some degree of accuracy, the meter range should be switched to $R \times 10$, $R \times 100$, and $R \times 1000$, depending on the actual resistance to be measured. These three ranges, $R \times 10$, $R \times 100$, and $R \times 1000$, will usually be found on the meter. In the $R \times 10$ range, any reading made on the basic scale must be multiplied by 10; in the $R \times 100$ range, any reading must be multiplied by 100; in the $R \times 1000$ range the reading must be multiplied by 1000.

After switching from one ohms range to another, check the Zero and Ohms controls and readjust if necessary.

Digital ohmmeters display resistance values directly. Digital meters contain range values as well as Low Ohms and High Ohms settings. Digital meters have an out-of-range indicator, usually a blinking display of the highest meter reading. For example, a 3½ digit meter displays four digits, the highest reading of which is 1999. When the ohms function of the meter is set initially, the display will blink 1999, indicating a reading (in this case infinity) is out of range. Touching the test leads together should produce a steady 000 reading. This indicator will vary depending on your type of DMM. Some DMMs will show a single 1 or O.L. on the left side of the display. As with analog meters, follow the manufacturer's instructions in adjusting the zero control of digital meters.

SUMMARY

1. The unit of resistance is the ohm.

2. The body of a fixed carbon resistor is color coded to specify its resistance value, tolerance, and reliability.

3. Twelve colors are contained in the color chart. These give the values of the significant figures of resistance, the tolerance, and reliability. Refer to Table 2–1 for the resistor color chart.

4. Resistor color codes use either four or five bands.

In the four-band code,

- First band is the first significant figure of the resistance.
- Second band is the second significant figure of the resistance.
- Third band is the number of zeros (or multiplier).
- Fourth band is the percent tolerance.

In the five-band code used for high-precision resistors,

- First band is the first significant figure of the resistance.
- Second band is the second significant figure of the resistance.
- Third band is the third significant figure of the resistance.
- Fourth band is the number of zeros (or multiplier).
- Fifth band is the percent tolerance.

In the five-band code used for military electronics (MILSTD),

- First band is the first significant figure of the resistance.
- Second band is the second significant figure of the resistance.
- Third band is the number of zeros (or multiplier).

- Fourth band is the percent tolerance.
- Fifth band is the percent reliability in defects per 1000 items.

5. High-wattage wirewound resistors are not color coded but have the resistance and wattage value printed on the body of the resistor.

6. Variable resistors are of two types, the rheostat and the potentiometer.

7. A rheostat is a two-terminal device whose resistance value may be varied between the two terminals.

8. A potentiometer is a three-terminal device. The resistance between the two end terminals is fixed. The resistance between the center terminal and either end terminal can be varied.

9. An ohmmeter or the ohms function of a VOM or electronic multimeter is used to measure resistance and continuity.

10. The scale of an analog ohmmeter is nonlinear.

11. An ohmmeter or the ohms function of a VOM or EVM has several ohms ranges ($R \times 1$, $R \times 10$, $R \times 100$, etc.).

12. Digital meters display resistance values directly and indicate out-of-range values by a blinking display or some other type of overrange indication.

SELF TEST

Check your understanding by answering the following questions:

1. A color code is used to indicate the _____ of a carbon resistor.

2. If the color red appears in either the first or second color band on a resistor, it stands for the number (significant figure) _____.

3. If the color yellow appears in the third band on a four-band color-coded resistor, it stands for _____ zeros.

4. If the color _____ appears on the fourth band of a four-band color code on a resistor, it indicates a tolerance value of 10 percent.

5. A resistor coded brown, black, black, gold has a value of _____ Ω and a tolerance of _____ percent.

6. (True/False) A high-wattage resistor is color coded in the same way as a low-wattage resistor. _____

7. A potentiometer has _____ terminals.

8. The fifth band on a MILSTD resistor is colored red, which means the resistor has a _____ of 0.1 percent.

9. A four-band resistor whose value is 120 Ω and whose tolerance is 20 percent is color coded _____.

10. If a resistor measures *infinite* ohms, the resistor is _____-circuited.

MATERIALS REQUIRED

Instruments:
- Digital multimeter (DMM) and volt-ohm-milliammeter (VOM)

Resistors:
- 10 assorted resistance values and tolerances; ½ W (color coded)
- 10 kΩ potentiometer

Miscellaneous:
- Length of hookup wire about 12 in. long
- Wire cutters

PROCEDURE

1. Your instructor will give you 10 resistors of various values and tolerances. Examine each one and determine its resistance and tolerance according to its color code. Record the color bands, the coded resistance value, and tolerance in Table 2–3 (p. 17).

2. Refer to Experiment 1 and the operator's manuals on the use of a digital multimeter and a volt-ohm-milliammeter to measure resistance. Zero the ohmmeter. Using the coded resistance value as a guide, select an appropriate meter range and measure the resistance of each of the 10 resistors. Record your readings under "Measured Value" in Table 2–3 (p. 17).

3. For each of the measured resistance values in step 2, calculate the percent accuracy using the equation:

$$\% \text{ Accuracy} = \frac{\text{Coded Value} - \text{Measured Value}}{\text{Coded Value}} \times 100$$

Record this % accuracy in Table 2–3.

4. **a.** Measure and record the resistance of a short length of hookup wire. $R =$ _____ Ω.

b. Select one of the resistors in step 1 and connect the wire in step 4a across it as shown in Figure 2–6. By connecting the wire across the leads of the resistor, the resistor has been short-circuited. Measure the resistance across the resistor-hookup wire combination. $R =$ _____ Ω.

5. Connect a hookup wire across the leads of your ohmmeter. Note the reading of the meter (you need not record this reading). Using your wire cutter, cut the hookup wire in half; the result is an open circuit. Note the reaction of the meter after the wire is cut.

6. **a.** Examine the potentiometer given to you. Place it on the lab table with the shaft pointing up [see Figure 2–4(b)]. The terminals of your potentiometer will be A, B, and C, as in the figure. Rotate the shaft completely counterclockwise (CCW). Connect the ohmmeter leads to terminals A and B and measure the resistance between terminals A and B (R_{AB}). Record your reading in Table 2–4 (p. 17).

   b. Next, measure the resistance between terminals A and C (R_{AC}). Record your reading in Table 2–4.

   c. Finally, measure the resistance between terminals B and C (R_{BC}) and record the reading in Table 2–4.

7. **a.** With the ohmmeter connected to terminals B and C, turn the shaft completely clockwise (CW). Note the reaction of the ohmmeter as the shaft is turned. Measure the resistance between terminals B and C and record it in Table 2–4.

   b. Connect the ohmmeter across terminals A and C. Measure the resistance between A and C and record it in Table 2–4.

   c. Measure the resistance between terminals A and B and record it in Table 2–4.

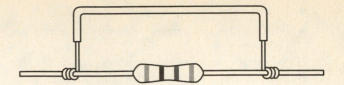

Figure 2–6. Resistor short-circuited by a piece of hookup wire.

8. Calculate the value of $R_{AC} + R_{BC}$ for steps 5 and 6 and record it in Table 2–4.

9. Connect a length of hookup wire from terminal B to terminal C. Measure the resistance between terminals A and C. $R_{AC} =$ _____ Ω. Disconnect the hookup wire.

10. connect the hookup wire from terminal A to terminal C. Measure the resistance between terminals B and C. $R_{BC} =$ _____ Ω. Disconnect the hookup wire.

Optional Activity

Use the Internet to find a source for purchasing resistors. If needed, have your instructor give you a list of WWW addresses.

ANSWERS TO SELF TEST

1. ohmic value	6. false
2. 2	7. 3
3. four	8. reliability
4. silver	9. brown; red; brown; no color
5. 10; 5	10. open

TABLE 2–3. Resistor Measured Resistance versus Color-Coded Values

	Resistor									
	1	*2*	*3*	*4*	*5*	*6*	*7*	*8*	*9*	*10*
First color band										
Second color band										
Third color band										
Fourth color band										
Coded value, Ω										
Tolerance, %										
Measured value, Ω										
Accuracy (%)										

TABLE 2–4. Potentiometer Measurements

Step	Setting of Potentiometer Control	R_{AB} Ω	R_{AC} Ω	R_{BC} Ω	$R_{AC} + R_{BC}$ Calculated Value
6	Completely CCW				
7	Completely CW				

QUESTIONS

1. When using an analog VOM, at which end of the ohms scale are resistance measurements more reliable, the zero end or the infinity end? Explain.

_____

_____

_____

_____

_____

2. What is the color code for each of the following carbon resistors?

    (a) 0.27 Ω, ½ W, 5% _____

    (b) 2.2 Ω, ¼ W, 10% _____

    (c) 39 Ω, ⅛ W, 10% _____

(d) 560 Ω, ½ W, 5% _____

(e) 33 kΩ, 1 W, 20% _____

3. Can inserting a resistor in a circuit produce an effect similar to a short circuit? Explain.

_____

_____

_____

_____

4. Can inserting a resistor in a circuit have an effect similar to an open circuit? Explain.

_____

_____

_____

_____

5. Explain the significance of the reaction noted in step 6a of the procedure.

_____

_____

_____

_____

6. An ohmmeter has the following ranges: $R \times 1$, $R \times 10$, $R \times 100$, $R \times 1$ k. At the $R \times 10$ range, the pointer indicates 1500 Ω. If the range were changed to $R \times 1$, would the pointer move toward the zero end or toward the infinity end? Would the new measurement be more or less accurate than at the $R \times 10$ range? Explain.

_____

_____

_____

_____

7. When measuring resistance with a DMM, what does a flashing display indicate?

_____

_____

_____

_____

MEASUREMENTS OF DC VOLTAGE

BASIC INFORMATION

In Experiment 1 you became familiar with the switches, controls, and features of a dc variable voltage-regulated power supply. You also investigated the capabilities and features of a dc voltmeter. In Experiment 2 you were introduced to resistors and resistance. In this experiment you combine the resistors to form a simple circuit. You power the circuit with a dc power supply and measure voltages across the circuit.

Voltage Measurement

Recall from Experiment 1 that the two types of voltmeters are analog meters and digital meters. Both are used in this experiment.

The analog meter will be an electronic transistorized multipurpose type. Measuring voltage is one function of this meter. The meter, known as a volt-ohm-milliammeter or VOM, can measure dc and ac voltage, dc and ac current, and resistance. (Most commercial electromechanical VOMs cannot measure ac current.) The digital meter will also be a multipurpose instrument called a digital multimeter or DMM. The DMM can measure the same quantities as the VOM though often with greater ranges. Modern DMMs have a wide variety of special functions. These include the ability to measure frequency, capacitance, and temperature. Sockets are often provided to allow the testing of semiconductor diodes and transistors.

Because of their low power requirements, light weight, and compact design, electronic VOMs and digital multimeters are designed for portable, battery-operated use. Older vacuum-tube meters use 120 V ac as well as heavier battery supplies. Some solid-state meters can operate on either ac or dc.

Meter Controls. Analog meters use movable pointers (also called needles) and calibrated scales to indicate the measured quantities. The meter controls and switches are located on the front panel of the meter just below or to one side of the housing containing the pointer, calibrated scale, and meter movement. The controls include an on-off switch, a function switch, and a range switch. In many cases the on-off switch is incorporated in the function switch, which may be a rotary selector switch. The range switch is also commonly a rotary selector switch, often spanning different types of ranges, depending on the setting of the function switch. For example, a particular setting of the range switch may be la-

OBJECTIVES

1 To measure dc voltage across a circuit

2 To operate a dc power supply

3 To measure the range of output voltages of the power supply

beled 10 V/10 mA/$R \times 10$ k. If the function switch is set to resistance (often marked simply R, Ω, or OHMS), then the range setting stands for $R \times 10$ k. On the other hand, if the function switch is set for voltage, then the same range setting would indicate a 10-V range. Some analog meters also have a reverse polarity switch, which can keep the pointer on scale in case the polarity of the voltage being measured is not connected to the correct positive and negative terminals of the meter. There are usually two output terminals on electronic meters. In most instruments, test leads with banana plugs are used with the jacks of the instrument.

Digital meters usually have controls and functions similar to those on analog meters. The digital meters, however, use numerical displays that provide direct reading of measurements. Many DMMs have autoranging, which eliminates the need for a separate range switch (though most DMMs have one, despite having autoranging capabilities).

Another feature unique to digital meters is automatic polarity. If the positive lead happens to be connected to a negative voltage, a negative sign (or some other similar sign) will be displayed, indicating a reverse voltage. Digital meters are usually specified by the number of digits they display accurately. Thus a 3½-digit meter displays four digits, but the lead digit is never greater than 1. For example, the maximum display of the 3½-digit meter is 1999. If a reading exceeds 1999, the display will either blank out or continue blinking 1999. If the value just exceeds 19.99 (if it is, for example, 20) the display will show only three digits (not 20.00), with the lead digit blank. When the value being measured exceeds the maximum value of the range, the display will either blank out, display O.L., blink 1999, or simply display a lead 1. The display itself usually is made up of seven-segment LEDs or LCDs. Although LCD displays cannot be read in the dark or in very dim light, their circuitry requires much less power. For that reason most newer digital displays use LCDs.

Calibrated Scales. The use of calibrated scales on analog meters requires a good working knowledge of scale reading. Figure 3–1 shows a typical calibrated scale of a VOM.

In this set of scales, each of the four arcs represents a different quantity. The outermost arc, unlike other arcs, is numbered 0 through ∞ (infinity) from right (0) to left (∞). This is the ohmic scale, used to measure resistance. The next inner arc is marked DC. Note that all the small divisions on this scale are the same width. Such a scale is called a *linear scale*. The next two arcs are marked AC with the smallest scale marked 6 V AC. These four scales are sufficient for all the functions and ranges that this meter is capable of measuring.

The resistance scale is obvious. If the function switch is set to measure resistance, and the range switch is set on any value, a short circuit would cause the pointer to swing

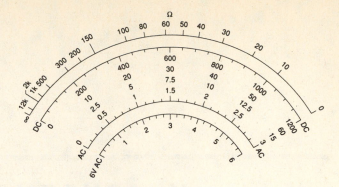

Figure 3–1. The calibrated scales of a typical volt-ohm-milliammeter. This meter can measure dc and ac voltage, dc current, and resistance.

to the right and stop at the 0 mark. An open circuit would cause the pointer to remain at rest, pointing to ∞.

In the case of dc voltage measurements, a set of numbers next to each major line division is associated with the range switch. This meter has maximum dc voltage ranges of 1200 V/600 V; 300 V, 60 V, 15 V, and 3 V. Notice that there is no 300 V mark at the end of the scale though there are marks for 1200, 60, 15, and 3 V. If the range switch were on 300 V and the voltage were 175 V, the pointer would simultaneously indicate 700 on the 1200-V scale, 35 on the 60-V scale, 8.75 on the 15-V scale, and 1.75 on the 3-V scale. By using multipliers ¼, 5, 20, and 100, respectively, the actual voltage can easily be determined.

If the function switch were on the 60-μA dc current range and the current were 25 μA, the pointer would simultaneously indicate 500 on the 1200 scale, 25 on the 60 scale, 6.25 on the 15 scale, and 1.25 on the 3 scale. Again, the current can easily be determined by using multipliers ¹⁄₂₀, 1, 4, and 20, respectively.

Similarly, voltages can be read on the ac scales using the same multipliers, with the exception of the 6-V scale, from which measurements are read directly.

Direct Current Power Supply $\cdots\cdots\cdots$

A variable voltage-regulated dc power supply is used in this experiment. The power supply has a low voltage range (0 to 15 V). A description of a typical supply follows; the one you will use may differ in some details.

The front panel controls include an on-off switch (often with a pilot light), a rotary dial used to adjust the output voltage, and two or three output jacks. If two jacks are provided, one is marked +, the other −. More often three output jacks are provided: +, −, and GND. The output voltage is provided across the + and − terminals, whereas GND is the equipment ground. Normally there is no voltage between GND and + or −. Some power supplies have separate + and − output controls. Such supplies will provide the + output between the + and GND jacks and the

− output across the − and GND jacks. The total output voltage is available across the + and − jacks. Such power supplies may be specified as having ± 15-V outputs.

Another control found on some supplies is labeled current. The current control is usually a rotary dial. This control sets a limit on the current output of the power supply. Although the power supply may have a 1-A rated output, the current control can limit the current to 0.5 A or any amount between 0 and 1 A. Such a control can be useful in preventing excessive currents from entering an experimental circuit.

Laboratory power supplies are usually equipped with a voltmeter on their front panel. The meter indicates the voltage at the terminals of the supply. Laboratory power supplies are line-operated equipment and are supplied with a cord and plug for 120-V operation.

When connecting a dc power supply to a circuit, you must note and follow the polarity requirements of the circuit (if any). Never make circuit changes while the power supply is connected to the circuit. If many and frequent circuit changes are necessary, as when performing the different steps in an experiment, use an external switch to disconnect the circuit from the supply.

In most cases it is preferable to bring the supply up to the required voltage level slowly rather than connect the entire voltage immediately. By bringing the voltage up slowly, any problems with the circuitry can be detected early and possible damage to components and meters avoided.

SUMMARY

1. Analog meters, whether electronic or electromechanical, use a pointer and a calibrated scale.

2. Calibrated scales are divided into small divisions. The divisions of the scale used to measure dc values are uniform. The scale is said to be linear.

3. The same set of divisions on a scale may be used to measure different quantities (for example, voltage and current) within different ranges (for example, 300 V, 60 V, 15 V, 3 V, 60 μA, 3 mA, 30 mA, 300 mA). By using the proper multipliers, the exact measurement can be obtained from any scale indication of the pointer.

4. Digital meters are often autoranging, so the range switch (if one is provided) need not be set before taking a measurement. Analog meters sometimes contain reverse polarity switches to read the measurements on scale in case the incorrect polarity has been connected to the meter. Digital meters have an automatic polarity feature that displays a negative sign when the meter is connected in reverse polarity.

5. Power supplies are rated according to their maximum voltage and current capacities.

6. A voltage-regulated power supply will maintain a set value of voltage even when the load current varies or the line voltage fluctuates, within specified limits.

7. Some supplies can produce separate + and − output voltages. Supplies of this type have three-jack outputs, +, −, and GND. The total voltage of + to GND and − to GND is available across the + and − terminals.

8. Some supplies have a current limit control. This control is used to prevent excessive currents from being drawn by the circuit.

SELF TEST

Check your understanding by answering the following questions.

1. The output voltage of dc power supplies used in the laboratory is _____, enabling them to supply a range of different voltages.

2. A power supply whose output voltage remains constant at a preset value despite changes in load current or line voltage fluctuations is voltage-_____.

3. An analog VOM contains a number of _____ across which the pointer moves to indicate electrical quantities.

4. (True/False) A digital meter specified as 3½ digits has a maximum display of 1999. _____

5. The range switch on an analog voltmeter is set at 60 V. If the pointer indicates 75 on the 0–100 calibrated scale, the actual voltage being measured is _____ V.

6. (True/False) The 0 position is on the extreme right, whereas the ∞ position is on the extreme left of the calibrated scale of the typical electromechanical ohmmeter. _____

7. The divisions of the calibrated scale of an electronic voltmeter are equal across the entire scale. The scale is therefore _____.

8. Most electromechanical multimeters cannot measure ac _____.

MATERIALS REQUIRED

Power Supply:
- Variable 0–15 V dc, regulated

Instruments:
- Digital multimeter (DMM)
- Volt-ohm-milliammeter (VOM) (analog; electronic)

Resistors:
- 2 220-Ω, 20-W

Miscellaneous:
- 2 SPST switches

Power supply controls and features differ from one manufacturer to another. Familiarize yourself with the supply you will be using by carefully studying the operating manual with the power supply available in front of you. Do not turn the power supply **on** until instructed to do so in the procedure.

A. Voltage Measurements

A1. With the power supply switch **off,** plug the supply into a 120-V ac source (if it is not already connected). Turn the voltage control completely counterclockwise (this should be the 0-V position). If the supply has a current control, turn this to the 0-A position also.

A2. Connect the VOM across the power-supply output terminals. Connect the + terminal of the meter to the + terminal of the supply and the − terminal of the meter (often labeled COM) to the − terminal of the power supply. Set the VOM to the highest voltage range.

A3. Turn **on** the power supply. Slowly increase the voltage until the power supply voltmeter measures 15 V. Maintain this voltage throughout part A. Check the power-supply voltage periodically and adjust if necessary. Record the VOM voltage range setting and the voltage measured by the VOM in Table 3–1. If the pointer does not indicate a measurable quantity, mark the space in Table 3–1 "no indication."

A4. Set the voltage range control to the next lower voltage range provided it is equal to or greater than 15 V. Record the range and the voltage measured by the VOM in Table 3–1. Again, if no readable indication is given by the pointer, mark "no indication" in the table.

A5. Set the voltage range control to the next lower voltage range provided it is equal to or higher than 15 V. Record the range and VOM measurement in Table 3–1 as in previous steps.

A6. Continue to set the voltage range control to the other ranges for all ranges equal to or greater than 15 V. Turn **off** the power supply. Set the voltage output control to 0 V.

A7. Connect a DMM across the power supply output and repeat steps A3 to A6. Record all values in Table 3–1.

B. Variable Voltage-Regulated Power Supply Operation

B1. With the power supply **off** and switches S_1 and S_2 **open,** connect the circuit of Figure 3–2.

B2. Turn **on** the power supply and adjust the output voltage to 15 V as measured by the VOM. If the power supply has a current control, set it to its maximum value. Record the VOM voltage in Table 3–2 in the first row.

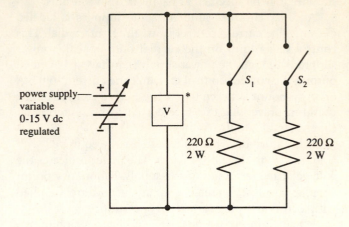

*Use VOM as a voltmeter.

Figure 3-2. Circuit for procedure step B1.

B3. Close S_1. *Do not adjust any controls on the power supply.* Record the VOM voltage in Table 3–2 in the second row.

B4. With S_1 closed, **close** S_2. *Do not adjust any controls on the power supply.* Record the VOM voltage in row 3 of Table 3–2.

B5. With S_1 and S_2 closed, decrease the power supply voltage to 10 V as measured by the VOM. Record this value in row 4 of Table 3–2.

B6. **Open** S_1 and S_2. *Do not adjust any controls on the power supply.* Record the VOM voltage in row 5 of Table 3–2.

B7. With S_1 and S_2 **open,** reduce the output voltage of the power supply to 1 V. Record the VOM voltage in row 6 of Table 3–2.

B8. **Close** S_1 and S_2. *Do not adjust any controls on the power supply.* Record the VOM voltage in row 7 of Table 3–2. After this step has been completed, **open** S_1 and S_2, turn **off** the power supply, and disconnect the VOM.

B9. Connect a DMM in place of the VOM and repeat steps B2 through B8. At the conclusion of the last step, **open** S_1 and S_2, turn **off** the power supply, and disconnect the circuit.

ANSWERS TO SELF TEST

1. variable
2. regulated
3. calibrated scales
4. true
5. 45
6. true
7. linear
8. current

Name _____ Date _____

TABLE 3–1. Voltage Measurements

Power-Supply Voltmeter Reading, V	VOM		DMM	
	Voltage Range, V	Measured Voltage, V	Voltage Range, V	Measured Voltage, V
15				

TABLE 3–2. Power Supply Operation

Row	Switch S_1	Switch S_2	VOM Voltage, V	DMM Voltage, V
1	Open	Open		
2	Closed	Open		
3	Closed	Closed		
4	Closed	Closed		
5	Open	Open		
6	Open	Open		
7	Closed	Closed		

QUESTIONS

1. List four precautions that should be observed when measuring dc voltage.

_____

_____

_____

_____

_____

_____

_____

_____

_____

_____

2. Electromechanical multimeters are usually capable of measuring ac and dc voltage and resistance. What other measurements can they make? What measurements are usually not made with this type of instrument?

_____

_____

_____

_____

_____

_____

3. What voltage ranges did your digital multimeter have? What was its digital rating? What is the highest readout it can display?

_____

_____

_____

_____

_____

_____

4. If your power supply has a current control, how can it be used to minimize the possibility of a short circuit?

_____

_____

_____

_____

_____

_____

_____

5. What possible damage can an analog meter suffer if the polarity connections are reversed?

_____

_____

_____

_____

_____

_____

_____

6. What advantages might an analog meter have over a digital meter in measuring voltages and current in a laboratory experiment?

_____

_____

_____

_____

_____

_____

EXPERIMENT

4

CONDUCTORS AND INSULATORS

BASIC INFORMATION

Conductors and Insulators

The useful effects of electricity result from the movement of electric charges in a circuit. This movement of electric charges is called *current*. Electric charges move easily through paths called *conductors,* but it is very difficult for these charges to move through *insulators.* Conductors are materials that permit current to flow easily with little electrical pressure (voltage) applied, whereas insulators are materials that permit very little or no current to flow. Copper is an example of a good conductor; rubber is a good insulator.

A complete (closed) circuit or path for direct current consists of a voltage source such as a battery or power supply, a load such as an electric lamp, and connecting conductors (copper wire) (Figure 4–1, p. 26). In such a closed circuit there is electric current, and if sufficient current flows, the lamp will light. If rubber cords instead of copper wires were used to tie the battery to the lamp, the lamp would not light, because rubber is an insulator and does not permit current to flow through it in the circuit.

The tungsten filament in the electric lamp is a conductor, but it is not as good a conductor as the connecting copper wires. Tungsten is used for the filament because its properties cause it to glow and give off light when it is heated by an electric current.

It is apparent that not all materials are equally good conductors of electricity. Specific materials are used for electrical components because of their unique characteristics as conductors and nonconductors.

Wire Conductors

The ability of a conductor to allow current flow is called *conductance,* denoted by the symbol G. The unit of conductance is the *siemen.* The opposition to current flow is called *resistance,* denoted by the symbol R. The unit of resistance is the *ohm.*

The degree of resistance of a material is called its *resistivity.* Different materials have different values of resistivity depending on their molecular structure.

The resistance of a material depends not only on its resistivity but also on its dimensions and temperature. For example, a round copper wire of a given diameter and length will have a certain value of resistance

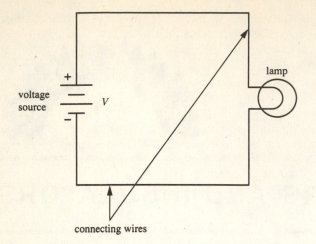

Figure 4–1. A complete (closed) electric circuit showing a voltage source, *V*, connecting wires, and a lamp (the electric load).

measured in ohms. If the length of this wire is doubled, its resistance will double. Its resistivity will remain the same as long as the temperature is constant. Resistance also depends on the cross-sectional area of a conductor. The resistance of a copper wire of the same length as the original wire but with double the cross-sectional area has one-half the resistance of the original wire. That is, resistance decreases as the cross-sectional area of a conductor increases; resistance increases as the length of the conductor increases.

For a given length and diameter of wire, the higher the resistivity, the higher the resistance. Stated another way, good conductors have low resistivities.

A formula that relates length, cross-sectional area, and resistivity to resistance is

$$R = \frac{\rho l}{A} \qquad (4\text{–}1)$$

where

R = resistance in ohms

l = length in feet

A = cross-sectional area in circular mils

ρ = resistivity at a given temperature

The resistivity of copper at 20°C (usually considered room temperature) is 10.37. Other materials often used as low-resistance conductors are silver, gold, and aluminum. The resistivity of silver is 9.85 and of aluminum is 17.00. In some applications, high-resistance conductors are necessary. High-resistance materials include nickel, iron, tungsten, a combination of nickel and chrome known as nichrome, and carbon.

An example will show how the resistance formula is used.

Problem. Find the resistance of 100 ft of copper wire having a diameter of 0.1 in.

Solution. A diameter of 0.1 in. is equal to 100 mils (1 mil = 0.001 in.). An area in circular mils is found by squaring the diameter in mils. Thus 100 mils in diameter yields a cross-sectional area d^2 of 10,000 circular mils. Substituting this value in formula (4-1), we have

$$R = \frac{\rho l}{A}$$

$$R = \frac{10.37 \times 100}{10,000}$$

$$R = 0.1037 \ \Omega$$

Wire Size

Round wire used as electrical conductors is specified by gage sizes. In the United States, the American Wire Gage (AWG) is used as the standard measure. The higher the AWG number, the smaller the diameter of the wire. For example, no. 12 copper wire (a common size for house wiring) has a much larger diameter than no. 22 wire (a common hookup wire for electronic circuits). The largest gage size is 0000, or, as it is often written, 4/0. Wire sizes larger than 4/0 are usually specified by giving their cross-sectional area in circular mils.

Electrical reference books often contain tables of wire sizes and properties. Included in these tables are AWG sizes, diameters, cross-sectional areas, and resistance per 1000 ft of wire.

The amount of current a conductor is able to carry before being damaged is usually much more than the conductor is *permitted* to carry by code regulations or law for reasons of safety to personnel and equipment. For example, a no. 12 copper wire with plastic insulation is limited by the National Electrical Code® to carrying 20 A of current when used as building wire under certain conditions. Yet the same no. 12 wire, uninsulated, when properly used in circuits in the laboratory, can safely carry considerably more than 20 A without melting or creating a safety hazard.

Wire Components

Some components are made by winding a length of wire around a form. For example, ceramic high-power resistors consist of a wire winding around a ceramic core. Iron-core inductors (chokes) are wound around a form through which an iron core is placed. Transformers consist of several windings around some specified core. If an ohmmeter is placed across each winding of these components, it will measure the resistance of that winding. Therefore, one way of determining if a wire winding is continuous, that is, if it is not broken, is to measure its resistance across the two end terminals. The normal resistance is usually specified by the manufacturer. If a winding measures infinite resistance, it is open. Because the resistance of these windings is often very low, the idea of the *continuity test* is to see that a complete electrical path exists rather than to measure resistance. A wide swing of the meter needle or a reading of very low resistance on a DMM indicates continuity of the electrical path.

Carbon Resistors as Conductors

In a previous experiment we learned how to determine the resistance of carbon resistors using a color code. Carbon is also considered a conductor of electricity. A complete circuit that contains a dc power source, connecting wires, and a resistor (Figure 4–2) will permit current to flow.

The physical construction of a resistor will determine its resistance. Here, as in circular wire, the resistance of the resistor varies inversely with the cross-sectional area of the carbon element in the resistor and directly with the length of the element. By controlling the diameter and length of the inner carbon element, we can control the resistance of R.

Comparing a resistor with a copper wire, we note that the resistance of the wire is *distributed* throughout its length. In a small carbon resistor, the resistance is lumped between the two terminals.

In the circuit of Figure 4–2 the current can be controlled by controlling the value of R, in ohms. With V held constant, the higher the resistance of R, the lower the current will be. That is, as the resistance of R increases, the opposition to current increases and the current decreases.

Insulators

There are some materials whose resistivity is so high that they permit *very* little current to flow. These materials are called *insulators*, or *nonconductors*. Rubber was previously mentioned as an insulator. Wood, glass, paper, mica, most plastics, and air are other examples of insulators.

If the leads of an ohmmeter are connected across an insulator, say a rubber cord, the meter will register infi-nite resistance. An infinite resistance will not permit direct current to flow in a circuit. Thus, Figure 4–3 is an open circuit, because direct current will not flow through the insulator.

Insulating materials are used to cover copper wires in electricity and electronics. The purpose of insulating electrical wires is to prevent them from making electrical contact (shorting) with any other conductor or component that they may touch accidentally. Wires on which the insulation has melted, cracked, or become frayed are safety hazards and should not be used.

Insulators may break down and become conductors if the voltage across them is high enough. Thus an air gap between two voltage terminals that are 1 cm apart will break down if the voltage across these terminals equals or exceeds 30,000 V. A descriptive term frequently used for this phenomenon is *arc-over*.

Semiconductors

The basic active materials of many modern electronic components are silicon and germanium. These two elements, in their pure state, are insulators. However, when certain impurities, such as gallium or antimony, are added, they can be made somewhat conductive, and they are then called *semiconductors*. The resistivity of semiconductors depends on the nature and amount of impurity injected. In general, however, the resistivity of semiconductors is said to be between that of insulators and conductors. Semiconductor devices are also referred to as *solid-state* devices.

The Human Body as a Conductor

The human body is a conductor of electricity, as evidenced by the electric shocks that people have experienced when touching the positive and negative terminals of a

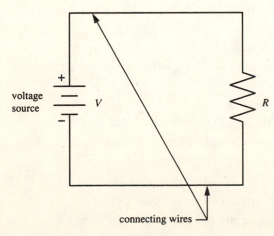

Figure 4–2. A complete electric circuit containing *V*, a dc power source, connecting wires, and *R*, a resistor.

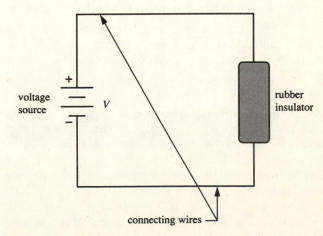

Figure 4–3. An open circuit. The rubber insulator does not permit direct current to flow through the circuit.

voltage source. Electric current will flow through the body, and the higher the voltage, the higher the current. Electric shock is *dangerous* and can be *fatal*. That is why the student of electricity and electronics should follow the safety precautions taught in order to avoid shock.

The resistance of the body is not a constant value. Its value at a specific point in time can be measured with an ohmmeter. The method to follow is given in the Procedure section.

SUMMARY

1. Conductors are materials that permit current to flow through them with ease. Silver, copper, and aluminum are examples of good conductors.

2. Insulators are materials that do not permit current to flow through them. Rubber, glass, mica, and most plastics are examples of insulators.

3. Conductors vary in their conductivity. Thus, copper is a better conductor than aluminum.

4. The resistance of wire varies directly with its length and inversely with its cross-sectional area. Resistance also depends on the material from which the conductor is made.

5. The American Wire Gage (AWG) is used to measure wire size.

6. Wire tables list AWG wire sizes, cross-sectional area of wire, resistance of 1000 ft of wire, and other electrical and physical properties of wire.

7. Insulators can withstand voltages up to a critical voltage. At the critical voltage or higher, insulators break down and permit current flow by arc-over.

8. The breakdown voltage of an insulator depends on its material.

9. The resistivity of a good conductor is very low and that of an insulator very high.

10. Semiconductors are materials with resistivity that lies between that of a conductor and insulator. Silicon and germanium to which impurities have been added are semiconductors.

11. The human body is a conductor of electricity. Body contact with electrical wires must be avoided to prevent shock.

SELF TEST

Check your understanding by answering the following questions:

1. List five materials that are good electrical conductors: _____ ; _____ ; _____ ; _____ ; _____ .

2. List five materials that are electrical insulators: _____ ; _____ ; _____ ; _____ ; _____ .

3. For current to flow in a circuit, it must have (a) _____ ; (b) _____ ; and (c) _____ .

4. A circuit in which current flows is called a _____ circuit.

5. Current will not flow in a(n) _____ circuit.

6. No. 22 AWG round copper wire has a _____ (larger/smaller) diameter than no. 12 wire.

7. The resistance of no. 10 wire is _____ (higher/lower) than that of no. 12.

8. A 50-ft piece of no. 22 wire has _____ the resistance of a 100-ft piece of the same wire.

9. An insulator may _____ if it is subjected to a voltage higher than its breakdown voltage.

10. (True/False) Semiconductors will, under the proper conditions, permit electric current to flow. _____

11. The diameter of a round copper wire is 0.025 in. The circular mil area of this wire is _____ .

12. The diameter of a round copper wire is 45 mils. Its resistivity is 10.37. The resistance of 1000 ft of this wire is _____ .

MATERIALS REQUIRED

Instruments:
- Digital multimeter or analog VOM

Resistors:
- 1 carbon resistor (resistance value not important)

Capacitors:
- 1 0.01-μF, 25 V dc

Semiconductors:
- 1 silicon rectifier, 1N4001 or equivalent

Miscellaneous:
- 12-in. length of #40 AWG solid copper wire
- 6-in. length of #40 nichrome wire
- 12-in. length of #40 nichrome wire
- 2-in. square of solid rubber, ¼-in. thick
- 2-in. square of wood, ¼-in. thick
- 2-in. square of acrylic plastic, ¼-in. thick
- Coil of wire

PROCEDURE

CAUTION: Be sure to hold the insulated probes on the meter leads when you measure resistance. This becomes *particularly important* when you are using the *higher* ranges of the ohmmeter. The reason is that the ohmmeter will read your body resistance if you hold the metal tips of the meter leads. This measurement of body resistance will affect the resistance reading of the object you are measuring, particularly if the object has high resistance.

1. In this experiment you will measure the resistance of each of the 10 objects listed in Table 4–1 (p. 31). Check your ohmmeter first by zeroing it as necessary. The test leads of your meter should be color coded with the common lead, or minus, black, and the plus lead red.

2. Connect the red test lead of the meter to one lead of the resistor and the black test lead to the other lead of the resistor. Measure the resistance and record it in column A of row 1 (Resistor).

3. Reverse the red and black test leads of the meter so that they are connected across the resistor again. Measure the resistance again and record the value in column B of row 1.

4. Indicate in column C, row 1, whether the resistor behaved like a conductor or an insulator.

5. Repeat steps 2 through 4 for the capacitor and record in row 2 the resistance values and whether the capacitor behaved as a conductor or an insulator.

6. Repeat steps 2 through 4 for the other eight objects listed in Table 4–1.

NOTE: Measure the resistance across the 2-in. dimension for the wood, rubber, and acrylic samples.

Body Resistance

7. Set the ohmmeter on the 2 MΩ or $R \times 10,000$ range and check that it is properly zeroed.

8. Measure your body resistance by grasping the metal tip of one lead with one hand and the metal tip of the other lead with your other hand. Record the value: Body resistance = _____ Ω.

ANSWERS TO SELF TEST

1. silver; copper; gold; aluminum; tungsten; and the like
2. wood; glass; paper; mica; air; and the like
3. (a) voltage source; (b) conductive load; (c) connecting conductive wiring
4. closed
5. open
6. smaller
7. lower
8. one-half
9. arc-over (or breakdown)
10. true
11. 625
12. 5.12 Ω

Name ——————————————————————— Date ———————————

TABLE 4–1. Resistance of Conductors and Insulators

Object	Resistance, Ω		Insulator or Conductor	Object	Resistance, Ω		Insulator or Conductor
	A	B	C		A	B	C
1. Resistor				6. Coil			
2. Capacitor				7. Wood			
3. 12-in. length #40 copper wire				8. Plastic			
4. 12-in. length #40 nichrome wire				9. Rubber			
5. 6-in. length #40 nichrome wire				10. Silicon rectifier 1N4001			

QUESTIONS

1. How were you able to determine whether a material was a conductor of direct current? (Refer to your actual test results.)

———

———

———

———

2. Compare columns A and B of Table 4–1. Did the polarity of the test leads affect the resistance measurements of the materials? If so, explain the effect; refer to actual test results.

———

———

———

———

3. Is the silicon rectifier a good conductor or a good insulator? How do your test results confirm your answer?

———

———

———

———

———

———

4. Formula (4–1) shows that the resistance of a wire varies directly with length for a given cross-sectional area of wire. Which measurement or measurements in this experiment, if any, confirm this relationship?

_____

_____

_____

5. Calculate the resistivity of nichrome wire using Formula (4–1) and the test results in Table 4–1. Use standard wire tables to determine the cross-sectional area of the #40 nichrome wire. Refer to a standard engineering handbook to compare your answer with standard published values. Explain any possible difference between your calculated value and the standard value.

_____

_____

_____

_____

6. Explain how to zero an analog ohmmeter at both ends of its scale.

_____

_____

_____

_____

7. Was the ohmmeter used in this experiment able to measure accurately the resistance of the insulating materials being tested? Explain your answer by referring to actual test results in Table 4–1.

_____

_____

_____

_____

8. Is the ohmic resistance of your body higher or lower than that of the surrounding air? What evidence from this experiment proves your answer?

_____

_____

_____

9. The ends of bare copper wire are often scraped with a knife or rubbed with fine sandpaper before being connected to a circuit or being soldered. What is a possible reason for this practice?

_____

_____

_____

EXPERIMENT 5

SWITCHES AND SWITCHING CIRCUITS

BASIC INFORMATION

In Experiment 4 you examined and measured the resistance of various conductors and insulators. Conductors were also used to complete connections between the voltage source and load. This allowed current to flow through the circuit. Electric charges easily flow through the conductors due to the low resistance of the conductors. Insulators on the other hand permitted little or no current to flow.

A switch is a device used to control an electric current path. When a switch is said to be closed its contacts complete the circuit and current is able to flow. When the switch is opened the circuit is broken and no current flows. Switches can operate mechanically or electronically. In this experiment we will concentrate on mechanical switches.

When a mechanical switch is closed a set of contacts physically connect forming a very low resistance path. When the switch is opened the contacts separate and the gap between the contacts forms a very high resistance.

As shown in Figure 5–1 (p. 34) when the switch is closed current flows and the entire voltage source is placed across the load. When the switch is opened no current flows and the voltage source is placed across the open switch.

A single switch can be used to control more than one path in a circuit. The switch in Figure 5–2 (p. 34) will allow current to flow through load A when contact 1 is closed. In this position load B will not have current. When the switch is moved to contact 2, load B will have current while load A will not. Switches are available to control three or more current paths in a similar manner.

Switch Types

Switches are usually specified by the terms *poles* and *throws*. The pole is the terminal in the switch that connects to the point in the circuit that is being controlled. The pole is sometimes referred to as the *common*. The throw is the number of ways the switch can be connected to other parts of the circuit. For example, in Figure 5–2, there is only one connection to the voltage source. This switch is called single pole. However the switch can be moved to either contact 1 or contact 2. This switch is a double throw. The complete specification for this switch is single-pole double-throw. Common switch types are shown in Figure 5–3 (p. 34). Abbreviations are usually used to refer to these switch types. The switch in Figure 5–2 would therefore be referred to as SPDT.

OBJECTIVES

1 ▸ To identify and test common types of switches

2 ▸ To construct switching circuits

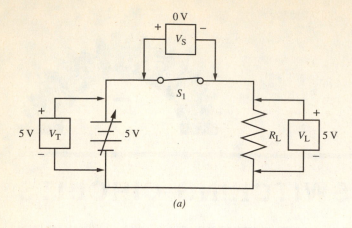

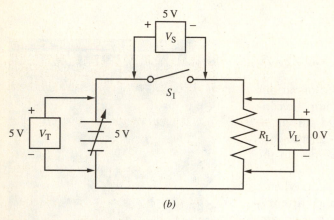

Figure 5–1. (a) Switch (S_1) in closed position. (b) Switch (S_1) in open position.

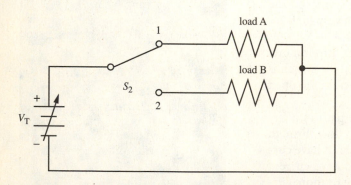

Figure 5–2. Switch used to control which part of a circuit will have a complete current path.

The switches described above maintain their positions once they are operated. An open switch will remain open until it is physically closed. Another type of switch is called a momentary contact switch. This type of switch is typically operated with a push button (Figure 5-4). If the normal condition of this type of switch is to be open (designated as normally-open or NO) pushing the button will close the switch. Once the button is released, the switch will spring back to its NO position. Similarly if the contacts of a switch are normally closed (NC), pressing the

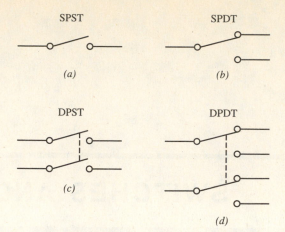

Figure 5–3. Types of switches. (a) single-pole single-throw; (b) single-pole double-throw; (c) double-pole single-throw; (d) double-pole double-throw.

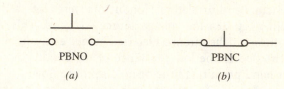

Figure 5–4. Types of push-button switches. (a) normally open, (b) normally closed.

button will open the contacts. Again, once the button is released the switch will revert to its NC position.

Mechanical switches such as those described above are manufactured with a wide variety of small handles, push buttons, and rotary dials. The principles of operation, however, are essentially the same for all of these types.

Switch Ratings

Switches are generally specified by their current and voltage ratings. The current rating is the maximum safe current that the switch can carry over a period of time. While the contacts of the switch can probably carry more current than the rating, the excess current would overheat the switch and damage its operating parts. Heat is one of the major factors that limit current ratings in electrical equipment. The voltage rating is the maximum safe voltage that can be applied across the switch contacts when the switch is opened. If the applied voltage is too high, current will try to jump across the switch contacts when the switch is opened (a process called *arcing*). Arcing can actually destroy the metal of the contacts. A house lighting circuit rated at 120 V should never be controlled by an automobile switch rated at 12 V.

Other factors are often considered when specifying switches such as the type of electrical system in which they

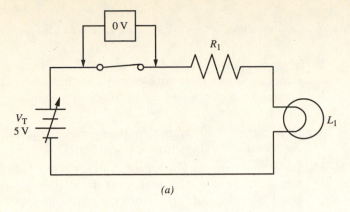

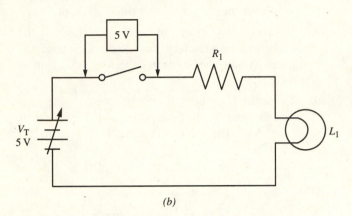

Figure 5–5. When testing a circuit in which voltage has been applied, take a voltage reading across the terminals. (*a*) When the switch is closed, the voltage drop should read close to 0 V. (*b*) In the open position, the voltmeter will read some value. In the circuit shown, the voltmeter will read the applied voltage.

are used and the type of load they will control. These factors will be discussed in later experiments.

Testing Switches ·

Since many switches are used in electric and electronic circuits, the electronics technician must be able to troubleshoot these devices. Most switch terminals are not labeled. This is especially true of the small switches used in electronic and electrical products. To test these switches the technician can use an ohmmeter to identify the poles and throws of the switch. A rough sketch of the switch poles and throws can then be drawn and used as an aid for further testing. The resistance across closed contacts should be very close to zero while the resistance across open contacts is almost infinite (Figure 5-5).

Ohmmeters must not be used in circuits to which a voltage has been applied. To test a switch in such a circuit a voltmeter can be used. If the voltage across the contacts of the switch is almost zero, the switch contacts are closed. If there is a voltage reading across the contacts, the switch is open or in some cases, defective. In addition, other parts of the circuit may have an effect on the voltage reading.

SUMMARY

1. A switch is used to control an electric current path.
2. A closed switch will have nearly zero ohms of resistance between its contacts.
3. An open switch will have almost infinite ohms of resistance between its contacts.
4. Switches are classified by the number of poles and throws.
5. Push-button momentary contact switches can be normally open (NO) or normally closed (NC).
6. Switches are specified by their current and voltage ratings.
7. An ohmmeter can be used to identify a switch's terminals.
8. A closed switch should have zero voltage across its contacts.

S E L F   T E S T

Check your understanding by answering the following questions.

1. When a switch is closed, its contacts will have approximately _____ ohms of resistance.
2. (True/False) When a switch is closed, the entire voltage source will be placed across the switch. _____
3. Switches are often classified according to their number of _____ and _____.
4. A DPST switch will have _____ connecting terminals.
5. Switches are usually specified by their _____ and _____ ratings.
6. An _____ can be used to determine a switch's terminals.
7. In a circuit in which a voltage is applied, a closed SPST switch will have _____ volts across its terminals.

MATERIALS REQUIRED

Power Supplies:
■ 2 variable 0–15-V dc, regulated

Instruments:
■ DMM or VOM

Resistors:
■ 1-1kΩ 1/2-W resistor

Miscellaneous:
■ 1 each SPST, SPDT, DPST, DPDT, PBNO, PBNC switch
■ 2 12-V incandescent lamps

PROCEDURE

Part A. Identifying and Testing Switches (record all results in Table 5–1 [p. 34])

A1. Examine the single-pole single-throw (SPST) switch and draw a sketch showing all terminals. Label each terminal on your diagram with a number.

A2. Draw the schematic symbol for the switch. Show the terminal numbers used in step A1.

A3. The switch has two operating positions, a and b. Using an ohmmeter check for continuity between the switch terminals in position a. Record the terminals that are connected together as indicated by the continuity test. If there is no continuity between the terminals, record the word "open."

A4. Operate the switch in its second position (position b) and perform the continuity test as in step A3. Record the results in the table.

A5. Repeat steps A1 through A4 for the single-pole double throw (SPDT), double-pole single-throw (DPST), and double-pole double-throw (DPDT) switches.

A6. The push-button normally-open (PBNO) switch is a momentary contact push-button switch. Draw a sketch of the switch and label each of its terminals with a number.

A7. Draw the schematic symbol for the switch and show the terminal numbers from step A6.

A8. Using an ohmmeter, check for continuity across the terminals of the PBNO switch. In the column labeled "Normal Position" record the terminals that are connected together as indicated by the test. If there is no continuity, record the word "open."

A9. Press the push button and hold while checking the continuity across the terminals. Record the results as in step A8 in the column labeled "Push Button Pressed."

A10. Release the push button and again check the continuity across the terminals. Record the results as in step A8 in the column labeled "Push Button Released."

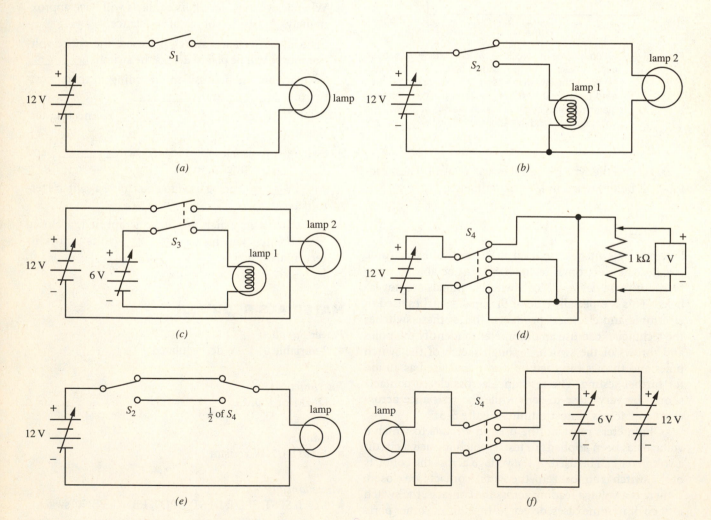

Figure 5–6. Circuits for part B of procedure.

A11. Repeat steps A6 through A10 for the push-button normally-closed (PBNC) switch.

A12. Using the SPST switch, measure and record the resistance across the terminals for position a. Set the switch to position b. Measure and record the resistance across the terminals.

Part B. Constructing Switching Circuits

Using the switches provided in Part A, construct each circuit shown in Figure 5–6(a) through 5–6(f). When you complete a circuit, have your instructor approve the connections before constructing the next circuit.

Part C. Circuit Testing

C1. With the power off, construct the circuit shown in Figure 5–7. Have your instructor approve the circuit before turning on power.

C2. With switch S_1 open, turn on the power supply and adjust the output to 12 V.

C3. Measure the voltage across the switch terminals and across the lamp. Record your measurements in Table 5–2 (p. 40).

C4. Close S_1 and repeat step C3.

C5. Replace the SPST switch with a PBNO switch. Have your instructor approve the circuit before applying power.

C6. Turn on the power supply and adjust the voltage to 12 V. Measure and record the voltage across the switch and across the lamp.

C7. Press the push button and hold while measuring the voltage across the switch terminals and across the lamp. Record your data in the table.

C8. Release the push button and measure the voltages as in step C6. Record your data in the table.

C9. Replace the PBNO switch with a PBNC switch. Have your instructor approve the circuit before turning on power. Repeat steps C6 through C8.

ANSWERS TO SELF TEST

1. zero
2. false
3. poles; throws
4. four
5. voltage; current
6. ohmmeter
7. zero

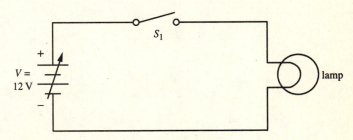

Figure 5–7. Circuit for part C of procedure.

EXPERIMENT

5

Table 5–1. Identifying and Testing Switches

Switch Type	Sketch	Schematic Symbol	Continuity Test		Resistance	
			a	b	a	b
Single-pole single-throw (SPST)						
Single-pole double-throw (SPDT)						
Double-pole single-throw (DPST)						
Double-pole double-throw (DPDT)						

Momentary Contact			Continuity Test		
			PB Normal	PB Pressed	PB Released
Push-button normally-open (PBNO)					
Push-button normally-closed (PBNC)					

TABLE 5–2. Circuit Testing

Switch S1	Position	Supply Voltage V_A	Voltage Across Switch S1	Voltage Across Lamp V_L
SPST	Open			
	Closed			
PBNO	Normal			
	Push button pressed			
	Push button released			
PBNC	Normal			
	Push button pressed			
	Push button released			

QUESTIONS

1. Explain in your own words what is meant by a switch's throw. Describe the action of the double throws in part B of this experiment.

_____

_____

_____

2. What part of a switch is sometimes referred to as the common? How is the common usually connected in a circuit?

_____

_____

_____

3. Examine your data for part C steps 3 and 4. Explain any differences between the voltages across the switch and across the lamp when the switch is open and when it is closed.

_____

_____

_____

4. Compare the behavior of the circuit with a SPST switch in part C steps 3 and 4 to that of the same circuit with a PBNO switch in steps 6 through 8.

_____

_____

_____

_____

MEASUREMENT OF DIRECT CURRENT

BASIC INFORMATION

Resistance, Voltage, and Current

Previous experiments used a circuit component called a *resistor,* whose resistance could be measured directly with an ohmmeter. The *value of the resistor,* expressed in ohms, is not dependent on the circuit in which it is connected. The value of the resistor is a specified amount of resistance within a given range, called the *tolerance* of the resistor.

Similarly, the voltage of a voltage source such as a battery or power supply can exist independently of any circuit. Its value can be measured with a voltmeter.

On the other hand, electric current cannot exist by itself. Current is defined as the movement of electric charges, but to have movement there must be voltage and a path along which the charges can move. A voltage source by itself cannot produce current. The electric circuit provides the complete path. Current is restricted to this closed path (Figure 6–1 p. 42).

The amount of current in a circuit is dependent on the amount of voltage applied by the voltage source and on the nature of the conductive path. If the path offers little opposition, the current is greater than it would be in a circuit where there is more opposition to current. Opposition to direct current is called *resistance* (measured in ohms). Current, then, can be controlled by the amount of resistance in a circuit.

Measuring Direct Current

Direct current in a circuit can be measured by means of a dc ammeter. In measuring current the circuit must be physically broken or opened and the meter inserted in series with the circuit. Suppose, for example, it is required to measure current in the circuit of Figure 6-1. The circuit is first broken at X (Figure 6–1). The ammeter is then inserted in series with the circuit at the two open leads A and B (Figures 6–2 and 6–3 p. 42).

When using an analog ammeter, the polarity of the meter terminals must be followed. The common (COM) or negative (−) lead of the meter must be connected to the point of lower potential. The positive (+) lead (sometimes called the "hot" lead) must be connected to the point of higher potential. When properly connected, the pointer or "needle" of the analog ammeter will move in an arc in a clockwise (left-to-right) direction. If the correct polarity is not followed, the pointer will swing sharply to the left off scale, where it will hit a stop pin or else move to the

OBJECTIVES

1 To measure current in a circuit

2 To measure the effect of resistance in controlling current

3 To measure the effect of voltage in controlling current

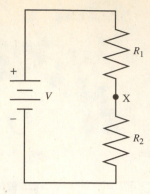

Figure 6-1. Current in the circuit is restricted to the closed path.

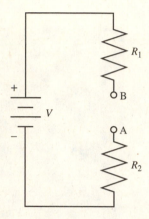

Figure 6-2. The circuit is open between points A and B.

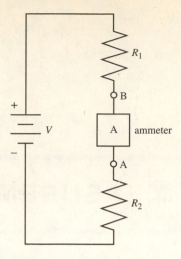

Figure 6-3. The circuit is completed through the ammeter in order to measure current.

extreme left of the scale. In either case the pointer can be bent or broken and the meter seriously damaged.

The terminals of digital meters may also be marked for polarity. However, the consequences of not following the correct polarity are less serious than with analog meters. Most digital meters when connected with reverse polarity will display a correct reading with a minus sign or some other appropriate symbol. Reversing the leads will produce the same reading without the minus sign.

Some ammeters are designed to make accurate measurements of low amperage. These meters are given special names to denote the ranges for which they are designed. The milliammeter is used to measure currents in the thousandths of an ampere range (*milli* means one-thousandth). The microammeter is used to measure currents in the millionths of an ampere range (*micro* means one-millionth).

CAUTION: An ammeter must never be connected across (that is, in parallel) with any component. Ammeters must always be connected in series with the conductors carrying current to a component in order to measure current in the component. Failure to follow this rule can result in serious damage to the meter.

SUMMARY

1. Opposition to current is a characteristic of a resistor. *Resistance can exist by itself.*

2. Voltage is a characteristic of an electromotive force (emf) source. *Voltage can exist by itself.*

3. *Current cannot exist by itself.* For current to exist, a voltage source and a closed path (circuit) are required.

4. In measuring direct current in a circuit, the circuit must be *broken* and an ammeter must be connected in *series* with the circuit.

5. In connecting an ammeter in a circuit, meter polarity must be observed.

6. Proper polarity is observed if the pointer of an analog ammeter moves up scale, from left to right, when the meter is connected in a circuit. Serious damage to the meter can result if the correct polarity is not followed.

7. The correct polarity should be followed when connecting digital meters. Failure to follow correct polarity, however, is not likely to damage the meter. Digital meters will usually display a minus sign or other distinctive symbol to indicate a reversal of polarity.

8. The amount of current in a circuit can be controlled by the amount of applied voltage and by the amount of resistance in the circuit.

S E L F   T E S T

Check your understanding by answering the following questions:

1. To measure current in a circuit, it is first necessary to _____ the circuit.

2. The ammeter is then connected _____ with the circuit.

3. Current can exist only in a _____ circuit.

4. _____ must be followed in connecting an ammeter in a circuit.

5. The _____ lead of a milliammeter is the common; the _____ lead of a milliammeter is frequently called the "hot" lead.

6. An ammeter must never be connected _____ a component. It must always be in _____ with the component.

7. The current in a circuit varies with the (a) _____ and (b) _____.

MATERIALS REQUIRED

Power Supplies:
- Variable 0–15-V dc, regulated

Instruments:
- 0–10-mA milliammeter
- VOM or DMM

Resistors (5%, ½-W):
- 3 1 kΩ

Miscellaneous:
- SPST switch

PROCEDURE

1. In this experiment you will observe the effect of resistance on current in a circuit. You will first measure resistance; then you will wire the resistance into a circuit and measure current.

   NOTE: Never measure the resistance of a component while it is connected in a circuit, especially if there is current in the circuit.

2. Select one 1-kΩ resistor and measure its actual resistance. Record the value in Table 6–1 (p. 47).

3. With the power supply off, connect the power supply, the meter, switch S_1, and the 1-kΩ resistor (labeled R_1), as shown in Figure 6–4. The switch must be in the **open** position. Note carefully that the negative terminal of the power supply must be connected to the negative terminal or lead of the meter. Have the instructor check the circuit.

4. After the circuit has been approved, turn on the power supply and adjust its output to 6 V.

5. Now, **close** the switch S_1. The meter will read the current in the circuit. Record the value in Table 6–1.

6. With the switch still closed and the meter reading current in the circuit, disconnect the resistor from the circuit. What effect did this have on the meter reading? Record in Table 6–1 the current read by the meter with the resistor out of the circuit. **Open** the switch.

7. Connect one lead of each of two 1-kΩ resistors together and measure the resistance across the two unconnected leads, as in Figure 6–5(a). Record this value in Table 6–1.

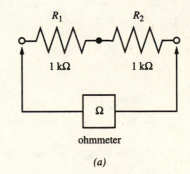

(a)

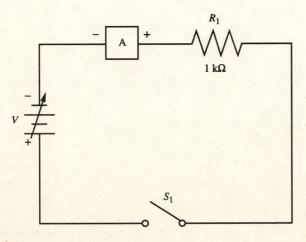

Figure 6–4. Circuit for procedure step 3.

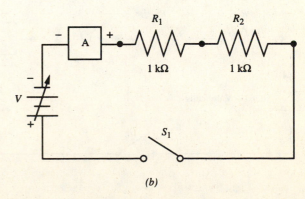

(b)

Figure 6–5. (a) Measuring resistance across two resistors in series. (b) Circuit for procedure step 7.

8. Connect the two joined resistors (labeled R_1 and R_2) in the circuit with the meter, switch, and power supply, as shown in Figure 6–5(b). **Close** the switch. Record the current read by the meter in Table 6–1. **Open** the switch. Disconnect R_1 and R_2 from the circuit.

9. Connect all three 1-kΩ resistors by adding the third 1-kΩ resistor (R_3) to the combination of R_1 and R_2. Measure the resistance across the three-resistor combination, as in Figure 6–6(a). Record the value in Table 6–1.

10. Connect R_1, R_2, and R_3 in the circuit, as shown in Figure 6–6(b). **Close** the switch and record the current read by the meter in Table 6–1. **Open** the switch.

Current in a Series Circuit

A series circuit exists when the power source and all resistors are connected through a single conducting path, as in Figure 6–6(b).

11. Disconnect the milliammeter and reconnect it between R_1 and R_2, as shown in Figure 6–6(c). Trace the path from the power supply terminals to make sure the meter polarity is correctly connected. Close the switch and in Table 6–1 record the current read by the meter. **Open** the switch.

12. Disconnect the milliammeter and reconnect it between R_2 and R_3, as shown in Figure 6–6(d). Again

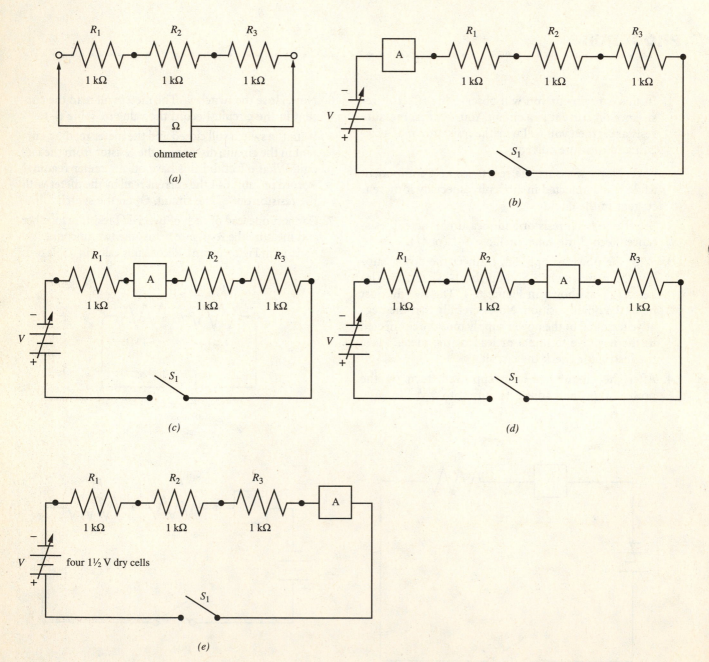

Figure 6–6. (a) Measuring resistance across three resistors in series. (b) Circuit for procedure step 10. (c)–(e) Changing the position of the ammeter in a series circuit.

observe meter polarity carefully. **Close** the switch. In Table 6–1, record the current read by the meter. **Open** the switch.

13. Disconnect the meter and reconnect it between R_3 and the switch, as shown in Figure 6–6(*e*). Observe meter polarity. **Close** the switch. In Table 6–1 record the current read by the meter. **Open** the switch.

Controlling Current by Voltage

14. Disconnect the resistors from the circuit in step 13. Separate the three resistors and reconnect one of the resistors into the circuit as in Figure 6–4. Adjust the power supply voltage, *V* to 8 V.

15. Close the switch and read the meter. Record the current value in Table 6–2 (p. 47). **Open** the switch.

16. Adjust the power supply voltage, V to 6 V.

17. Close the switch and read the meter. Record the current value in Table 6–2. **Open** the switch.

18. Adjust the power supply voltage, V to 4 V.

19. Close the switch and read the meter. Record the current value in Table 6–2. **Open** the switch.

20. Adjust the power supply voltage, V to 2 V.

21. Close the switch and read the meter. Record the current value in Table 6–2. **Open** the switch.

22. Reduce the power supply voltage, V to 0 V.

23. Close the switch and read the meter. Record the current value in Table 6–2. Open the switch and disconnect all components.

Optional Activity ·

This activity requires the use of electronics simulation software. Using the software construct the simulated circuit of Figure 6–5(*b*). It is not necessary to show the switch in your circuit diagram.

Adjust the power supply voltage, *V* to 12 V and maintain that voltage throughout this activity. Vary the value of resistance in five steps of 100 Ω above 1 kΩ. Record the current for each value of resistance.

Next, vary the value of resistance in five steps of 100 Ω below 1 kΩ. Record the current for each value of resistance.

Describe and analyze your results in a brief report.

ANSWERS TO SELF TEST

1. break (or open)
2. in series
3. closed
4. polarity
5. negative; positive
6. across; series
7. (a) applied voltage; (b) resistance

EXPERIMENT 6

TABLE 6–1. Controlling Current by Resistance. Voltage Source Constant

Step	Resistance, Ω	Circuit Condition	Current, mA	Meter Location
2,5		Closed		Between R_1 and V
6	✕	Open		Between R_1 and V
7		Open	✕	✕
8		Closed		Between R_1 and V
9		Open	✕	✕
10		Closed		Between R_1 and V
11		Closed		Between R_1 and R_2
12		Closed		Between R_2 and R_3
13		Closed		Between R_3 and V

TABLE 6–2. Controlling Current by Voltage. Resistance Constant

Steps	Voltage of Source, V	Current, mA
14, 15	8 V	
16, 17	6 V	
18, 19	4 V	
20, 21	2 V	
22, 23	0 V	

QUESTIONS

1. Under what conditions will there be current in a circuit? Refer to your test results as recorded in Table 6–1 to support your answer.

_____

_____

_____

_____

2. What precautions must be observed when connecting the milliammeter in the circuits of this experiment?

_____

_____

_____

3. Step 1 of the procedure cautions against measuring resistance of a component while it is connected in a circuit, especially if there is current in the circuit. Why is this caution so important?

_____

_____

_____

_____

_____

_____

4. What conclusions can you draw from the results of procedure steps 10 through 13?

_____

_____

_____

_____

_____

_____

5. Examine the results of steps 5, 8, and 10 as recorded in Table 6–1. With the voltage constant, what can you conclude about the relationship between current I in the circuit and the resistance R in their circuit? Using the symbols V for voltage, I for current, and R for resistance, express this relationship as a mathematical formula.

_____

_____

_____

_____

_____

_____

6. Examine the results of steps 14 through 21 as recorded in Table 6–2. With resistance constant, what can you conclude about the relationship between current I in the circuit and the voltage V applied to the circuit? Using the symbols V for voltage, I for current, and R for resistance, express this relationship as a mathematical formula.

_____

_____

_____

_____

_____

OHM'S LAW

BASIC INFORMATION

Experiment 6 established that there is a definite relationship between current, voltage, and resistance in a circuit. It was found that in a closed circuit containing voltage *V* and resistance *R* there is a current *I*. It was further found that if the voltage remains constant, the current decreases as the resistance increases. If the resistance remains constant, the current increases as the voltage increases.

These results are important, but they are only descriptive of some relationship that exists between current, voltage, and resistance. When working with circuits it is necessary to have a more exact statement of the relationship in the form of a mathematical formula. The formula not only shows the change but makes it possible to predict how much of a change will occur.

Using the method in Experiment 6 it is possible to develop a mathematical formula for the relationship between *I*, *V*, and *R*. To do this, it is necessary to make many precise measurements of *V* and *R* in a circuit. By applying mathematical methods to the experimental results, a formula can be written that will fit the measured quantities. More measurements can be taken to verify or modify the formula, if necessary.

For example, the circuit of Figure 7–1 was used to study the relationship between *I* and *V* for a constant value of *R*. A voltmeter was used to measure the voltage of the circuit, and an ammeter was used to measure the current. The voltage was varied from 10 to 50 V dc in steps of 10 V with a 10-Ω resistor in the circuit. The results are tabulated in Table 7–1, p 50.

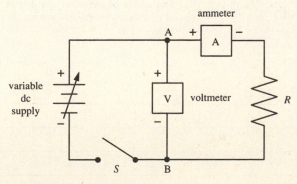

Figure 7–1. Circuit for verifying Ohm's law.

An examination of the data in Table 7–1 show an exact relationship between I and V in that the ratio V/I for each step is equal to 10. As a formula the ratio can be written as

$$\frac{V}{I} = 10, \quad \text{or} \quad \frac{V}{10} = I$$

Since the value of resistance was 10 Ω, it might be concluded that the ratio of V/I is always equal to R; that is,

$$\frac{V}{I} = R \tag{7–1}$$

or

$$\frac{V}{R} = I \tag{7–2}$$

Of course, to verify this relationship for a more general case, the preceding experiment would need to be repeated many times using different voltages and different resistances. For each result, the formula $V/R = I$ would need to be confirmed exactly.

It should be noted that the preceding experiment was described using ideal conditions. That is, the resistance of the meters was ignored, and it was assumed that the entire resistance of the circuit was concentrated in the 10-Ω resistor. In a practical experiment, meter and circuit resistance have to be taken into account if exact measurements are to be obtained.

Ohm's Law

Using the assumptions given for the preceding experiment and the actual results from Experiment 6, it is possible to state a relationship between V, I, and R.

Formula (7–2) states that current is directly proportional to voltage and inversely proportional to resistance. Stated another way, with R constant, as V is increased, I will increase; with V constant, as R is increased, I will decrease; where I is in amperes, V is in volts, and R is in ohms.

Formula (7–2) is known as *Ohm's law*, because it was first formulated by Georg Simon Ohm. The law is one of the fundamental relationships on which electricity and electronics is based.

Ohm's law can also be written as

$$V = I \times R$$

All three forms of Ohm's law are exactly the same. The choice of the formula to be used in any particular problem is determined by which values are known (V, I, or R) and which values must be found.

Measurement Errors

In the preceding discussion it was assumed that all the measurements made experimentally were 100 percent accurate. In practice, this is never so. Errors do occur and for several reasons.

One possible error results from reading the scale of an analog meter incorrectly. This can be corrected by exercising greater care and by taking the average of a number of the same measurements. Interpolating incorrectly between calibrated markers on a scale may be another source of error. A digital meter eliminates these particular errors.

Parallax is another source of error that can easily be corrected. It occurs when a meter reading is taken from an off center position—that is, when the line of sight between the viewer and the meter pointer is not perpendicular to the meter scale. Figure 7–2 illustrates the error of parallax. When the viewer is in position P_1, the line

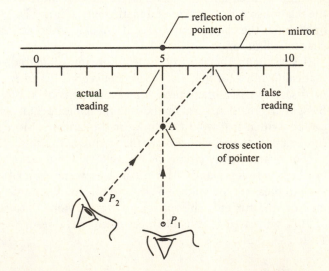

Figure 7–2. Parallax errors occur when the line of sight of the viewer and the meter pointer are not perpendicular to the meter scale.

TABLE 7–1. Developing a Formula for *I* when *R* = 10 Ω

R	$10\ \Omega$				
V (volts)	10	20	30	40	50
I (amperes)	1	2	3	4	5

between the eye and the meter pointer A is perpendicular to the meter scale. This gives a correct reading of 5. However, if the viewer is in position P_2, the reading will be 7, an error due to parallax. To eliminate errors of parallax, a mirror strip is sometimes placed just below the meter scale. The correct reading position is the one in which the pointer is positioned directly above its reflected image in the mirror.

Meter reading errors can be eliminated by using meters with a numerical readout—that is, digital meters.

There are other errors that are not so obvious and that cannot be corrected so easily. For example, there are inherent errors in the instruments used. The instrument manufacturer usually specifies the percentage of instrument error. For more accuracy, laboratory-precision instruments are required. These are highly precise instruments whose inherent error is held to a fraction of a percent.

Another source of error results from the process of inserting an instrument in a circuit to make a measurement. If the instrument alters circuit conditions in any way, incorrect readings may be obtained. Insertion errors called meter loading are discussed in greater detail in later experiments.

The fact that errors do occur is mentioned here because in this experiment the objective is to develop the formula for Ohm's law from experimental data. You can expect your data to contain some errors of measurement.

SUMMARY

1. The relationship between the voltage V applied to a closed circuit by some source such as a battery, the total resistance R, and the current I in that circuit is given by the formula $I = V/R$.

2. The relationship between the voltage drop V across a resistor R and the current I in that resistor is given by the formula $I = V/R$.

3. The formula $I = V/R$ is a mathematical statement of Ohm's law.

4. To verify Ohm's law experimentally, many measurements must be made and the results of the measurements must be substituted in formula (7–2) to verify the formula.

5. One set of data is obtained by measuring I while holding the measured value of V constant and varying the measured value of R. The data obtained should fit the formula $I = V/R$.

6. Another set of data is obtained by measuring I while holding the measured value of R constant and varying the measured value of V. The data obtained should also fit the formula $I = V/R$.

7. Measurement errors do occur, and these must be considered in attempting to establish the accuracy of a formula such as Ohm's law.

8. Among the measurement errors that may occur are (a) incorrect reading of the meter scale, (b) incorrect meter readings due to parallax, (c) errors resulting from the accuracy of the instrument used, and (d) errors introduced by inserting the instrument in the circuit (insertion or loading errors).

Check your understanding by answering the following questions:

1. The current in a fixed resistor is _____ proportional to the voltage across that resistor.

2. If the voltage across a resistor is held constant, the current in that resistor is _____ proportional to its resistance.

3. The formula that gives the mathematical relationship between I, V, and R in a closed circuit is $I =$ _____.

4. The formula in question 3 is called _____.

5. If the voltage across a 10-kΩ resistor is 125 V, the current in the resistor is _____ A.

6. If the voltage across a resistor is 60 V, and the current in the resistor is 50 mA, the value of the resistor is _____ Ω.

7. If the current in a 1.5-kΩ resistor is 0.12 A, the voltage across the resistor is _____ V.

8. In reading an analog meter scale, the line of sight between the viewer and the meter pointer should be _____ to the meter scale.

9. The error of parallax may be eliminated by placing a _____ just below the meter scale. The correct reading position occurs when the pointer and its _____ coincide.

MATERIALS REQUIRED

Power Supplies:
- Variable 0–15 V dc, regulated

Instruments:
- 0–10 mA milliammeter
- DMM or VOM

Resistors:
- 1 100 Ω ½-W, 5%
- 1 5-kΩ 2-W potentiometer

Miscellaneous:
- SPST switch

PROCEDURE

Part A

Adjust the potentiometer so that the resistance between terminals A and B [Figure 7–3(a)] measures 1000 Ω. Always measure the resistance of the potentiometer when it is disconnected from the circuit.

A1. Connect the circuit in Figure 7–3(b). Make sure power is **off** and switch S_1 is **open** before wiring the circuit. Have your instructor check and approve the circuit before proceeding to step A2.

A2. Turn power **on. Close** switch S_1 to apply power to the circuit. Slowly increase the voltage until the voltmeter reads 2 V. Read the milliammeter and record the value in Table 7–2 (p.55) in the "2 V" column.

A3. Adjust the voltage again until the voltmeter reads 4 V. Record the milliammeter reading in the "4 V" column of Table 7–2.

A4. Adjust the voltage again until the voltmeter reads 6 V. Record the milliammeter reading in the "6 V" column of Table 7–2.

A5. Adjust the voltage again until the voltmeter reads 8 V. Record the milliammeter reading in the "8 V" column of Table 7–2. **Open** switch S_1. Turn **off** the power.

A6. Calculate the value of V/I for each of your test values of voltage and current in Table 7–2. Record your answers in the "V/I" row of the table.

A7. From step A6 you should be able to deduce the relationship between V and I. Write the relationship, or formula, in Table 7–2 where indicated.

A8. Using the formulas in step A7, calculate the value of I when $V = 5.5$ and $V = 9.0$ V. Record your answers in the "Formula Test" section of Table 7–2 in the "I calculated" row.

A9. Turn power **on; close** S_1. Adjust the voltage until the voltmeter reads 5.5 V. Record the milliammeter reading in Table 7–2 in the "I measured" row. Increase the voltage to 9.0 V and record the milliammeter reading in the "I measured" row again. **Open** S_1. Turn **off** power to the circuit. Remove the potentiometer from the circuit.

Part B

Adjust the potentiometer so that the resistance across AB is 2 kΩ.

B1. Restore the potentiometer to the circuit of Figure 7–3. Turn power **on. Close** S_1 to apply power to the circuit. Increase the voltage until the voltmeter reads

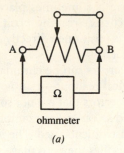

ohmmeter

(a)

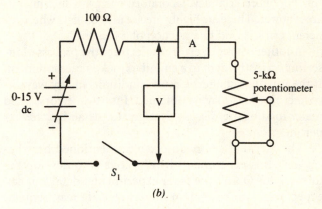

Figure 7–3. (a) Measuring the resistance across the terminals of a potentiometer. (b) Circuit for procedure step A1.

4 V. Record the milliammeter reading in Table 7–3 (p. 55) in the "4 V" column.

B2. Repeat step B1 using voltages of 6 V, 8 V, and 10 V in order. Record all milliammeter readings in Table 7–3. **Open** S_1; turn **off** power.

B3. Calculate the value of V/I for each of your test values of voltage and current in steps B1 and B2. Record your answers in the "V/I" row in Table 7–3 for each V and I.

B4. Write a formula for the V and I relationship indicated in step B3. Record the formula in Table 7–3 where indicated.

B5. Using the formulas in step B4, calculate the value of I when $V = 6$ V and $V = 12$ V. Record your answers in the "Formula Test" section of Table 7–3 in the "I calculated" row.

B6. Turn **on** power; **close** S_1. Adjust the voltage until the voltmeter reads 6 V. Record the milliammeter reading in Table 7–3 in the "I measured" row. Increase the voltage to 12 V and record the milliammeter reading in the "I measured" row of Table 7–3. **Open** S_1 and turn power **off.** Remove the potentiometer from the circuit.

Part C

Repeat the procedures for parts A and B using a potentiometer resistance of 3 kΩ. The test voltages are 6 V, 8 V, 10 V, and 12 V. The "Formula Test" voltages are 7 V and 13.5 V. Record all data for part C in Table 7–4 p.55.

Part D

Repeat the procedures of parts A and B using a potentiometer resistance of 4 kΩ. The test voltages are 8 V, 10 V, 12 V, and 14 V. the "Formula Test" voltages are 8 V and 14.5 V. Record all data for part D in Table 7–5.

ANSWERS TO SELF TEST

1. directly
2. inversely
3. *V/R*
4. Ohm's law
5. 12.5 mA
6. 1.2 k
7. 180
8. perpendicular
9. mirror strip; reflected image

Name _____ Date _____

TABLE 7–2. Part A: Measurements to Verify Ohm's Law

R	1 kΩ				Formula Relating, V, I, and R	Formula Test		
V, volts	2	4	6	8	When R = 1 kΩ, $\dfrac{V}{I} =$ $I = \dfrac{V}{}$	V, volts	5.5	9.0
I, mA						I measured mA		
V/I						I calculated mA		

TABLE 7–3. Part B

R	2kΩ				Formula Relating, V, I, and R	Formula Test		
V, volts	4	6	8	10	When R = 2 kΩ, $\dfrac{V}{I} =$ $I = \dfrac{V}{}$	V, volts	6	12
I, mA						I measured mA		
V/I						I calculated mA		

TABLE 7–4. Part C

R	3 kΩ				Formula Relating, V, I, and R	Formula Test		
V, volts	6	8	10	12	When R = 3 kΩ, $\dfrac{V}{I} =$ $I = \dfrac{V}{}$	V, volts	7	13.5
I, mA						I measured mA		
V/I						I calculated mA		

TABLE 7–5. Part D

R	4 kΩ				Formula Relating, V, I, and R	Formula Test		
V, volts	8	10	12	14	When R = 4 kΩ, $\dfrac{V}{I} =$ $I = \dfrac{V}{}$	V, volts	8	14.5
I, mA						I measured mA		
V/I						I calculated mA		

QUESTIONS

1. From your data in Table 7–2, 7–3, 7–4, and 7–5, what can you conclude about the relationships between the current I, voltage V, and resistance R of a circuit? Discuss these relationships in your own words.

_____

_____

_____

2. Represent the relationships discussed in Question 1 as mathematical formulas.

_____

_____

3. Referring to your data in Tables 7–2 through 7–5, discuss any experimental errors in your measurements.

_____

_____

_____

4. Explain in your own words how parallax introduces errors in reading analog meters.

_____

_____

_____

5. On a separate sheet of $8\frac{1}{2} \times 11$ graph paper, plot a graph of current I versus voltage V for the data in each of Tables 7–2 through 7–5. Use the horizontal (x) axis for voltage and the vertical (y) axis for current. Label each of the four graphs with the table number from which the data came.

6. Is there any similarity among the four graphs plotted in Question 5? If so, discuss the similarity. If not, discuss the differences in the graphs.

_____

_____

_____

7. Using the graphs plotted in Question 5, find the following directly on the graph and record the answers here:

   (a) With $R = 1$ kΩ and $I = 5$ mA, find V: _____ V.

   (b) With $R = 3$ kΩ and $V = 9$ V, find I: _____ A.

8. Using the graphs in Question 5, discuss how it is possible to find the current in a circuit with $R = 2$ kΩ and $V = 20$ V.

_____

_____

_____

SERIES CIRCUITS

BASIC INFORMATION

Finding R_T for Series-Connected Resistors · · · · · · · · ·

In electric circuits there may be one or more resistors connected in series, in parallel, or in a series-parallel arrangement. The technician must be able to analyze such circuits in order to be able to determine and predict the effect of a resistor or combination of resistors in controlling current.

With the knowledge you have already gained it is possible to study the effect of series-connected resistors R_1, R_2, R_3, etc., on current through experiments.

Figure 8–1 shows a voltage source V applied across a resistor R_1. You have already determined that if the values of V and R_1 are given, you can predict the current in this circuit by substituting the values of V and R_1 in Ohm's law:

$$I = \frac{V}{R} \tag{8–1}$$

Now, knowing the values of V, R_1, R_2, R_3, etc., is it possible to predict the current I in the circuit containing two resistors R_1 and R_2 connected in series [Figure 8–2(a)]? Three resistors R_1, R_2, and R_3 connected in series [Figure 8–2(b) p. 58]? Any number of resistors connected in series?

What are the facts on which an educated guess may be made? In Experiment 6 it was established that the current I is everywhere the same in a series circuit. This is self-evident, since there is one and only one path for current in a series circuit. Moreover, in Experiment 6, measurement showed that the current I in a circuit decreased as more resistance was added in series, if the applied voltage was kept constant. It would appear,

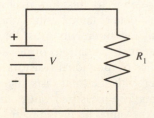

Figure 8–1. Voltage V applied across resistor R_1.

OBJECTIVES

1 To verify experimentally what the total resistance R_T is in a circuit in which the resistors R_1, R_2, R_3, etc., are connected in series

2 To develop a formula, based on experimental results, that gives the total resistance R_T of resistors connected in series

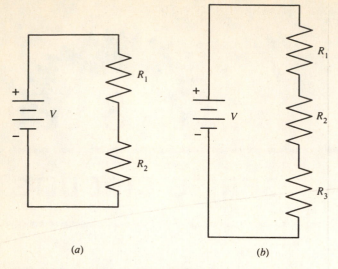

Figure 8–2. Voltage V applied across a series circuit. (a) Two resistors, R_1 and R_2, connected in series. (b) Three resistors, R_1, R_2, and R_3, connected in series.

then, that the effect of adding resistors in series is to increase the opposition to current in a circuit. Again, this conclusion now seems evident, for since there is only one path for current, each resistor must contribute to the control of current. It is the cumulative effect, the equivalent or total resistance R_T of the series combination, that is controlling current.

Resistance R_T for a specific circuit may be found by measuring the current I_T in that circuit and solving for R_T by substituting the measured values I_T and V in Ohm's law. Thus,

$$R_T = \frac{V}{I_T} \qquad\qquad (8\text{–}2)$$

Formula (8–2) suggests that a *single resistor* R_T may be substituted for two or more resistors connected in series, since R_T has the same effect in controlling current as the series-connected resistors.

An experimental procedure for finding a formula by which R_T may be determined is suggested by the preceding discussion. Take a group of resistors whose values are known or whose values may be determined by measurement with an ohmmeter. Connect two of these resistors in series. Apply a measured voltage V across the circuit and measure the current I_T. Calculate the equivalent resistance R_T of this combination, using Ohm's law. Suppose, for the sake of discussion, that $R_T = R_1 + R_2$ in this case. Now, connect two other resistors in the circuit and experimentally determine their equivalent resistance. Repeat this procedure for two more resistors. Study the results. Is there a pattern you can recognize? Can it be stated by a formula—for example, $R_T = R_A + R_B$? If it can, experimentally determine if the formula still holds for three resistors in series, four resistors in series, or any number of resistors in series.

Measuring Resistance · · · · · · · · · · · · · · · · ·

Because the value of resistance is constant whether or not it is in a circuit, another method can be used to find the relationship between R_T and R_1, R_2, R_3, etc.

A group of resistors can be connected in series and an ohmmeter used to measure the resistance across the two end leads. Because the values of the individual resistors are known (or can be easily measured), a formula can be developed directly from the measured resistance values.

From an experimental standpoint it would be worthwhile to use both methods, to verify the results of each method.

SUMMARY

1. The effect of two or more resistors connected in series in a closed circuit containing a voltage source V is to offer more resistance to current than can any one of these resistors acting alone in the circuit.

2. Two methods can be used to find experimentally the relationship between total resistance R_T and individual resistors connected in series.

3. *Method* 1. The effective or total resistance R_T of series-connected resistors R_1, R_2, etc, in a circuit containing a voltage source V may be determined experimentally by measuring the applied voltage V and the current I. Then R_T may be found by substituting the measured values of V and I in Ohm's law. Thus,

$$R_T = \frac{V}{I}$$

Two or more series-connected resistors may be replaced by a resistor R_T determined by this method, since the effect of R_T in controlling current is the same as the effect of the series-connected resistors.

4. *Method* 2. The value R_T of series-connected resistors may also be measured directly with an ohmmeter. However, in using an ohmmeter to measure total resistance in a series circuit, the power source must be disconnected from the resistors.

5. A general formula for determining the value R_T of series-connected resistors may be found from experimental data obtained using either method 1 or method 2. But measurements must be made involving many different combinations of resistors. In all cases the resistance of individual resistors must be measured, together with the total resistance of each series combination. The measured values must then be compared with the values predicted by the general formula to see if they are equal.

Check your understanding by answering the following questions:

1. The current in a circuit containing four series-connected resistors measures 0.025 A. If the voltage applied to the circuit is 10 V, the total resistance R_T of the four series-connected resistors is _____ Ω.

2. The total resistance R_T of series-connected resistors is _____ (greater, less) than the resistance of any one of these resistors taken by itself.

3. In a series-connected closed circuit there is just _____ path for current.

4. Current in a series circuit is _____ everywhere.

5. An unknown number of resistors of equal value are connected in series across a 10-V source. The current in the circuit measures 0.2 A. One of the resistors is removed and the circuit reconnected. This time the current measures 0.25 A. How many resistors were in the original circuit? _____

MATERIALS REQUIRED

Power Supply:
- Variable 0–15 V dc, regulated

Instruments:
- DMM or VOM
- 0–10 mA milliammeter

Resistors (5%, ½-W):
- 1 330-Ω
- 1 470-Ω
- 1 1.2 k-Ω
- 1 2.2 k-Ω
- 1 3.3 k-Ω
- 1 4.7 k-Ω

Miscellaneous:
- SPST switch

PROCEDURE

Part A: Finding R_T for Series-Connected Resistors by the Ohmmeter Method

A1. Obtain six resistors with the nominal (rated) values of 330, 470, 1.2 k, 2.2 k, 3.3 k, and 4.7 kΩ. Measure the actual resistance of each resistor and record the values in Table 8–1 (p. 61).

A2. Using the symbols R_1, R_2, R_3, R_4, R_5, and R_6 write a formula for finding the total resistance of each combination in Table 8–2 (p. 61). For example, if the measured value of R_1 is 300 Ω and of R_2 is 450 Ω and the measured resistance across the combination of R_1 and R_2 is 750 Ω, then we can conclude that

$$R_T = R_1 + R_2$$

A3. Using the measured values of resistance from Table 8–1, calculate the value of "Total Resistance" for each of the combinations in Figure 8–3(*a*) to (*g*) (p. 60). Record the results in Table 8–2 in the "Calculated Value" column.

A4. Connect the resistors in the seven combinations shown in Table 8–2 and Figure 8-3. Measure the resistance across AB for each combination and record the value in Table 8–2 in the "Measured Value" column.

Part B: Finding R_T for Series-Connected Resistors Using Ohm's Law

The six resistors in part A are used in this part in the seven combinations of Table 8–2.

B1. Connect the circuit of Figure 8–4 (p. 60). Power is **off** and switch S_1 is **open.** Connect the resistors in combination *a* across AB. Turn **on** power and adjust the voltage until the voltmeter reads 10 V.

B2. **Close** S_1. Current through the resistors will be indicated by the milliammeter. Check the voltmeter and adjust the power supply to maintain 10 V across the circuit. Record this voltage in Table 8–3 (p. 61) in the "Voltage Applied" column. Read the milliammeter. Record this value for combination *a* in the "Current Measured" column.

B3. **Open** S_1. Replace the resistors of combination *a* across AB with combination *b*. **Close** S_1. Check the voltmeter to see that 10 V is maintained across the circuit. Adjust the power supply if necessary. Record the voltage (10 V) and current for combination *b* in Table 8–3.

B4. For each of the other combinations *c* through *g*, follow the procedure of steps B2 and B3. Record *V* and *I* in Table 8–3. Remember to check the voltage to maintain 10 V across the circuit.

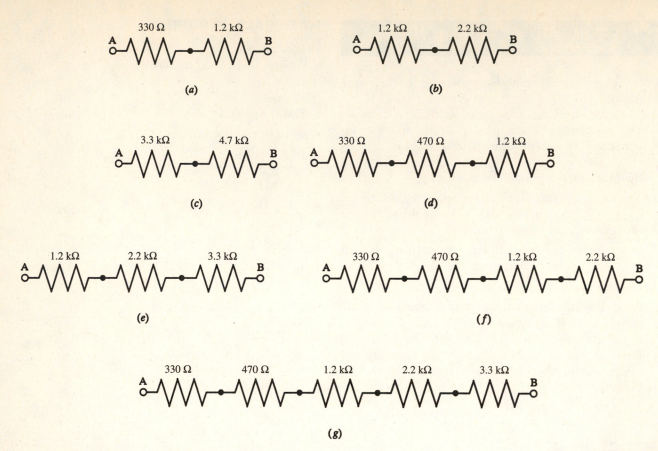

Figure 8–3. (a)–(g) Resistor combinations for procedure step A2.

B5. From the measured values of V and I in Table 8-3, calculate R_T using Ohm's law,

$$R_T = \frac{V}{I}$$

Record your answers in the "Ohm's Law" column.

B6. Transfer the calculated values of R_T from Table 8–2 to Table 8–3 in the column indicated for each of the combinations of resistors a to g.

B7. Using the symbols R_1, R_2, R_3, R_4, R_5, and R_6 write a formula that fits the results indicated in Table 8–3 for each combination a to g.

B8. Based on your results in parts A and B, write a general formula for any combination of series-connected resistors:

$$R_T =$$

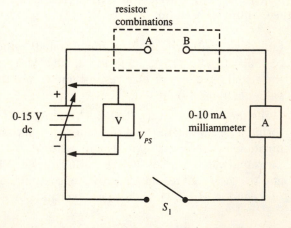

Figure 8–4. Circuit for procedure step B2 to be used with the resistor combinations of Figure 8-3(a) to (g).

ANSWERS TO SELF TEST

1. 400
2. greater
3. one
4. the same
5. 5

TABLE 8–1. Measured Value of Resistors in Part A

Resistor	R_1	R_2	R_3	R_4	R_5	R_6
Rated value, Ω	330	470	1.2 k	2.2 k	3.3 k	4.7 k
Measured value, Ω						

TABLE 8–2. Total Resistance of Series-Connected Resistors—Method 1

Combination	Resistor Rated Value, Ω						Total Resistance R_T, Ω		Formula for R_T
	R_1	R_2	R_3	R_4	R_5	R_6	Calculated Value	Measured Value	
a	330		1.2 k						
b			1.2 k	2.2 k					
c					3.3 k	4.7 k			
d	330	470	1.2 k						
e			1.2 k	2.2 k	3.3 k				
f	330	470	1.2 k	2.2 k					
g	330	470	1.2 k	2.2 k	3.3 k				

TABLE 8–3. Total Resistance of Series-Connected Resistors—Method 2

Combination	Voltage Applied (V), V	Current Measured (I), mA	Ohm's Law, $R_T = \dfrac{V}{I}$	Total Resistance, R_T, Ω Calculated Value from Table 8–2	Formula for R_T
a					
b					
c					
d					
e					
f					
g					

QUESTIONS

1. Explain in your own words the two methods you used in this experiment to find the total resistance of a series circuit.

_____

_____

_____

_____

2. Why was it necessary to measure the resistance of each resistor individually with an ohmmeter in order to satisfy the objectives of this experiment?

_____

_____

_____

_____

3. Examine your data in Tables 8–1 and 8–2. Do the results of the experiment suggest a general formula for finding the total resistance of a group of series-connected resistors? Explain the reason for your answer.

_____

_____

_____

_____

4. Compare the measured values and the calculated values of R_T in Table 8–2. If they are not exactly the same, how would you explain the differences?

_____

_____

_____

_____

5. Compare the value of the total resistance calculated using Ohm's law and the measured current in Table 8–3 with the calculated value of total resistance from Table 8–2. If they are not exactly the same, how would you explain the difference?

_____

_____

_____

_____

6. If the position of the resistors in the series R_1, R_2, R_3, R_4 were interchanged—for example, R_1, R_4, R_2, R_3, or R_3, R_1, R_2, R_4—what would be the effect on the total resistance of the entire group?

_____

_____

_____

DESIGNING SERIES CIRCUITS

BASIC INFORMATION

Designing a Series Circuit to Meet Specified Resistance Requirements

The law for total resistance of series-connected resistors can be applied to the solution of simple design problems. An example will indicate the techniques to be used.

Problem 1. A technician has a stock of the following resistors: four 56-Ω, five 100-Ω, three 120-Ω, two 180-Ω, two 220-Ω, and one each of 330-Ω, 470-Ω, 560-Ω, 680-Ω, and 820-Ω. A resistance value of 1 kΩ is needed for a circuit being designed. Find at least four combinations of resistors from those in stock, using the least possible number of components, that will satisfy the design requirement.

Solution. Experiment 8 developed the generalization that the total resistance R_T of series-connected resistors R_1, R_2, R_3, etc., is equal to the sum of their resistances. Stated as a formula,

$$R_T = R_1 + R_2 + R_3 + \cdots \qquad (9\text{--}1)$$

The technician can use Formula (9–1) to solve Problem 1. Thus,

$$1 \text{ k}\Omega = R_1 + R_2 + R_3 + \cdots$$

1. By inspection it is apparent that the 820-Ω resistor and the 180-Ω resistor connected in series will add to 1 kΩ. Hence this is one solution. It is also the solution that requires the least number of components, two.
2. Another solution is to connect the 680-Ω, 220-Ω, and 100-Ω resistors in series. Here three components are used.
3. Another solution is to connect a 560-Ω and two 220-Ω resistors in series. Again, three components are used.
4. A fourth solution is to connect a 470-Ω, a 330-Ω, and two 100-Ω resistors in series. Here four components are used.

There are other combinations that will add to 1 kΩ, so that the technician has a fairly wide choice. The restriction that the least number of components be used, however, does limit the choice.

OBJECTIVES

1 To design a series circuit that will meet specified resistance requirements

2 To design a series circuit that will meet specified voltage and current requirements

3 To design a series circuit that will meet specified current and resistance requirements

4 To construct and test the circuits to see that they meet the design requirements

Finally, the technician should connect the resistors and measure their total resistance with an ohmmeter to confirm the solution.

Designing a Series Circuit to Meet Specified Voltage and Current Requirements ·

Ohm's law and the law for total resistance of series-connected resistors can be be applied to the solution of this type of design requirement. Again, an example will illustrate the procedures to be used.

Problem 2. A technician has a 15-V battery and the same stock of resistors as in Problem 1. A circuit is to be designed in which current must be 10 mA. Show the circuit arrangement that can be used, including the values of all resistors.

Solution. Assume a closed series circuit is used. Two of the circuit conditions are known, voltage and current. Using Ohm's law, the total resistance in the circuit can be found. Thus,

$$R_T = \frac{V}{I} \qquad (9-2)$$

Substituting the known values of V and I in Formula (9–2), we get

$$R_T = \frac{15\ V}{10\ mA} = 1.5\ k\Omega$$

This is the circuit resistance that will hold circuit current at 10 mA. It is now necessary to find a combination of resistances from those in stock that will add to 1.5 kΩ. This is merely a process of trying different combinations to see which will give the required result. In this case the 820-Ω and 680-Ω resistors satisfy the conditions, since 820 + 680 = 1.5 kΩ.

As a final step, the technician should connect the circuit of Figure 9–1 and verify by measurement that there is indeed 10 mA of current in the circuit.

The procedure for solving this type of problem is as follows:

1. Solve for R_T by substituting the known values of V and I in the formula

$$R_T = \frac{V}{I}$$

2. Find the combination of resistances whose sum will add to the given value of R_T using the formula

$$R_T = R_1 + R_2 + R_3 + \cdots$$

3. Connect the circuit using the combination of resistors determined in step 2 and verify by measurement that the circuit conditions have been met.

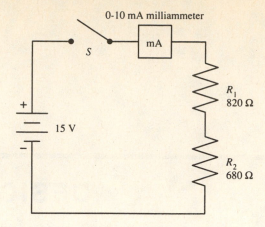

Figure 9–1. Circuit used to limit current to 10 mA.

Designing a Series Circuit to Meet Specified Current and Resistance Requirements ·

As in the preceding problem, Ohm's law and the formula for finding R_T of series-connected resistors are applied here. Again, a problem will illustrate the process.

Problem 3. Resistors R_1, R_2, and R_3 in Figure 9–2 are components of an electronic device that requires 50 mA to operate properly. What voltage should the technician connect to this series circuit to provide the necessary current?

Solution.

1. Find the total resistance R_T.

$$R_T = 33 + 47 + 56 = 136\ \Omega$$

2. Substitute the values $I = 50$ mA and $R_T = 136$ Ω in formula (9–3), which was derived from Ohm's law:

$$V = I \times R$$

$$V = (5\ mA)(136\ \Omega) = 6.8\ V \qquad (9-3)$$

3. Connect the circuit of Figure 9–2 and set the power-supply voltage V to 6.8 V. The milliammeter should read 50 mA.

SUMMARY

1. If it is required to make up a resistance value R_T from a group of series-connected resistors whose values are R_1, R_2, R_3, etc., the combination of resistors should satisfy the formula

$$R_T = R_1 + R_2 + R_3 + \cdots$$

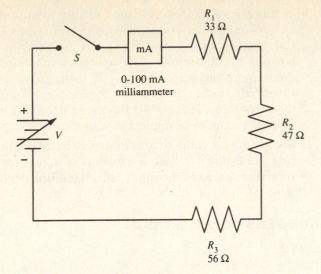

Figure 9–2. Adjusting the voltage V so that 50 mA flows in the circuit.

2. If it is required to design a series circuit that will meet specified voltage (V) and current (I) requirements, find the value R_T that will satisfy the given voltage and current by substituting V and I in the formula

$$R_T = \frac{V}{I}$$

Then select the resistors R_1, R_2, etc., so that

$$R_T = R_1 + R_2 + R_3 + \cdots$$

3. If it is required to design a series circuit that will meet specified current (I) and resistance (R_T) values, first select those resistors whose resistance sum is R_T.

$$R_T = R_1 + R_2 + R_3 + \cdots$$

Then solve for the unknown voltage V by substituting the given values of I and R_T in Ohm's law,

$$V = I \times R_T$$

4. After the circuit has been designed, connect it and measure the unknown quantity to see that it is in fact the required design value.

PROCEDURE

The six resistors used in this experiment will be identified as follows:

$R_1 = 330 \ \Omega$
$R_2 = 470 \ \Omega$
$R_3 = 1.2 \ \text{k}\Omega$

SELF TEST

Check your understanding by answering the following questions:

1. The formula that gives the total resistance of series-connected resistors is $R_T =$ _____.

2. The Ohm's law formula that gives the relationship between the applied voltage V, the current I, and the resistance R of a closed circuit is $V =$ _____.

3. To design a circuit powered by a battery of V volts that draws I amperes, it is necessary to find _____. This is done by substituting V and I in the formula _____.

4. To design a circuit that draws I amperes through a resistance of R ohms, it is necessary to find _____. This is done by substituting I and R in the formula _____.

5. The final step in the design of a circuit, after the values of V, R, and I have been determined, is to _____ the circuit and _____ the quantities involved.

MATERIALS REQUIRED

Power Supply:
- Variable 0–15 V dc, regulated

Instruments:
- DMM or VOM
- 0–10 mA milliammeter

Resistors (5%, ½-W):
- 1 330-Ω
- 1 470-Ω
- 1 1.2-kΩ
- 1 2.2-kΩ
- 1 3.3-kΩ
- 1 4.7-kΩ

Miscellaneous:
- SPST switch

$R_4 = 2.2 \ \text{k}\Omega$
$R_5 = 3.3 \ \text{k}\Omega$
$R_6 = 4.7 \ \text{k}\Omega$

1. Refer to Table 9–1 (p. 67). In the first row, $R_T = 2 \ \text{k}\Omega$. Select three resistors from R_1 through R_6 that, when

connected in series, total 2 kΩ. Record the rated values in the column for each resistor. For example, if you were asked to select two resistors that, when connected in series, total 1.67 kΩ, you would pick $R_2 = 470\ \Omega$ and $R_3 = 1.2\ \text{k}\Omega$. You would then write 470 in the R_2 column and 1.2 k in the R_3 column.

2. Connect the three resistors chosen in step 1 in series and measure the resistance of the combination. Record this value in row 1 in the "R_T Measured" column.

3. Choose as many resistors from the group of six as needed that will have a total resistance of 5.3 kΩ when connected in series. Record the rated values of the resistors in Table 9–1.

4. Connect the resistors in step 3 in series. Measure their total resistance and record the value in the 5.3 kΩ row in the "R_T Measured" column.

5. Repeat steps 3 and 4 for the remaining total resistances of 7.5 kΩ, 10 kΩ, and 11 kΩ. Record all values in Table 9–1. At the end of this step the "R_T Measured" column should be completely filled.

6. Design a series circuit that will produce a current of 5 mA when supplied by 10 V. The resistors chosen for your design must come from the group of resistors R_1 through R_6. Record the values chosen in Table 9–2 in the 10-V row.

7. With power **off** and switch S_1 **open,** connect the circuit of Figure 9–3 using the resistor combination found in step 6. Use the 0–10 mA milliammeter. After checking the circuit, turn **on** the power and **close** S_1.

8. Adjust the power supply until the voltmeter reads 10 V. Read the milliammeter and record the value in the "Circuit Current, Measured" column of the 10-V row.

9. Repeat steps 6 through 8 for each of the remaining combinations of V and I in Table 9–2. Record all resistor combinations and milliammeter readings in the table.

10. Design a circuit that will draw 4 mA. The only conditions are that the resistors used in each case must be from among the six R_1 to R_6 used in other parts

of this experiment. The voltage can vary from 0 to 15 V. Combination 1 must consist of two resistors. Combination 2 must consist of three resistors. Combination 3 must consist of four resistors. In choosing the resistors for your circuit, use the actual, or measured, value rather than the rated value. Record the measured values in Table 9–3. Also, record the design value of voltage to be applied to your circuit.

11. Construct each of the circuits in step 10 based on the circuit of Figure 9–3. Record the values of voltage and current measured by the voltmeter and milliammeter.

ANSWERS TO SELF TEST

1. $R_1 + R_2 + R_3 + \cdots$

2. $I \times R$

3. resistance; $R = V/I$

4. voltage V across R; $V = I \times R$

5. connect; measure

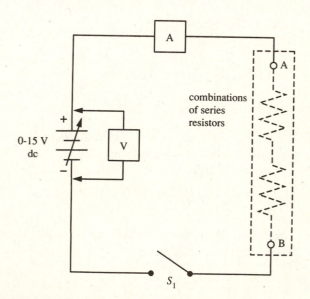

Figure 9–3. Circuit for procedure step 7.

TABLE 9–1. Measured versus Rated Values of Series-Connected Resistors

R_T Required, Ω	Rated Value of Resistors Whose Sum Will Satisfy R_T						R_T Measured, Ω
	R_1	R_2	R_3	R_4	R_5	R_6	
2 k							
5.3 k							
7.5 k							
10 k							
11 k							

TABLE 9–2. Circuit Design for Specified Values of *V* and *I*

V Applied, V	Circuit Current I, mA		Rated Value of the Design Resistor, Ω					
	Required	Measured	R_1	R_2	R_3	R_4	R_5	R_6
10	5 mA							
12	4 mA							
5.5	1 mA							
8	10 mA							
11.4	1 mA							

TABLE 9–3. Circuit Designed to Draw 4 mA

Combination	Measured Value of the Design Resistor, Ω						V Applied, Design Value, V	I measured, mA
	R_1	R_2	R_3	R_4	R_5	R_6		
1 (2 resistors)								
2 (3 resistors)								
3 (4 resistors)*								

QUESTIONS

1. Refer to your data in Table 9–1. Compare the R_T required value with the R_T measured value for each of the five resistances. Are they equal? If not, explain why. In each case, is the difference, if any, consistent with the tolerance of the individual resistors?

_____

_____

_____

2. Refer to your data in Table 9–2. Compare the required current with the measured current for each of the required current values. Are they equal? If not, explain why.

_____

_____

_____

_____

_____

_____

_____

3. Refer to your data in Table 9–3. Compare the measured current with the design value of 4 mA for each of the three resistor combinations. Are they equal? If not, explain why the measured value is not equal to 4 mA.

_____

_____

_____

_____

_____

_____

_____

_____

4. Three ½-W 5% resistors are connected in series. Their color-coded values are 1 kΩ, 5 kΩ, and 10 kΩ, respectively. What would be the possible range of resistance readings of an ohmmeter, assuming the meter has a 0% error? Show all calculations.

_____

_____

_____

_____

_____

_____

_____

_____

_____

VOLTAGE-DIVIDER CIRCUITS (UNLOADED)

BASIC INFORMATION

Series-Connected Voltage-Divider Circuits

Ohm's law finds immediate application in working with voltage-divider circuits. Resistive voltage dividers can be very simple circuits or complex arrangements of resistors servicing one or more loads. In this experiment we will be concerned with unloaded dividers, that is, with *circuits that are not required to deliver current to an external load.*

The simplest dc voltage divider consists of two resistors R_1 and R_2 connected in series, across which a dc voltage V is applied (Figure 10–1). Assume that V is 12 V and that the resistors R_1 and R_2 and 7.5 kΩ and 2.5 kΩ, respectively. The voltages V_1 across R_1 and V_2 across R_2, measured with a voltmeter, are 9 V and 3 V, respectively. The 12-V source has thus been divided by the circuit of Figure 10–1 to produce two lower voltages.

Figure 10–1 can be modified by the addition of one or more resistors to produce any number of lower voltages, measured across individual resistors or measured with respect to some common point such as C. The choice of resistors to produce specific voltages can be made by trial and error or by first analyzing the circuit. The trial-and-error method is too time consuming and inefficient. Carefully studying the circuit and calculating the values of the resistors that will produce a required result is both fast and effective.

The basic formulas of electricity are used in analyzing the problem and obtaining a solution. For example, in Figure 10–1, the current I may be found by substituting the values of V and R_T in formula (10–1).

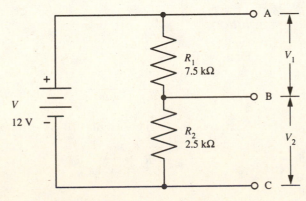

Figure 10–1. A dc voltage divider.

69

$$I = \frac{V}{R_T} \qquad \textbf{(10--1)}$$

where V is the applied voltage and

$$R_T = R_1 + R_2$$

Since $V = 12$ V and $R_1 + R_2 = 10$ kΩ,

$$I = \frac{12\ V}{10\ k} = 1.2\ mA$$

Now $\qquad V_1 = I \times R_1 \qquad \textbf{(10--2)}$

and $\qquad V_2 = I \times R_2$

Therefore, $\qquad V_1 = (1.2\ mA)(7.5\ k\ \Omega) = 9\ V$

and $\qquad V_2 = (1.2\ mA)(2.5\ k\Omega) = 3\ V$

A formula can be found to simplify our work. Consider Figure 10–2. It is required to find V_1, V_2, V_3, and V_4. Assume that I is the current in this circuit. Then

$$V = I \times R_T \qquad \textbf{(10--3)}$$

where $\qquad R_T = R_1 + R_2 + R_3 + R_4$

Since $\qquad V_1 = I \times R_1$

$$V_2 = I \times R_2 \qquad \textbf{(10--4)}$$

$$V_3 = I \times R_3$$

$$V_4 = I \times R_4$$

we can find the ratio of V_1, V_2, V_3, and V_4, to V. Thus

$$\frac{V_1}{V} = \frac{I \times R_1}{I \times R_T} = \frac{R_1}{R_T} \qquad \textbf{(10--5)}$$

and $\qquad V_1 = V \times \frac{R_1}{R_T} \qquad \textbf{(10--6)}$

Similarly, $\qquad V_2 = V \times \frac{R_2}{R_T}$

$$V_3 = V \times \frac{R_3}{R_T}$$

$$V_4 = V \times \frac{R_4}{R_T}$$

We have thus derived formula (10–6) for finding the voltage across any resistor in a series circuit. Stated in words, the voltage across any given resistor in a series circuit is equal to the ratio of the resistance of the given resistor to the total resistance of the series circuit multiplied by the total applied voltage. This formula applies to a

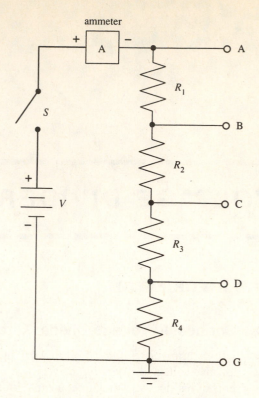

Figure 10–2. A dc voltage divider with four resistors.

series circuit containing any number of resistors. This formula is sometimes referred to as the Voltage Divider Rule.

As an example, let us apply the formula to Figure 10–1.

$$V_1 = V \times \frac{R_1}{R_T} = 12\ V \times \frac{7.5\ k}{10\ k} = 9\ V$$

Similarly, $\qquad V_2 = 12\ V \times \frac{2.5\ k}{10\ k} = 3\ V$

Here is another example to illustrate how the formula can be used to design a voltage divider.

Problem. Using a 25-V source, we wish to find the values of resistors R_1 through R_4 connected in a simple series circuit to provide 2.5, 5.0, 7.5, and 10.0 V across R_1, R_2, R_3, and R_4, respectively. Assume that the current I in this circuit must be limited to 1 mA.

Solution.

1. First, we find R_T:

$$R_T = \frac{V}{I} = \frac{25\ V}{1\ mA} = 25\ k\Omega$$

2. Next, we can rewrite formula (10–6) as follows:

$$R_1 = \frac{V_1}{V} \times R_T$$

$$R_2 = \frac{V_2}{V} \times R_T \qquad \textbf{(10--7)}$$

$$R_3 = \frac{V_3}{V} \times R_T$$

$$R_4 = \frac{V_4}{V} \times R_T$$

3. Substituting $V = 25$ V, $R_T = 25$ kΩ, $V_1 = 2.5$ V, $V_2 = 5.0$ V, $V_3 = 7.5$ V, and $V_4 = 10.0$ V in (10–7) gives

$$R_1 = \frac{2.5 \text{ V}}{25 \text{ V}} \times 25 \text{ k} = 2.5 \text{ k}\Omega$$

$$R_2 = \frac{5.0 \text{ V}}{25 \text{ V}} \times 25 \text{ k} = 5 \text{ k}\Omega$$

$$R_3 = \frac{7.5 \text{ V}}{25 \text{ V}} \times 25 \text{ k} = 7.5 \text{ k}\Omega$$

$$R_4 = \frac{10.0 \text{ V}}{25 \text{ V}} \times 25 \text{ k} = 10 \text{ k}\Omega$$

These are the required values of resistance. The circuit of Figure 10–2 can be connected using these values of R_1, R_2, R_3, and R_4, and V, and the required voltages can be verified by measurement with a voltmeter.

In analyzing the voltage-divider circuits of Figures 10–1 and 10–2, we have considered the voltages developed across individual resistors of the divider. Another view of the divider is relative to a common point. In Figure 10–1, point C is the common, or ground, return of the circuit. In Figure 10–2, point G is ground. Now consider Figure 10–2. What is the voltage at point A relative to ground? B to ground? C to ground? D to ground?

These voltages can be found by a modification of formula (10–6). First, the voltage from A to G is obviously the applied voltage V. Now, the voltage V_{BG} from B to ground is

$$V_{BG} = V \times \frac{R_2 + R_3 + R_4}{R_T}$$

From C to ground:

$$V_{CG} = V \times \frac{R_3 + R_4}{R_T} \qquad \textbf{(10–8)}$$

From D to ground:

$$V_{DG} = V \times \frac{R_4}{R_T}$$

Another method of calculating V_{BG} and V_{CG} in Figure 10–2 is to solve for the voltages V_1, V_2, V_3, and V_4 by formula (10–6). Then

$$V_{BG} = V_2 + V_3 + V_4$$

$$V_{CG} = V_3 + V_4 \qquad \textbf{(10–9)}$$

and $\qquad V_{DG} = V_4$

Variable Voltage-Divider Circuits (Unloaded)

Suppose we want to set up a divider, as in Figure 10–1, with an applied voltage of 10 V and a total resistance of 10k Ω, whose divider ratio is such that $V_1 = 6.9$ V and $V_2 = 3.1$ V. Solution of this circuit by the method just described yields the results

$$R_1 = 6.9 \text{ k}\Omega$$

$$R_2 = 3.1 \text{ k}\Omega$$

It would normally be expensive to get resistors of these exact values. To overcome this difficulty, we use a potentiometer. A potentiometer is a three-terminal variable resistor. The resistance between the two outer terminals is fixed at the rated value of the potentiometer. The center terminal, or arm, is connected to a slider that makes contact with the resistive material of the potentiometer. The arm can be rotated manually to select different values of resistance between the center terminal and either end terminal. Thus, if R_1 and R_2 of Figure 10–1 are replaced by a 10 kΩ potentiometer, the corresponding circuit is that of Figure 10–3.

As arm B moves toward A, resistance R_1 decreases, while the resistance R_2 increases. As arm B moves toward C, R_1 increases and R_2 decreases. When B is at A, $R_1 = 0$ and $R_2 = 10$ kΩ; when B is at C, $R_2 = 0$ and $R_1 = 10$ kΩ.

Thus, by manually adjusting the position of the slider, we can set the ratio R_1/R_2 and thus have a means for setting the voltage V_1 at any value between zero and the total voltage V across the potentiometer. In this process we have not changed the total resistance of the potentiometer (resistance from A to C).

In practice, if a potentiometer is used to obtain a voltage, a voltmeter is connected across the arm and one of the end terminals. The potentiometer is varied until the desired voltage is measured.

It is possible to limit the range of voltage variation by placing a potentiometer in series with one or more fixed resistors. Thus, in Figure 10–4 (p. 72) the voltage variation (range) from B to C is from 5 to 15 V.

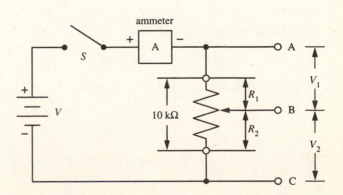

Figure 10–3. Potentiometer as a variable voltage divider.

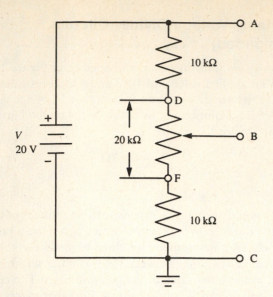

Figure 10–4. Limiting the range of variation of a voltage divider.

It should be noted that these results are true only if the voltage divider is unloaded—that is, if no current is drawn by any external circuit.

Variable voltage dividers are used as volume controls in radios, focus controls in television receivers, speed controls in electronic motor-control circuits, voltage regulator controls, and similar applications.

SUMMARY

1. The voltage across each resistor in a series-connected resistive voltage divider can be found by the formula

$$V_1 = V \times \frac{R_1}{R_T}$$

where V_1 is the voltage across R_1, V is the total voltage applied to the circuit, R_1 is a resistor in a series-connected circuit, and R_T is the total resistance of the circuit.

2. A somewhat longer method to determine the voltage across any resistor in a series-connected divider is as follows:

First calculate the total resistance,

$$R_T = R_1 + R_2 + R_3 + R_4$$

Next, solve for the current I in the circuit:

$$I = \frac{V}{R_T}$$

Knowing I, find the voltage drop across R_1 using Ohm's law.

$$V_1 = I \times R_1$$

3. If it is required to find the voltage to common (or ground) or to any reference point from any point in a series-connected voltage divider, the methods in 1 or 2 can be used.

4. Variable voltage-divider circuits can be formed by using a potentiometer across a voltage source.

5. The range of voltage variation of a voltage divider can be limited by connecting a potentiometer in series with voltage-dropping resistors.

6. The voltage relationships found by using the given formulas are for unloaded voltage dividers.

SELF TEST

Check your understanding by answering the following questions:

1. In Figure 10–1, if the positions of resistors R_1 and R_2 were reversed in the circuit, the voltage across the 7.5-kΩ resistor would be _____ V.

2. In Figure 10–2, assume $R_T = 15$ kΩ and $R_3 = 3$ kΩ. If the applied voltage V is 22.5 V, the voltage across R_3 is _____ V.

3. In Figure 10–2, $R_T = 15$ kΩ, $R_1 = 3.5$ kΩ, and $V = 30$ V. The voltage across B and G, V_{BG}, is _____ V.

4. In Figure 10–2, $R_T = 10$ kΩ, $V_{BC} = 2$ V, and $V = 8$ V. The value of R_2 is _____ Ω.

5. In the variable voltage divider (Figure 10–3), assume $V = 35$ V. The range of variation of V_{BC} is from _____ V (maximum) to _____ V (minimum).

6. In the variable voltage divider (Figure 10–4), the battery voltage V is 6 V. The values of the resistors are those shown in the figure. The range of V_{BC} is from _____ V (maximum) to _____ V (minimum).

7. In Figure 10–2, $R_1 = 1$ kΩ, $R_2 = 2.2$ kΩ, $R_3 = 680$ Ω, $R_4 = 220$ Ω, and $V = 16$ V.

$V_1 = $ _____ V, $\qquad$ $V_2 = $ _____ V,
$V_3 = $ _____ V, and $V_4 = $ _____ V.

8. For the same conditions as in question 7, $V_{CG} = $ _____ V, $V_{BG} = $ _____ V, and $V_{BD} = $ _____ V.

Resistors (5%, ½-W):
- 1-820-Ω
- 1-1 kΩ
- 1-2.2 kΩ
- 1-3.3 kΩ
- 1-10 kΩ, 2-W potentiometer

Miscellaneous:
- SPST switch

PROCEDURE

A. Fixed Voltage-Divider Measurements

A1. With power **off** and switch S_1 **open,** connect the circuit of Figure 10–5. The values for R_1 through R_4 are the rated values of the resistors.

A2. Connect the voltmeter across the power supply and adjust the power supply until the voltmeter reads 15 V. Maintain this voltage throughout steps A3 and A4.

A3. **Close** S_1. Measure the supply voltage and record the value in Table 10–1 (p. 75) The voltmeter should read 15 V; adjust the power supply if it does not read 15 V.) Connect the voltmeter across AB to read the volt-

age across R_1; this is voltage V_1. Similarly, connect the voltmeter across BC to read V_2, the voltage across R_2; across CD to read V_3, the voltage across R_3; and across DE to read V_4, the voltage across R_4. Record all measured values in Table 10–1.

A4. Connect the voltmeter across BE to measure voltage V_{BE}, the voltage across the series combination of R_2, R_3, and R_4. Similarly, connect the voltmeter across CE to measure V_{CE}; and across DE to measure V_{DE}. Record all measured values in Table 10–1. **Open** S_1.

A5. Using the rated values of the resistors and the supply voltage of 15 V in Figure 10–5, calculate the current I delivered by the power supply and V_1, V_2, V_3, V_4, V_{BE}, V_{CE}, and V_{DE}. Use the formulas discussed in the Basic Information section. Record your answers in Table 10–1.

A6. With the circuit still connected as in Figure 10–5, **close** S_1. Adjust the power supply so that the milliammeter reads 1.5 mA. Measure and record V_1, V_2, V_3, V_4, V_{BE}, V_{CE}, and V_{DE} in Table 10–1. **Open** S_1.

A7. Using the rated value of the resistors and the supply current of 1.5 mA in Figure 10–5, calculate the power supply voltage V_{PS} and V_1, V_2, V_3, V_4, V_{BE}, V_{CE}, and V_{DE}. Use the formulas discussed in the Basic Information section. Record your answers in Table 10–1.

B. Variable Voltage-Divider Measurements

B1. With power **off** and switch S_1 **open,** connect the circuit of Figure 10–6 (p. 74). Adjust the power supply voltage to 15 V and maintain this value throughout this experiment.

B2. **Close** S_1. Rotate the potentiometer shaft to the maximum clockwise (CW) direction so that the wiper arm is at A. Measure V_{AB}, V_{BC}, and I and record the values in Table 10–2 (p. 75).

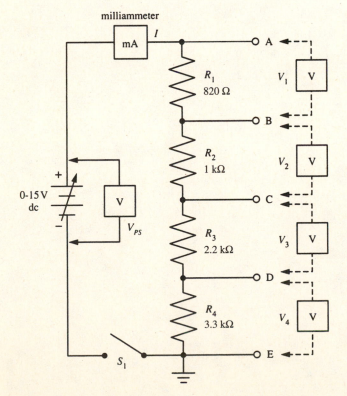

Figure 10–5. Fixed voltage-divider circuit for procedure step A1.

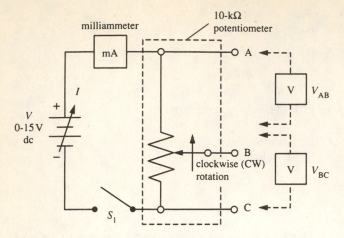

milliammeter

10-kΩ potentiometer

V 0-15 V dc

V_{AB}

V_{BC}

clockwise (CW) rotation

S_1

Figure 10–6. Variable voltage-divider circuit for procedure step B1.

B3. Rotate the shaft of the potentiometer until the arm is at the midpoint between A and C. Measure V_{AB}, V_{BC}, and I and record the values in Table 10–2.

B4. Rotate the shaft to the maximum counterclockwise (CCW) direction so that the arm is at C. Measure and record V_{AB}, V_{BC}, and I in Table 10–2.

B5. With the power-supply voltage V maintained at 15 V, adjust the potentiometer until the voltage across BC (V_{BC}) is 9 V. Measure and record V, V_{AB}, V_{BC}, in Table 10–3. Do not change the position of the potentiometer arm.

B6. **Open** S_1. Use an ohmmeter to measure the resistance across AB (R_{AB}), BC (R_{BC}), and AC (R_{AC}). Record these values in Table 10–3.

B7. Using $V = 15$ V and the total potentiometer resistance of 10 kΩ, calculate the values of R_{AB} and R_{BC} necessary to produce $V_{BC} = 9$ V.

C. Designing Voltage-Divider Circuits

Before performing this step, read the Optional Activity given below.

Design a voltage-divider circuit that will deliver a variable voltage from 0 to 11.5 V from a constant 15-V power supply. Select only the resistors and potentiometer in the materials list of this experiment. Draw a diagram of the circuit showing all component values. After approval by the instructor, wire the circuit and measure the voltages and current in the circuit. Tabulate your results.

HINT: See Figure 10–4 for method of using fixed and variable resistors in a voltage-divider circuit.

Optional Activity ·

This activity requires the use of electronic simulation software. Before constructing and testing the circuit in Part C1, simulate the circuit using the software. Show the meters necessary to test your design. Record the voltages and current obtained. Compare the results obtained using simulation with those from the circuit you constructed. Explain any discrepancies between the two results.

ANSWERS TO SELF TEST

1. 9
2. 4.5
3. 23
4. 2500
5. 35; 0
6. 4.5; 1.5
7. 3.9; 8.6; 2.7; 0.86
8. 3.5; 12.1; 11.2

EXPERIMENT

10

TABLE 10–1.   Fixed Voltage-Divider Measurements: Part A

Step		V	I (mA)	V_1	V_2	V_3	V_4	V_{BE}	V_{CE}	V_{DE}
A3, A4	Measured	15								
A5	Calculated	✕								
A6	Measured									
A7	Calculated	✕								

TABLE 10–2.   Variable Voltage-Divider Measurements: Part B

Step	Position of Arm	Measured Values				Calculated Values
		V	I (mA)	V_{AB}	V_{BC}	$V_{BC} + V_{AB}$
B2	Max CW (at A)	15				
B3	Midpoint	15				
B4	Max CCW (at C)	15				

TABLE 10–3.   Variable Voltage-Divider Values

V	I (ma)	V_{BC}	V_{AB}	R_{BC}	R_{AB}	R_{AC}	R_{BC}	R_{AB}
		Measured Values					Calculated Values	
15		9						

QUESTIONS

1. Refer to Table 10–1. Compare the measured values of V_1, V_2, V_3, and V_4 (step A3) with their respective calculated values (step A5). If any of the related values are not equal, explain any differences.

_____

_____

_____

_____

2. Refer to Table 10–1. Compare the measured values of V_1, V_2, V_3, V_4, V_{BE}, V_{CE}, and V_{DE} (step A6) with their respective calculated values (step A7). If any of the related values are not equal, explain any differences.

_____

_____

_____

_____

3. Use your data from Table 10–3.
   (a) Calculate the ratios V_{BC}/V_{AB} and R_{BC}/R_{AB}.
   (b) Are the ratios from (a) equal? Should they be? Why?
   (c) How are the measured values of R_{AB}, R_{BC}, and R_{AC} related?
   (d) Explain the effect on the current I (measured) as the potentiometer arm is varied.

_____

_____

_____

_____

_____

_____

_____

_____

_____

_____

_____

_____

_____

_____

4. Explain in your own words whether or not formula (10–9) was confirmed by this experiment. Refer to specific data in Tables 10–1, 10–2, and 10–3.

_____

_____

_____

_____

_____

_____

_____

_____

5. From the data in Table 10–2, what is true about the measured values of V_{AB} and V_{BC}, regardless of the position of the potentiometer's arm?

_____

_____

_____

_____

_____

_____

_____

_____

EXPERIMENT 11

CURRENT IN A PARALLEL CIRCUIT

BASIC INFORMATION

Branch Currents

In considering the series circuit in Experiment 6, it was found that a closed circuit is required for current, that current stops when the circuit is open, and that current in a series circuit is the same everywhere. How does a parallel circuit differ from a series circuit?

Figure 11–1 shows three resistors connected in parallel and a voltage V applied across the combination. If the line connecting the battery to the parallel network is broken at X or at Y and an ammeter is inserted in the circuit at X or Y (Figure 11–2, p.78), the ammeter will measure the total current delivered by the voltage source. This line current is drawn by the three resistors from the source.

A simple experiment suggests an important characteristic of a parallel circuit. If, in Figure 11–2, resistor R_1 is removed from the circuit, the line current measured by the ammeter decreases. If R_2 is then removed from the circuit, the line current decreases further. What remains is a simple series circuit consisting of V, R_3, and the ammeter. The line current is now the current drawn by R_3 from V; it may be computed directly by Ohm's law.

The results of this experiment indicate that in Figures 11–1 and 11–2 there are actually three conductive paths for current—namely, R_1, R_2, and R_3. When all three paths are closed, there is maximum line current. When path R_1 is broken, there is less line current because only two paths remain, R_2 and R_3, and as was shown before, when both R_1 and R_2 are open, only path R_3 remains. The individual paths are called *branches*, or *legs*, of the parallel circuit.

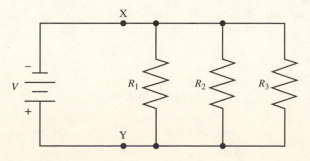

Figure 11–1. Voltage V applied across three resistors connected in parallel.

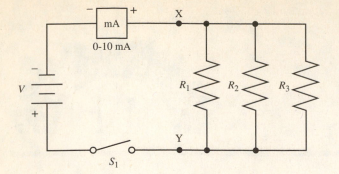

Figure 11–2. Measuring total current in a parallel circuit.

One characteristic of a parallel circuit containing only resistors is that the total current I_T in the circuit is greater than the current in any branch. It follows that each branch current in a parallel circuit containing only resistors is less than the total or line current I_T.

An example will illustrate this characteristic. Assume in Figure 11–3 that the voltage source V is 6.6 V. Note that 6.6 V appears across each branch resistor in the circuit— that is, the voltage across R_1 is 6.6 V, that across R_2 is 6.6 V, and that across R_3 is 6.6 V also. The individual currents in each branch, I_1 in R_1, I_2 in R_2, and I_3 in R_3, can be calculated by Ohm's law. Thus,

$$I_1 = \frac{V}{R_1} = \frac{6.6 \text{ V}}{2 \text{ k}\Omega} = 0.0033 \text{ A} = 3.3 \text{ mA}$$

$$I_2 = \frac{V}{R_2} = \frac{6.6 \text{ V}}{3 \text{ k}\Omega} = 0.0022 \text{ A} = 2.2 \text{ mA}$$

$$I_3 = \frac{V}{R_3} = \frac{6.6 \text{ V}}{10 \text{ k}\Omega} = \xi°{+}0.00066 \text{ A} = 0.66 \text{ mA}$$

In a parallel circuit, the total current is greater than the current in any branch. Therefore the total current I_T in this circuit must be greater than 3.3 mA.

Total Current in a Parallel Circuit · · · · · · ·

The only source of current for R_1, R_2, and R_3 in the parallel circuit of Figure 11–3 is the voltage source V.

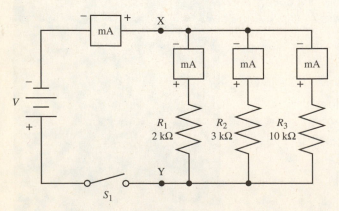

Figure 11–3. Circuit used to find the relationship between branch currents and total current in a parallel circuit.

Therefore, each of the currents I_1, I_2, and I_3 combine in the line to form the line, or total, current I_T.

Does the process of combining mean adding? This question can be answered by performing an experiment on the parallel circuit. By placing ammeters in each of the branches and one in the line, we can measure I_1, I_2, I_3, and I_T. If the experiment is performed carefully, the relationship between branch currents and line or total current will be found to be

$$I_T = I_1 + I_2 + I_3$$

which states that *the total current in a parallel circuit is the sum of the branch currents.*

The total current in the preceding example is

$$I_T = 3.3 \text{ mA} + 2.2 \text{ mA} + 0.66 \text{ mA} = 6.16 \text{ mA}$$

The total current, 6.16 mA, is also seen to be greater than any of the branch currents, 3.3 mA, 2.2 mA, or 0.66 mA.

SUMMARY

1. In a parallel circuit containing only resistors, the individual current in each branch is less than the total line current.

2. The total line current is greater than each individual branch current.

3. The total line current is equal to the sum of all the branch currents.

4. The voltage across each branch is the same.

SELF TEST

Check your understanding by answering the following questions:

1. The voltage across each branch of a parallel network must be _____.

2. In Figure 11–3, the voltage V_1 across R_1, the voltage V_2 across R_2, and the voltage V_3 across R_3 must all be _____ V.

3. In Figure 11–3, if $V = 10$ V, find the branch currents I_1, I_2, and I_3. $I_3 = $ _____ mA, $I_2 = $ _____ mA, and $I_3 = $ _____ mA.

4. The total current in the circuit of question 3 must be greater than the _____ branch current.

5. In Figure 11–3, for the branch currents listed in question 3, $I_T = $ _____ mA.

MATERIALS REQUIRED

Power Supply:
- Variable 0–15 V dc, regulated

Instruments:
- DMM or VOM (with at least 100 mA dc range)

Resistors (5% ½-W):
- 1-820-Ω
- 1-1 kΩ
- 1-2.2 kΩ
- 1-3.3 kΩ
- 1-4.7 kΩ

Miscellaneous:
- SPST switch

PROCEDURE

1. Measure the resistance of each of the five resistors supplied and record values in Table 11–1 (p. 81).

2. With the power supply **off** and switch S_1 **open,** connect the circuit of Figure 11–4(a) (p. 80). If you are using a single DMM to measure all quantities in this experiment, do not connect the ammeter to measure I_T at this point.

3. With power **on, close** S_1 and adjust the power supply V_{PS} to 10 V. Branches 1, 2, and 3 must be connected across the power supply and drawing current when voltage V_{PS} is adjusted. **Open** S_1.

4. Connect the ammeter in the circuit to measure I_T as shown. Do not change the adjustment on the power supply.

5. In the next steps it will be necessary to measure the current in each of the branches as well as the total current in the circuit. If only a single ammeter is available, it will be necessary to break the branch circuit to insert the meter in each case to measure I_1, I_2, and I_3. The main circuit will need to be broken to measure I_T. **Open** S_1 in each case before changing the meter and circuit connections. The following steps assume all necessary meter and circuit connections have been made to enable the currents and voltages to be measured.

6. **Close** S_1. Measure V_{PS}, I_T, I_1, I_2, and I_3. Record these values in Table 11–2 (p. 81). Calculate I_T (the sum of all the branch currents) and record your answer in Table 11–2.

7. Remove the 820-Ω resistor from branch 1. Adjust V_{PS} = 10 V. Measure I_T, I_2, and I_3. Record the values in Table 11–2. Calculate I_T as in step 6 and record your answer in Table 11–2.

8. Remove the 1-kΩ resistor from branch 2 so that only branch 3 remains in the circuit. Adjust V_{PS} to equal 10 V. Measure I_T and I_3. Record the values in Table 11–2. Calculate I_T as in step 6 and record your answer in Table 11–2. **Open** S_1 and turn the power **off.**

9. Connect the circuit in Figure 11–4(b). Reread the suggestions in steps 2 through 5. Voltage V_{PS} should remain 10 V with the three branches connected.

10. With power **on** and S_1 **closed,** measure V_{PS}, I_T, I_1, I_2, and I_3. Record all values in Table 11–2. Calculate I_T (the sum of the branch currents) and record your answer in Table 11–2.

11. Remove the 1-kΩ resistor from branch 1. Adjust V_{PS} = 10 V. Measure I_T, I_2, and I_3. Record the values in Table 11–2. Calculate I_T as in step 10 and record your answer in Table 11–2.

12. Remove the 2.2-kΩ resistor from branch 2 so that only branch 3 remains in the circuit. Adjust V_{PS} = 10 V. Measure I_T and I_3. Record the values in Table 11–2. Calculate I_T as in step 10 and record your answer in Table 11–2. **Open** S_1 and turn power **off.**

13. Connect the circuit in Figure 11–4(c). Reread the suggestions in steps 2 through 5. Voltage V_{PS} should remain 10 V with the three branches connected.

14. With power **on** and S_1 **closed,** measure V_{PS}, I_T, I_1, I_2, and I_3. Record the values in Table 11–2. Calculate I_T (the sum of the branch currents) and record your answer in Table 11–2.

15. Remove the 2.2-kΩ resistor from branch 1. Adjust V_{PS} = 10 V. Measure I_T, I_2, and I_3 and record the values in Table 11–2. Calculate I_T (as in step 14) and record your answer in Table 11–2.

16. Remove the 3.3-kΩ resistor from branch 2 so that only branch 3 remains in the circuit. Adjust V_{PS} = 10 V. Measure I_T and I_3 and record the values in Table 11–2. Calculate I_T (as in step 14) and record your answer in Table 11–2. **Open** S_1 and turn the power **off.**

ANSWERS TO SELF TEST

1. equal
2. equal to
3. 5; 3.3; 1
4. highest
5. 9.3

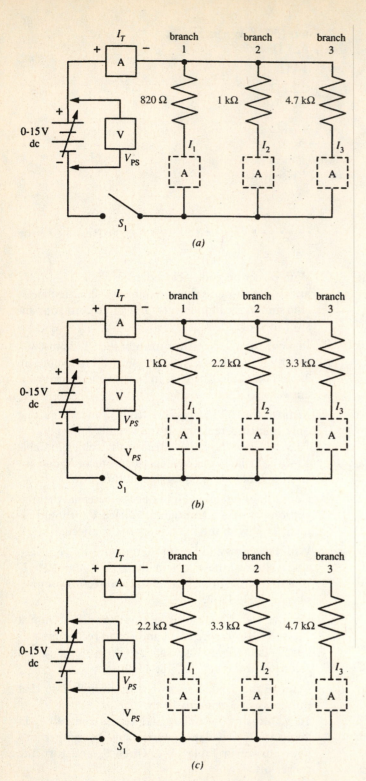

Figure 11–4. (*a*) Circuit for procedure step 2. (*b*) Circuit for procedure step 9. (*c*) Circuit for procedure step 13.

Name _____ Date _____

TABLE 11–1. Measured Values of Experimental Resistors

Resistor	R_1	R_2	R_3	R_4	R_5
Rated value, Ω	820	1 k	2.2 k	3.3 k	4.7 k
Measured value, Ω					

TABLE 11–2. Measured and Computed Values in Parallel Circuit

Step	Rated Value of Branch Resistors, Ω					Measured Values V	Measured Values mA				I_T Calculated (sum of branch I) mA
	R_1	R_2	R_3	R_4	R_5	V	I_T	I_1	I_2	I_3	mA
6	820	1 k			4.7 k						
7		1 k			4.7 k						
8					4.7 k						
10		1 k	2.2 k	3.3 k							
11			2.2 k	3.3 k							
12				3.3 k							
14			2.2 k	3.3 k	4.7 k						
15				3.3 k	4.7 k						
16					4.7 k						

QUESTIONS

1. Explain, in your own words, how your experimental results confirmed the two objectives of this experiment. Refer to the data in Tables 11–1 and 11–2 to support your discussion.

_____

_____

_____

_____

2. Why was it important to measure the values of the resistors (step 1)?

_____

_____

_____

3. Discuss the effect on the total current of parallel-connected resistors if:
   (a) the number of resistors in parallel is increased.
   (b) the resistance of each resistor is increased.
   Support your answers in (a) and (b) by referring to your experimental data in
   Table 11–2.

_____

_____

_____

_____

_____

_____

_____

_____

_____

_____

_____

4. Examine your data in Tables 11–1 and 11–2. What general relationship is indicated
   between the branch current and the total circuit current? State this relationship in
   your own words; then write the relationship as a mathematical formula.

_____

_____

_____

_____

_____

_____

_____

_____

5. Using Figure 11–3, what would happen to the current through R if R_3, became
   open?

_____

_____

_____

_____

_____

_____

_____

RESISTANCE OF A PARALLEL CIRCUIT

BASIC INFORMATION

Total Resistance in a Parallel Circuit

The resistance R_T to which voltage V is connected in Figure 12–1 limits the current in the circuit to I_T. If a single resistor could be found that would draw the same current I_T when connected across V, then the value of this resistor would be equivalent to the three parallel resistors. This equivalent resistor would also represent the total resistance R_T of the three parallel resistors.

Measuring Total Resistance

With V removed, an ohmmeter placed across the end points X and Y of a parallel circuit such as in Figure 12–2 (p. 84) would measure the total resistance of the three resistors.

CAUTION: Always disconnect resistors from their power source before making resistance measurements with an ohmmeter.

The Ohm's law formula for finding resistance is

$$R = \frac{V}{I}$$

If R is the total resistance across V, then I is the total current delivered by V to the resistance. This provides another method for measuring R_T.

If the circuit in series with V were broken and an ammeter inserted in the break, as in Figure 12–3 (p. 84), the ammeter would read the total current delivered by V. The value of V can be measured by connect-

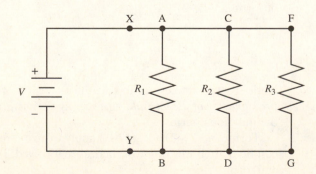

Figure 12–1. Voltage V applied across three resistors connected in parallel.

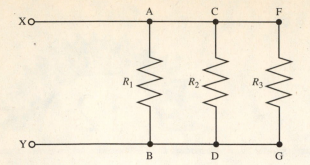

Figure 12–2. Finding the total resistance of three parallel resistors.

ing a voltmeter directly across the voltage source. Two of the three factors of Ohm's law, V and I, would then be known and the third factor, R, could be calculated using the formula

$$R_T = \frac{V}{I_T} \qquad (12–1)$$

In Experiment 11 it was found that the total current drawn by a parallel circuit is greater than the current in any branch. From Ohm's law and the fact that resistance is inversely proportional to current (that is, if voltage is held constant, current will decrease as resistance increases) a similar characteristic is true for parallel resistors.

In a parallel circuit the total resistance of the circuit is *less* than the *lowest* resistance in the parallel combination. For example, if three resistors with values of 47 Ω, 68 Ω, and 100 Ω were connected in parallel as in Figure 12–2, the total resistance would be *less* than 47 Ω. (The exact value and a method for finding it are discussed next.)

The Relationship Between Branch Resistance and Total Resistance · · · · · · ·

It seems logical that there should be some definite relationship between parallel resistances R_1, R_2, R_3, . . . and their total or equivalent resistance. And it should be possible to express this relationship in a formula.

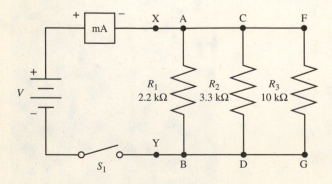

Figure 12–3. Measuring total current in a parallel circuit.

In the discussion that follows, we make a very important (but valid) assumption. We will assume that the resistance of the conductors in the circuits is zero and that the only resistance in the circuit is that of the resistors themselves.

In Figure 12–1 this assumption allows us to say that electrically points B, D, and G are the same as point Y and points A, C, and F are the same as point X.

This fact makes one condition of the circuit very obvious: *The voltage across each branch resistor is exactly the same*, and in the circuit of Figure 12–1, it is equal to V.

Using Ohm's law we can find the current in each branch of the circuit in Figure 12–1:

$$I_1 = \frac{V}{R_1}$$

$$I_2 = \frac{V}{R_2}$$

$$I_3 = \frac{V}{R_3}$$

The total current delivered to this circuit by V is

$$I_T = I_1 + I_2 + I_3$$

$$= \frac{V}{R_1} + \frac{V}{R_2} + \frac{V}{R_3}$$

We can rewrite formula (12–1) as

$$I_T = \frac{V}{R_T}$$

Substituting this in the previous formula we get

$$\frac{V}{R_T} = \frac{V}{R_1} + \frac{V}{R_2} + \frac{V}{R_3}$$

Canceling the Vs,—that is, dividing both sides of the formula by V—results in

$$\frac{1}{R_T} = \frac{1}{R_1} + \frac{1}{R_2} + \frac{1}{R_3} \qquad (12–2)$$

Formula (12–2) states the relationship between branch resistance and total resistance of a parallel circuit: *The reciprocal of the total resistance of a parallel circuit is equal to the sum of the reciprocals of the individual branch resistances.*

To find R_T we need to take the reciprocals of both sides of formula (12–2)

$$R_T = \frac{1}{\dfrac{1}{R_1} + \dfrac{1}{R_2} + \dfrac{1}{R_3}} \qquad (12–3)$$

Notice that it is necessary to take the reciprocal of the *entire right side of the formula*, not merely the reciprocals of the individual terms.

Formulas (12–2) and (12–3) apply to any parallel circuit no matter how many branches are involved. The general form is

$$\frac{1}{R_T} = \frac{1}{R_1} + \frac{1}{R_2} + \frac{1}{R_3} + \cdots$$

where the dots indicate that any number of reciprocal resistances can be added in a particular case.

Although formulas (12–2) and (12–3) were obtained using Ohm's law, their validity can be checked experimentally. Using the methods discussed before, the total resistance of a parallel circuit can be measured directly using an ohmmeter or else calculated using measured values of V and I_T and Ohm's law.

Measuring Individual Resistances in a Parallel Circuit

To verify formulas (12–2) and (12–3) using the circuit in Figure 12–3, we need to measure the individual branch resistances R_1, R_2, and R_3 in the parallel network. How can this be done? Obviously it *cannot* be done by placing the ohmmeter across each resistor in the network, because this procedure would give the measurement of R_T, not the branch resistor. We can measure R_1 by disconnecting it from the parallel network and measuring it outside of the circuit. Or we can disconnect one lead of R_1, say at point A, thus removing the effect of the network. We can then measure R_1 by placing an ohmmeter across it. A similar procedure can be followed to measure the resistances of R_2 and R_3.

SUMMARY

1. The voltage V across each branch (i.e., each resistor in Figures 12–1 and 12–3) of a parallel circuit is the same.

2. The total or equivalent resistance R_T of two or more resistors connected in parallel, as in Figure 12–3, can be found experimentally by measuring the total current I_T, measuring the voltage V across the parallel network, and substituting the measured values in the formula

$$R_T = \frac{V}{I_T}$$

3. Another method of finding the total resistance R_T of two or more parallel-connected resistors, as in Figure 12–2, is to place an ohmmeter across the parallel circuit. The meter will measure R_T.

4. Resistance should never be measured when there is power applied to the circuit. If the parallel resistance of R_1, R_2, and R_3 in Figure 12–3 is required, power must first be disconnected.

5. A formula that expresses the relationship between R_T and R_1, R_2, R_3, etc., of parallel-connected resistors is:

$$R_T = \frac{1}{\dfrac{1}{R_1} + \dfrac{1}{R_2} + \dfrac{1}{R_3} + \cdots}$$

6. Another way of writing the formula for R_T is

$$\frac{1}{R_T} = \frac{1}{R_1} + \frac{1}{R_2} + \frac{1}{R_3} + \cdots$$

7. To measure one of two or more resistors connected in parallel, say R_1 in Figure 12–2, disconnect one lead of R_1 from the circuit. Then measure the resistance of R_1.

SELF TEST

Check your understanding by answering the following questions:

1. In the circuit of Figure 12–1, $I_T = 20$ mA. $V = 5$ V. The total resistance R_T equals _____ Ω.

2. (True/False) For the conditions in question 1, it is possible for R_1 to equal 100 Ω. _____

3. For the conditions in question 1, the voltage V across $R_2 = $ _____ V.

4. (True/False) To measure the resistance of R_3 in Figure 12–2, simply place the ohmmeter leads across GF and read the resistance. _____

5. In Figure 12–2 the resistance of $R_1 = 25$ Ω, $R_2 = 33$ Ω, $R_3 = 75$ Ω. $R_T = $ _____ Ω.

MATERIALS REQUIRED

Power Supplies:
- Variable, 0–15 V dc, regulated

Instruments:
- DMM or VOM

Resistors (5%, ½-W):
- 1 820-Ω
- 1-1 kΩ
- 1-2.2 kΩ
- 1-3.3 kΩ
- 1-4.7 kΩ

Miscellaneous:
- SPST switch

PROCEDURE

A. Finding R_T by Formula

NOTE: For the calculations in Part A use the general form of the parallel resistance formula discussed in the BASIC INFORMATION section.

A1. Measure the resistance of each of the resistors supplied and record the value in Table 12–1 (p. 89)

A2. Connect the two resistors in parallel as in Figure 12–4(a). Calculate the total resistance R_T of this combination. Record the answer in Table 12–2 (p. 89). Using an ohmmeter measure the total resistance R_T of this combination. Record the value in Table 12–2.

A3. Connect the third resistor to the parallel combination as in Figure 12–4(b). Calculate the total resistance R_T of this combination. Record the answer in Table 12–2. Using an ohmmeter measure the total resistance R_T of this combination. Record the value in Table 12–2.

A4. Connect the fourth resistor to the parallel combination as in Figure 12–4(c). Calculate the total resis-

tance R_T of this combination. Record the answer in Table 12–2. Using an ohmmeter measure the total resistance R_T of this combination. Record the value in Table 12–2.

A5. Connect the fifth resistor to the parallel combination as in Figure 12–4(d). Calculate the total resistance R_T of this combination. Record the answer in Table 12–2. Using an ohmmeter measure the total resistance R_T of this combination. Record the value in Table 12–2.87

B. Finding R_T Using the Voltage-Current Method

B1. With power **off** and switch S_1 **open,** using the parallel combination of resistors in step A5, connect the circuit of Figure 12–5(a). Power **on,** S_1 **closed.** A constant voltage, $V_{PS} = 10$ V, will be applied to all circuits in part B. Measure V_{PS} (it should be 10 V) and I_T. Record values in Table 12–3 (p. 89).

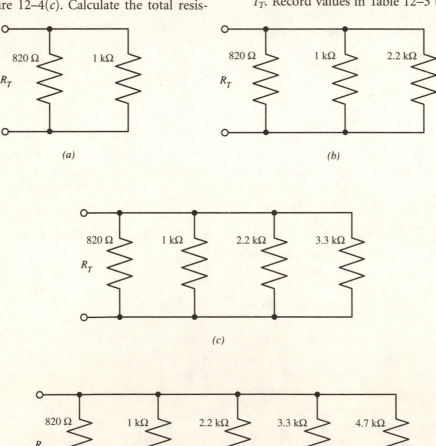

Figure 12–4. (a) Circuit for procedure step A2. (b) Circuit for procedure step A3. (c) Circuit for procedure step A4. (d) Circuit for procedure step A5.

B2. Remove the 4.7-kΩ resistor to obtain the circuit of Figure 12–5(b). Adjust V_{PS} to 10 V. Measure I_T and record values in Table 12–3.

B3. Remove the 3.3-kΩ resistor from the circuit of step B2 as in Figure 12–5(c). Adjust V_{PS} to 10 V. Measure I_T and record values in Table 12–3.

B4. Remove the 2.2-kΩ resistor from the circuit of step B3, leaving two resistors in parallel as in Figure 12–5(d). Adjust V_{PS} to 10 V. Measure I_T and record values in Table 12–3. Power **off**, S_1 **open**.

B5. For each circuit in steps B1 through B4, calculate R_T by using the Ohm's law formula discussed in the Basic Information section. Record your answers in Table 12–3.

ANSWERS TO SELF TEST

1. 250
2. false
3. 5
4. false
5. 12

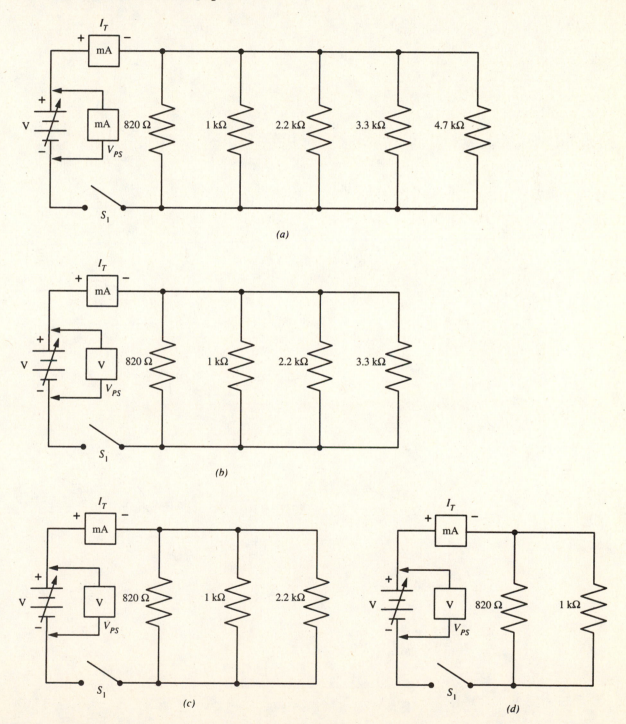

Figure 12–5. (a) Circuit for procedure step B1. (b) Circuit for procedure step B2. (c) Circuit for procedure step B3. (d) Circuit for procedure step B4.

RESISTANCE OF A PARALLEL CIRCUIT   **87**

Name _____ Date _____

TABLE 12–1. Measured Values of Experimental Resistors

Resistor	R_1	R_2	R_3	R_4	R_5
Rated value, Ω	820	1 k	2.2 k	3.3 k	4.7 k
Measured value, Ω					

TABLE 12–2. Part A: Finding R_T of Parallel-Connected Resistors by Formula and Measurement

Step	Rated Value, Ω					Calculated Value of R_T, Ω	Measured Value of R_T, Ω
	R_1	R_2	R_3	R_4	R_5		
A2	820	1 k					
A3	820	1 k	2.2 k				
A4	820	1 k	2.2 k	3.3 k			
A5	820	1 k	2.2 k	3.3 k	4.7 k		

TABLE 12–3. Part B: Finding R_T by the Voltage-Current Method

Step	Measured Values		Calculated Values R_T, Ω
	V_{PS}, V	I_T, mA	
B1			
B2			
B3			
B4			

QUESTIONS

1. Explain, in your own words, the relationship between branch resistances and the total resistance of a parallel circuit.

_____

_____

_____

_____

2. Write the relationship discussed in Question 1 as a mathematical formula.

_____

_____

3. Discuss the effect on total resistance of a parallel circuit, if:
   (a) the number of parallel resistors is increased.
   (b) the resistance of each resistor is increased.
   Support your answers by referring to your experimental data.

   _____

   _____

   _____

   _____

   _____

   _____

4. Discuss three methods used in finding the total resistance of parallel-connected
   resistors.

   _____

   _____

   _____

   _____

   _____

   _____

5. Parts A and B use similar circuits. For each comparable combination of resistors
   in parts A and B, compare the calculated values. Discuss the possible reasons for
   differences, if any.

   _____

   _____

   _____

   _____

   _____

   _____

6. In reference to Figure 12–2, what would the total resistance, R_T, between points
   X and Y equal if R_3 became shorted?

   _____

   _____

   _____

   _____

   _____

   _____

DESIGNING PARALLEL CIRCUITS

BASIC INFORMATION

Designing a Parallel Circuit to Meet Specified Resistance Requirements

The formulas for total resistance of parallel-connected resistors can be applied to the solution of simple design problems. An example will indicate the techniques to be used.

Problem 1. A technician has a stock of the following color-coded resistors: four 68–Ω, five 82–Ω, two 120–Ω, three 180–Ω, two 330–Ω, and one each of 470–Ω, 560–Ω, 680–Ω, and 820–Ω. A circuit being designed needs a 37–Ω resistance. Find a combination of resistors, using the least possible number of components, that will satisfy the design requirement. Assume the measured values of the resistors are the same as the color-coded values.

Solution. Because the value of 37 Ω is less than the resistance of the smallest resistor in stock, a parallel arrangement will be required. (Recall from Experiment 12 that the total resistance of a parallel circuit is less than the resistance value of the smallest resistor.) The formula for total resistance R_T of a parallel circuit is

$$\frac{1}{R_T} = \frac{1}{R_1} + \frac{1}{R_2} + \frac{1}{R_3} + \cdots \tag{13–1}$$

If two resistors in parallel satisfy the design requirements, this formula can be rewritten as

$$\frac{1}{R_T} = \frac{1}{R_1} + \frac{1}{R_2}$$

Upon simplification this becomes

$$\frac{1}{R_T} = \frac{R_1 + R_2}{R_1 R_2}$$

Solving for R_T yields

$$R_T = \frac{R_1 \times R_2}{R_1 + R_2} \tag{13–2}$$

That is, the total resistance of two parallel resistors is equal to the product of their resistances divided by the sum of their resistances. Formula

OBJECTIVES

1 To design a parallel circuit that will meet specified voltage, current, and resistance requirements

2 To construct and test the circuit to see that it meets the design requirements

(13–2) is a general formula that can be used to find the total resistance of all two-resistor parallel combinations. Using this formula where it applies can save considerable calculating time.

Assume that two resistors will meet the design requirements of this problem and that the resistors are 68 Ω and 82 Ω. Substituting these values in formula (13–2) gives

$$R_T = \frac{68 \times 82}{68 + 82} = \frac{68 \times 82}{150} = 37.2 \ \Omega$$

It is evident then that the two values selected meet the problem requirements, because when connected in parallel, their total resistance is very close to 37 Ω. Trial and error will show that no other combination of two resistors in stock will produce a value closer to 37 Ω.

It is not always possible to select the two required resistors so easily. Another method can be used. Suppose we assume that 68 Ω is one of the resistors. We wish to find another resistor R_X that, when connected in parallel with 68 Ω, will produce an equivalent or total resistance of 37 Ω. Substitute the known values of R_T and R_1 in formula (13–2):

$$37 = \frac{68 \times R_X}{68 + R_X}$$

This formula can be rewritten to solve for R_X

$$R_X = \frac{37 \times 68}{68 - 37}$$

$$= 81.2 \ \Omega$$

The 82–Ω resistor in stock could therefore satisfy (within limits) the requirements of the circuit.

Problem 2.
Assume the same stock as in Problem 1 is available. The technician must design a circuit requiring a 60-Ω resistor.

Solution.
The 60-Ω requirement is exactly half the value of the 120-Ω resistor in stock. Assume that two resistors will satisfy the design requirements, and one of the resistors is 120 Ω. Substitute the known values in formula (13–2):

$$R_T = \frac{120 \times R_2}{120 + R_2} = 60$$

Solving for R_2, we have

$$R_2 = \frac{60 \times 120}{120 - 60} = \frac{7200}{60}$$

$$= 120 \ \Omega$$

Thus, the two 120-Ω resistors in parallel will produce an equivalent resistance of one-half the value of one of the resistors; two 68-Ω resistors in parallel are equivalent to 34 Ω; two 1 kΩ resistors in parallel are equivalent to 500 Ω, and so on. What would be the equivalent resistances of three, four, or more equal resistors in parallel?

Formula (13–1) will be used to test the condition of three equal resistors connected in parallel.

$$\frac{1}{R_T} = \frac{1}{R} + \frac{1}{R} + \frac{1}{R} = \frac{3}{R}$$

$$R_T = \frac{R}{3}$$

Similarly, applying formula (13–1) to four, five, and six equal resistors results in total resistances equal to R/4, R/5, and R/6, respectively. We can draw a general rule and formula from these calculations: *The total resistance of n equal resistors connected in parallel is the value of a single resistance R divided by the number of equal resistors*, or

$$R_T = \frac{R}{n} \qquad\qquad (13–3)$$

The technician could also use three 180-Ω resistors connected in parallel, because their total resistance is also 60 Ω.

$$R_T = \frac{180}{3} = 60 \ \Omega$$

However, here three components are used rather than the two 120-Ω resistors.

Problem 3.
In the circuit of Figure 13–1 a technician measures the resistance between points A and B and finds it is 180 Ω. The circuit requires a resistance between A and B of 45 Ω. How can the technician modify the circuit to meet the required value, assuming the same stock of resistors as in problem 1 is available?

Solution.
Assume that there is a resistor R_X that, when connected in parallel with 180 Ω, will bring the resistance R_{AB} down to 45 Ω. Substitute the known values in formula (13–2).

$$45 = \frac{180 \times R_X}{180 + R_X}$$

Solving for R_X, we have

$$R_X = \frac{45 \times 180}{180 - 45} = \frac{8100}{135}$$

or

$$R_X = 60 \ \Omega$$

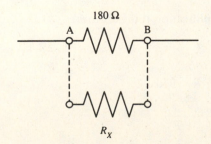

Figure 13–1. When resistor R_X is connected across the 180-Ω resistor, the total resistance across A–B must be 45 Ω.

What is required, then, is a 60-Ω resistor. From problem 2 it is evident that two 120-Ω resistors, connected in parallel with the 180-Ω resistor, will yield the required result:

$$R_{AB} = 45 \; \Omega$$

To verify the design we will find the total resistance of a parallel circuit consisting of three branch resistors, two 120-Ω and one 180-Ω.

$$\frac{1}{R_T} = \frac{1}{120} + \frac{1}{120} + \frac{1}{180}$$

$$\frac{1}{R_T} = \frac{3 + 3 + 2}{360} = \frac{8}{360}$$

$$R_T = \frac{360}{8} = 45 \; \Omega$$

The procedure, then, for designing a parallel circuit having a specified resistance value R_T from a group of known resistors is to apply formulas (13–1), (13–2), and (13–3) to the resistors at hand and find a combination that equals or is closest to the required value. After this is done the technician should connect the resistors in parallel and measure their total resistance with an ohmmeter to confirm the solution.

Designing a Parallel Circuit to Meet Specified Resistance and Current Requirements · · · · · · · · · · · · · · · · · · ·

Ohm's law and the formulas for total resistance in a parallel circuit are applied in the solution of this type of problem. Again, an example illustrates the procedure.

Problem 4. In the circuit of Figure 13–2(a), find the value of V_V to which the variable dc power supply must be set to obtain a current of 20 mA.

Solution. In order to find the applied voltage V, it is first necessary to determine the total resistance R_T in the circuit. When that is known, V can be found by using the Ohm's law formula $V = I \times R_T$.

1. *Procedure for finding R_T:* The 680-Ω resistor between A and B is connected in parallel with the two series-connected resistors, 1.2 kΩ and 820 Ω. The first step is to replace the 1.2 k and 820-Ω resistors with a single resistor R_1 whose resistance value is equal to the sum of the two resistors. Thus,

$$R_1 = 1200 + 820 = 2020 \; \Omega$$

The resulting equivalent circuit is then Figure 13–2(b), whose resistance and current characteristics are the same as those of Figure 13–2(a).

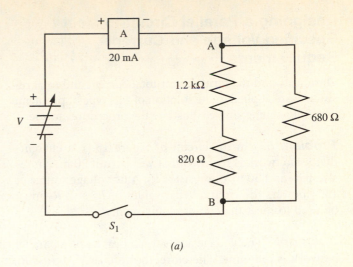

(a)

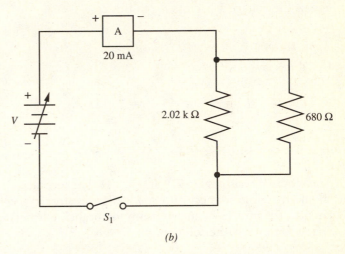

(b)

Figure 13–2. Finding the voltage that will draw 20 mA from the supply.

The total resistance can now be calculated by use of formula (13–2).

$$R_T = \frac{680 \times 2020}{680 + 2020} = 509 \; \Omega$$

2. *Procedure for finding V:*

$$V = I_T \times R_T$$
$$= 20 \; \text{mA} \times 509 \; \Omega = 10.2 \; \text{V}$$

The required voltage is approximately 10.2 V.

The technician must now connect the circuit of Figure 13–2(a) and apply a measured 10.2 V from the dc supply. The measured current should be 20 mA.

In actual practice the technician would have connected the circuit of Figure 13–2 and adjusted the output of the dc supply until the ammeter measured 20 mA. The technician could read the dc voltage, which should be 10.2 V, on the power supply voltmeter.

Designing a Parallel Circuit to Meet Specified Voltage and Current Requirements ·

Ohm's law and the formulas for total current and total resistance are applied in the solution of this type of problem. The following examples illustrate the procedures to use.

Problem 5. In the circuit of Figure 13–3, a current-divider network is required that will permit 20 mA in resistor R_1 and 30 mA in resistor R_2. A dc voltage source of 15 V powers the circuit. What values of R_1 and R_2 must be used to meet the circuit requirements?

Solution. The voltage across each branch of a parallel network is the same. Therefore, the voltage V_1 across R_1 equals the voltage V_2 across R_2. In this case,

$$V_1 = V_2 = 15 \text{ V}$$

By Ohm's law,

$$R_1 = \frac{V_1}{I_1} = \frac{15 \text{ V}}{20 \text{ mA}} = 750 \text{ } \Omega$$

$$R_2 = \frac{V_2}{I_2} = \frac{15 \text{ V}}{30 \text{ mA}} = 500 \text{ } \Omega$$

The required resistors are 750 and 500 Ω. The technician should therefore connect 750- and 500-Ω resistors in the circuit of Figure 13–3 and measure the branch currents I_1 and I_2 to verify the solution.

Problem 6. A technician must design a circuit that will deliver 0.25 A from a 5-V supply. Using the appropriate formula, determine what combination of resistors, using the least number of components, will satisfy the design requirements. The stock of resistors that the technician has on hand includes two 40-, two 60-, three 150-, three 180-, three 680-, and one each of 330- and 470-Ω resistors.

Solution. The total resistance that will satisfy the requirements of the problem is

$$R_T = \frac{5 \text{ V}}{0.25 \text{A}} = 20 \text{ } \Omega$$

A series combination will not give the required solution, but a parallel arrangement may. By inspection, it is apparent from the stock of resistors that two 40-Ω resistors in parallel will give the required value of 20 Ω. Although other combinations are possible (Can you find them?), this is the only solution requiring just two resistors.

SUMMARY

1. If a circuit requires a resistance value R_T that is less than the lowest resistance value of a stock of resistors, it may be possible to approximate the required value by paralleling two or more of the available resistors. The procedure is to choose a resistor R_1 whose resistance is slightly higher than the required R_T, then find R_2 from the formula for total resistance

$$R_T = \frac{R_1 \times R_2}{R_1 + R_2}$$

If R_2 is in stock, the two resistors have been found. Otherwise it may be necessary to repeat the process. Finally, if two resistors whose parallel combination is R_T cannot be found, it may be necessary to use three or four resistors in parallel.

2. The equivalent resistance R_T of n equal-valued resistors R_1, connected in parallel, is

$$R_T = \frac{R_1}{n}$$

3. If it is necessary to calculate the voltage V that will produce I amperes in a circuit containing R ohms, the Ohm's law formula can be used:

$$V = I \times R$$

4. It is possible to find experimentally the voltage V necessary to produce I amperes in a circuit containing R ohms. A circuit is connected containing the variable dc source in series with an ammeter and the resistance R. The voltage is varied until the required I is measured. The voltage V, measured at the terminals of the dc source, is the required voltage.

5. If it is required to calculate the resistance R that will draw I amperes from a voltage source V, use the formula

$$R = \frac{V}{I}$$

If the value of R is available in stock, that solves the problem. Otherwise, use series or parallel resistors in a combination that will satisfy the required R.

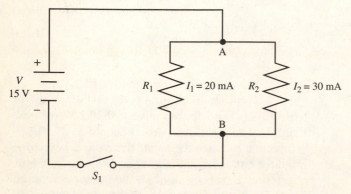

Figure 13–3. Finding the values of R_1 and R_2 to meet the given voltage and current requirements.

Check your understanding by answering the following questions:

1. What two resistors in parallel will yield a total resistance R_T of 30 Ω? Give at least two solutions.
   (a) _____ Ω in parallel with _____ Ω;
   (b) _____ Ω in parallel with _____ Ω.

2. Three resistors, 22-, 33-, and 47-Ω, are connected in parallel. Their total resistance will be (more/less) _____ than 22 Ω.

3. A 10-Ω resistance can be constructed from _____ 50-Ω resistors connected in parallel.

4. In the circuit of Figure 13–4, the voltage V that will deliver 150 mA to the resistors is _____ V.

5. In a circuit with two parallel resistors similar to Figure 13–3, there is a total current of 3 mA. The applied voltage, V, is 15 V. The total resistance R_{AB} between points A and B is _____ Ω.

MATERIALS REQUIRED

Power Supply:
- Variable 0-15 V dc, regulated

Instruments:
- DMM or VOM
- 0–100 mA milliammeter

Resistors (½-W, 5%)
- 1 820-Ω
- 1-1 kΩ
- 1-2.2 kΩ
- 1-3.3 kΩ
- 1-4.7 kΩ
- 1-5.6 kΩ

Miscellaneous:
- SPST switch

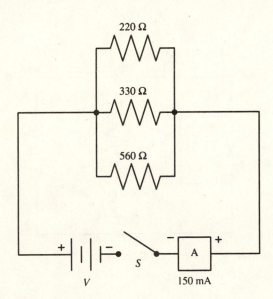

Figure 13–4. Circuit for self-test question 4.

PROCEDURE

1. Measure the resistance of each of the resistors supplied for this experiment and record its value in Table 13–1 (p. 97).

2. Using the formulas for finding the R_T of a parallel circuit, find the parallel combinations of two or three resistors that will produce the values of R_T in Table 13–2 (p. 97). Calculate R_T using the rated values of the supplied resistors only. Record the rated values in Table 13–2.

3. Connect each of the parallel combinations in Table 13–2 and measure the R_T with an ohmmeter. Record the measured values in Table 13–2.

4. Using the resistors supplied, design a two-resistor parallel circuit that will draw approximately 20 mA when 15 V is applied across the parallel combination. If more than one combination appears to be suitable, choose the circuit closest to drawing the 20 mA. In

Table 13–3 (p. 97), record the rated values of the resistors used.

5. With power **off** and switch S_1 **open,** connect the circuit in step 4 as in Figure 13–5. Turn **on** power; **close** S_1. Adjust V_{PS} to 15 V and measure I_T. Record the value in Table 13–3.

6. Using the formulas in the Basic Information section, calculate the voltage required to deliver 20 mA to a circuit consisting of a 1-kΩ resistor in parallel with a 2.2 kΩ resistor. Record your answer in Table 13–3.

7. Connect the circuit of step 6 to a variable voltage supply as in Figure 13–6. Adjust V_{PS} until the ammeter reads 20 mA. Measure V_{PS} and record the value in Table 13–3.

8. Design a three-branch parallel circuit with branch currents as shown in Figure 13–7. Choose the resistors from the six supplied for this experiment. Calculate the voltage required to deliver the branch currents specified. Record your answers in Table 13–4.

9. With power **off** and S_1 **open,** connect the circuit of Figure 13–7 using your designed resistor values. Turn **on** the power; **close** S_1. Adjust V_{PS} until the specified I_T is being delivered. Measure I_1, I_2, I_3, I_T, and V_{PS} and record the values in Table 13–4.

10. Design a parallel circuit containing four branches for which $I_T = 10$ mA (approximately). The individual branch currents are not critical. The voltage applied to the circuit cannot exceed 15 V. Draw a diagram of the circuit showing a meter for measuring I_T and a switch for opening and closing the circuit. Show all calculations for finding the resistance value of each branch, and the value of V_{PS}. Record your answers in Table 13–5 (p. 98).

11. Before performing this step, read the Optional Activity below. With power **off** and S_1 **open,** connect your circuit of step 10. Adjust V_{PS} to your design value.

Measure I_T and record the values of V_{PS} and I_T in Table 13–5. Turn power **off; open** S_1.

Optional Activity

This activity requires the use of electronic simulation software. Build a simulated circuit of your design in step 10. Show the instruments necessary to measure the total current and the power supply voltage. Measure I_T and V. In a brief report analyze and describe any discrepancies with the results in step 11.

ANSWERS TO SELF TEST

1. (a) 60, 60; (b) 50, 75
2. less
3. 5
4. 16.1
5. 5 k

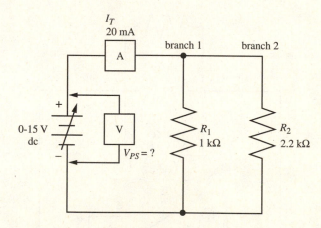

Figure 13–6. Circuit for procedure step 7.

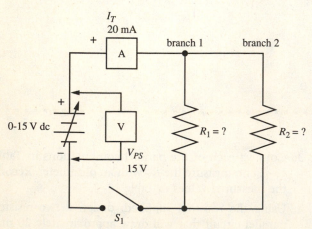

Figure 13–5. Circuit for procedure step 5. Values of R_1 and R_2 are determined in step 4.

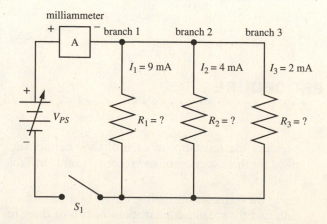

Figure 13–7. Circuit for procedure steps 8 and 9. Values of R_1, R_2, and R_3 are determined in step 8.

TABLE 13–1.   Measured Value of Experimental Resistors

Resistor	R_1	R_2	R_3	R_4	R_5	R_6
Rated value, Ω	820	1 k	2.2 k	3.3 k	4.7 k	5.6 k
Measured value, Ω						

TABLE 13–2.   Designing a Parallel Resistive Network for a Required R_T

R_T Required, Ω	Combination of Parallel Resistors (Rated value, Ω)			Measured Value R_T, Ω
	Branch 1 R	Branch 2 R	Branch 3 R	
374				
417				
530				
825				
1068				
1320				
1440				
2555				

TABLE 13–3.   Designing a Circuit That Will Yield I_T with V Known

Steps	V_{PS}, V			I_T, mA		Parallel Resistors, Ω	
	Given Value	Calculated Value	Measured Value	Required Value	Measured Value	Branch 1 R	Branch 2 R
4, 5	15	✕		20			
6, 7	✕			20		1k	2.2k

TABLE 13–4.   Designing a Current-Divider Circuit

Branch Current, mA						Rated Value of Resistor, Ω			Voltage, V	
Required			Measured			Branch 1 R	Branch 2 R	Branch 3 R	Calculated	Measured
I_1	I_2	I_3	I_1	I_2	I_3					
9.0	4.0	2.0								

TABLE 13–5.   Design Circuit: Given I_T and R, Find V

Steps	Rated Value of Resistors, Ω				V_{PS}, V (Design Value)	I_T, Measured, mA
	Branch 1 R	Branch 2 R	Branch 3 R	Branch 4 R		
10, 11						

QUESTIONS

1. Did all the resistors used in your design fall within their tolerance ratings? Refer to your measurements in Table 13–1 and the resistor color code to support your answer.

_____

_____

_____

_____

2. Why was it necessary to measure the resistance of the resistors used in your design?

_____

_____

_____

3. Did the measurement of I_T in step 5 confirm that the values of the design resistors found in step 4 would satisfy the specifications of the problem? If not, explain the difference.

_____

_____

_____

_____

4. Was the measured voltage, V_{PS}, in step 9 the same as that calculated in step 8? If not, discuss possible reasons for the difference.

_____

_____

_____

5. On a separate sheet of paper, design a three-branch parallel circuit that divides the current such that the current in the first branch is double the current in the second branch and three times more than the current in the third branch. The total current in the circuit, I_T is 110 mA. The total resistance of the parallel circuit, R_T, is approximately 305 Ω. Find the value of each parallel resistor. (The resistors can have any whole-number values.) Also find the value of the applied voltage. Show all formulas and calculations. Draw and label the circuit diagram.

RESISTANCE OF SERIES-PARALLEL CIRCUITS

BASIC INFORMATION

Total Resistance of a Series-Parallel Circuit · · · · · · ·

Figure 14–1 shows a series-parallel arrangement of resistors. In this circuit, R_1 is in series with the parallel circuit between points B and C, which is then in series with R_3. What is the total resistance between points A and D? Obviously we can measure R_T with an ohmmeter, or R_T can be found by the voltage-current method described in Experiment 8. Is it possible, however, to write a formula by which R_T of a series-parallel circuit can be computed without measurement?

Again refer to Figure 14–1. If R_1 and R_3 were removed and if the parallel circuit between points B and C were left standing alone, we could compute the total resistance R_{T2} between B and C by the formula for parallel resistors. We can, therefore, replace the parallel circuit in Figure 14–1 with its equivalent resistance R_{T2}, and the circuit will now take the form of Figure 14–2. The total resistance R_T of the series circuit in Figure 14–2 is the same as the total resistance R_T between points A and D in Figure 14–1.

This suggests that to find the total resistance of a series-parallel network, replace the parallel circuits first by their equivalent resistances and

OBJECTIVES

1 To verify experimentally the rules for finding the total resistance R_T of a series-parallel circuit

2 To design a series-parallel network that will meet specified current requirements

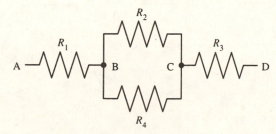

Figure 14–1. A series-parallel resistor circuit.

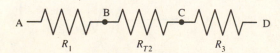

Figure 14–2. Equivalent circuit for Figure 14–1.

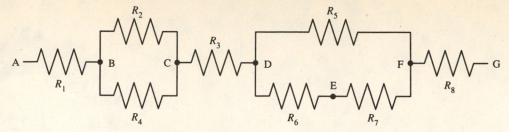

Figure 14–3. Two parallel circuits in a series-parallel combination.

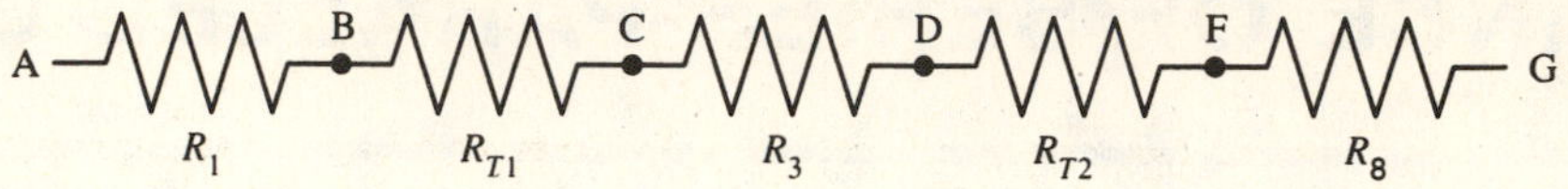

Figure 14–4. Equivalent resistance for Figure 14–3.

treat the resultant network like a simple series circuit. In the case of Figure 14–1, the total resistance R_T between points A and D is

$$R_T = R_1 + R_{T2} + R_3$$

The circuit in Figure 14–3 is similar to the circuit in Figure 14–1 except that there are two parallel circuits in Figure 14–3. To find the total resistance of the circuit, first find the equivalent resistance of each of the parallel sections. This results in a series circuit having five components, as in Figure 14–4.

The equivalent resistance R_{T1} of the parallel circuit between points B and C can be found using the formula from Experiment 13.

$$R_{T1} = \frac{R_2 R_4}{R_2 + R_4}$$

The same formula can be used to find the equivalent resistance R_{T2} between points D and F. Note, however, that one branch of this parallel section has two resistors in series. The total resistance of the branch is therefore $R_6 + R_7$. Applying the formula for two branch parallel circuits (as before), we have

$$R_{T2} = \frac{R_5 (R_6 + R_7)}{R_5 + (R_6 + R_7)}$$

The total resistance R_T of the resulting series circuit (Figure 14–4) can be found by adding all resistances in series, including the two equivalent resistances

$$R_T = R_1 + R_{T1} + R_3 + R_{T2} + R_8$$

The method just discussed is applicable to all series-parallel networks. However, in solving more complex series-parallel circuits, the initial step usually involves identifying series and parallel combinations of compo-

nents. This is not always easy because the circuit diagrams are rarely as straightforward as those in Figures 14–1 through 14–4. Often a choice must be made in isolating one particular section of an overall complex network and then concentrating on solving that section.

An example is used to demonstrate this method of solving a complex circuit. Other methods for solving complex networks are discussed elsewhere in this book.

Problem 1. Find the current I_T drawn by the circuit in Figure 14–5.

Solution. The circuit in Figure 14–5 can be redrawn as in Figure 14–6(a). These two circuits are exactly the same electrically. Notice that R_2 and R_3 are in series and that the $R_2 + R_3$ combination is across—that is, in parallel with—R_5. The circuit in Figure 14–6(b) shows the result of finding the equivalent resistance of R_5 in parallel with $(R_2 + R_3)$. Again, a series circuit consisting of R_4

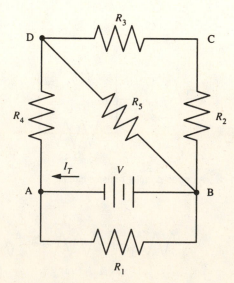

Figure 14–5. Circuit for problem 1.

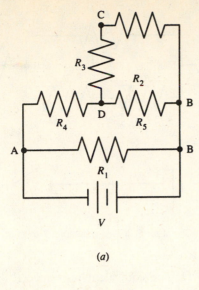

(a)

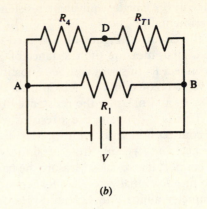

(b)

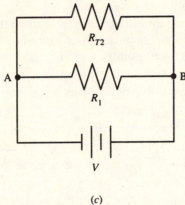

(c)

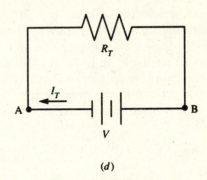

(d)

Figure 14–6. Equivalent circuit simplification for Figure 14–5.

in series with R_{T1} can be simplified by finding the total resistance $R_{T2} = R_4 + R_{T1}$.

Figure 14–6(c) shows the circuit after that simplification. Finally, the parallel branches R_{T2} and R_1 can be combined using the formula for two parallel resistances,

$$R_T = \frac{R_1 R_{T2}}{R_1 + R_{T2}}$$

We are left with the equivalent circuit in Figure 14–6(d). As far as the voltage source is concerned, this is the circuit it is feeding, and I_T is the current drawn by R_T. Therefore,

$$I_T = \frac{V}{R_T}$$

A numerical example shows how the simplifying process works step by step.

Problem 2. Assume that the following values apply to the circuit of Figure 14–5: $V = 12$ V, $R_1 = 500$ Ω,

$R_2 = 680$ Ω, $R_3 = 320$ Ω, $R_4 = 1$ kΩ, and $R_5 = 1$ kΩ. Find I_T.

Solution. The steps described in problem 1 are followed in this problem, substituting the known values.

$$R_2 + R_3 = 680 + 320 = 1 \text{ k}\Omega$$

$$R_{T1} = \frac{R_5 (R_2 + R_3)}{R_5 + (R_2 + R_3)} = \frac{1 \text{ k} \times 1 \text{ k}}{1 \text{ k} + (680 + 320)} = 500 \ \Omega$$

The equivalent resistance R_{T1} is in series with R_4; therefore,

$$R_{T2} = R_4 + R_{T1} = 1 \text{ k} + 500 = 1.5 \text{ k}\Omega$$

Finally,

$$R_T = \frac{R_1 R_{T2}}{R_1 + R_{T2}} = \frac{500 \times 1.5 \text{ k}}{500 + 1.5 \text{ k}} = 375 \ \Omega$$

$$I_T = \frac{12 \text{ V}}{375 \ \Omega} = 0.032 \text{ A, or } 32 \text{ mA}$$

Verifying the Simplification Process · · · · ·

The process of combining series resistances into a single equivalent resistance and, similarly, of combining parallel branches into a single equivalent resistance, can be tested experimentally.

Again, the circuit of Figure 14–5 is used to describe the method. First, measure the resistance of R_1 through R_5. Next, connect R_2 in series with R_3. Substitute a resistor whose value is equal to $R_2 + R_3$. Connect this new resistor across R_5 and measure the resistance of the parallel circuit. This is R_{T1}. Again, use a resistor whose value is R_{T1} and connect it in series with R_4. Measure the series combination to find R_{T2}. Substitute a resistor whose value is R_{T2} and connect it across R_1. Measure the parallel combination to find R_T. Finally, connect R_T across V in series with an ammeter, adjust V to 12 V, and measure I_T.

SUMMARY

1. In a series-parallel network like that in Figure 14–1, the total resistance R_T of the network measured across the end terminals A and D can be found by replacing each parallel combination by its equivalent resistance R_{T2}, leaving an equivalent series circuit (Figure 14–2). The total resistance R_T can then be calculated by using the series-resistance formula:

$$R_T = R_1 + R_{T2} + R_3 + \cdots$$

2. In determining the equivalent resistance R_{T2} of one of the parallel combinations, such as R_2, R_4 in Figure 14–1, the formula for parallel resistors is used.

3. A parallel branch may sometimes have two or more series resistors, such as branch R_6–R_7 in Figure 14–4. In that case R_6 and R_7 are combined like series resistors, and their equivalent series resistance $R_{6-7} = R_6 + R_7$ is substituted in the circuit.

4. In a parallel network the voltage across each branch is the same.

SELF TEST

Check your understanding by answering the following questions:

1. In Figure 14–1, $R_1 = 280\ \Omega$, $R_2 = 120\ \Omega$, $R_3 = 330\ \Omega$, and $R_4 = 470\ \Omega$. The value of R_{T2} measured from B to C is _____ Ω.

2. For the same conditions as in question 1, the value of R_T measured from A to D is _____ Ω.

3. In Figure 14–3, $R_1 = 470\ \Omega$, $R_2 = 56\ \Omega$, $R_3 = 33\ \Omega$, $R_4 = 68\ \Omega$, $R_5 = 120\ \Omega$, $R_6 = 20\ \Omega$, $R_7 = 100\ \Omega$, and

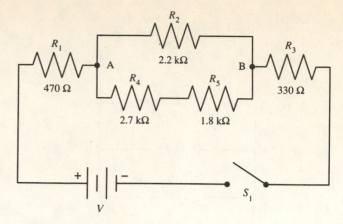

Figure 14–7. Circuit for Self Test questions 7 through 9.

$R_8 = 100\ \Omega$. The equivalent resistance R_{T2} between points B and C is _____ Ω.

4. For the same conditions as in question 3, the equivalent resistance R_{6-7} in the lower branch of the parallel network between points D and F is _____ Ω.

5. For the same conditions as in question 3, the equivalent resistance R_{T3} of the parallel network between points D and F is _____ Ω.

6. For the same conditions as in question 3, the total resistance R_T between points A and G is _____ Ω.

7. In Figure 14–7, the total resistance R_T in the circuit is _____ Ω.

8. In Figure 14–7, if $V = 25$ V, the total current in the circuit is _____ mA.

9. If in Figure 14–7 the current in R_2 is 67.5 mA the voltage V_{AB} across the series combination of R_4 and R_5 is _____ V.

MATERIALS REQUIRED

Power Supply:
- Variable 0–15 V dc, regulated

Instruments:
- DMM or VOM
- 0–10 mA milliammeter

Resistors (½-V, 5%):
- 1 330–Ω
- 1 470–Ω
- 1 560–Ω
- 1-1.2 kΩ
- 1-2.2 kΩ
- 1-3.3 kΩ
- 1-4.7 kΩ
- 1-10 kΩ

Miscellaneous:
- SPST switch

PROCEDURE

1. Measure the resistance of each of the resistors supplied and record its value in Table 14–1 (p. 105).

2. Connect the resistors as in Figure 14–8(a). Measure the resistance between A and D (R_T) and the resistance between B and C (R_{BC}). Record the values in Table 14–2 (p. 105).

3. Complete the first row of Table 14–2 under "Calculated Value" in the following manner:

   R_T (a)   Calculate the total resistance between A and D by adding the measured values of R_1, R_{BC}, and R_3.

   R_{BC} Use the formulas from the Basic Information section to calculate the resistance of the paral-

lel combination of R_2 and R_4. Use measured values from Table 14–1 in your calculations.

   R_T (b)   Calculate the total resistance between A and D by adding the measured values of R_1 and R_3 to the calculated value of R_{BC}.

4. Connect the resistors as in Figure 14–8(b). Complete the second row of Table 14–2 using the procedures in steps 2 and 3.

5. Connect the resistors as in Figure 14–9(a). Measure the resistance across A and G (R_T), across B and C (R_{BC}), and across D and F (R_{DF}). Record the values in the third row of Table 14–2.

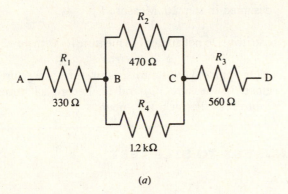
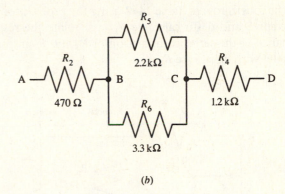

(a)   (b)

Figure 14–8. (a) Resistor combination for procedure step 2. (b) Resistor combination for procedure step 4.

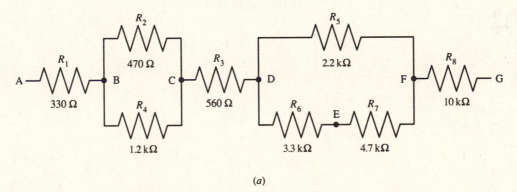

(a)

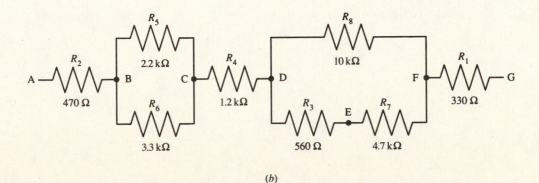

(b)

Figure 14–9. (a) Resistor combination for procedure step 5. (b) Resistor combination for procedure step 7.

RESISTANCE OF SERIES-PARALLEL CIRCUITS   **103**

6. Complete the third row of Table 14–2 under "Calculated Value" in the following manner:

R_T (a) Calculate the total resistance between A and D by adding the measured values of R_1, R_{BC}, R_3, R_{DF}, and R_8.

R_{BC} Use the formulas from the Basic Information section to calculate the resistance of the parallel combination of R_2 and R_4. Use measured values from Table 14–1 in your calculations.

R_T (b) Calculate the total resistance between A and G by adding the measured value of R_1, the calculated value of R_{BC}, the measured value of R_3, the calculated value of R_{DF}, and the measured value of R_8.

7. Connect the resistors as in Figure 14–9(b). Complete the fourth row of Table 14–2 using the procedures in steps 5 and 6. To calculate R_{DF}, calculate the resistance of the series combination of R_3 and R_7 in parallel with resistance R_8.

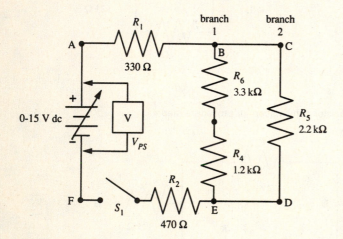

Figure 14–10. Circuit for procedure step 8.

8. With power **off** and switch S_1 **open,** connect the circuit of Figure 14–10. Power **on.** Adjust V_{PS} to 15 V and maintain this voltage for the next step.

9. **Close** S_1. Measure the voltage across R_1, R_2, branch 1 (points B and E) (V_{BE}), branch 2 (V_{CD}), points A and E (V_{AE}), and points B and F (V_{BF}). Record the values in Table 14–3.

10. Using the rated values of the resistors supplied, design a four-resistor series-parallel circuit as in Figure 14–1, such that the circuit will draw a total current I_T of approximately 5 mA when connected across a 10-V dc supply. On a separate sheet of paper, draw the circuit showing all resistor values, an ammeter to measure I_T, and a switch S_1 to **open** and **close** the circuit. Show all calculations used to find the circuit resistances.

11. With power **off,** and S_1 **open,** connect the circuit you designed in step 10. Measure R_T.

CAUTION: Do not attempt to measure R_T with power **on.**

After approval by the instructor, turn power **on** and **close** S_1. Measure V_{PS} and I_T. Record all values in Table 14–4. Power **off;** S_1 **open.**

ANSWERS TO SELF TEST

1. 96
2. 706
3. 30.7
4. 120
5. 60
6. 694
7. 2278
8. 11
9. 148.5

EXPERIMENT 14

TABLE 14–1.   Measured Values of Resistors

Resistor	R_1	R_2	R_3	R_4	R_5	R_6	R_7	R_8
Rated value, Ω	330	470	560	1.2 k	2.2 k	3.3 k	4.7 k	10 k
Measured value, Ω								

TABLE 14–2.   Ohmmeter Method for Determining R_T in a Series-Parallel Network

Steps	Measured Value, Ω			Calculated Value, Ω			
	R_T	R_{BC}	R_{DF}	R_T (a)	R_{BC}	R_{DF}	R_T (b)
2, 3 [Figure 14–8(a)]			✕			✕	
4 [Figure 14–8(b)]			✕			✕	
5, 6 [Figure 14–9(a)]							
7 [Figure 14–9(b)]							

TABLE 14–3.   Branch Voltage in a Series-Parallel Network

V Applied	V_1 (across R_1)	V_2 (across R_2)	Branch 1 V_{BE}	Branch 2 V_{CD}	V_{AE}	V_{BF}

TABLE 14–4.   Design Problem

Design Values			Measured Values		
V_A	I_T	R_T	V_{PS}	I_T	R_T
10 V	5 mA				

QUESTIONS

1. Explain, in your own words, the rules for finding the total resistance of a series-parallel circuit.

_____

_____

2. Explain why it is essential to disconnect power from a circuit before measuring resistance in the circuit with an ohmmeter.

_____

_____

3. When measuring the value of a resistor in a series-parallel circuit, one lead of the resistor should be disconnected from the circuit. Explain why.

_____

_____

_____

_____

4. What does your data in Tables 14–1 and 14–2 confirm about the total resistance of a series-parallel circuit? Refer to specific measurements in the tables to support your conclusions.

_____

_____

_____

_____

_____

_____

5. What measurements will you need to make to find the current in each resistor of a series-parallel circuit?

_____

_____

_____

6. Refer to Figure 14–7 (neglect the resistor values). If I_2 is the current through R_2, $I_{4,5}$ is the current through resistors R_4 and R_5, and I_T is the total current supplied by V, what is the relationship between $I_2 R_2$, $I_3 R_3$, $I_{4,5} (R_4 + R_5)$, and $I_T R_T$, where R_T is the total resistance of the series-parallel circuit? (State the relationship as a mathematical formula.)

_____

_____

_____

_____

7. Refer to question 6. Do your data in Tables 14–2 and 14–3 support your answer? If so, cite specific measurements.

_____

_____

_____

_____

_____

DIRECT-CURRENT ANALOG METER PRINCIPLES

BASIC INFORMATION

Multimeters

The proper measurement of voltage, current, and resistance values is essential for an electronic technician. Technicians make these measurements on a daily basis using meters called multimeters. These meters may be of an analog or digital variety as shown in Figure 15–1 (p. 108). An analog multimeter is commonly referred to as a volt-ohm-millammeter or VOM while the digital meter is often called a digital multimeter or DMM. Understanding the basic characteristics of these multimeters is vital to making safe and accurate measurements. This experiment will focus on the analog direct-current meter used to measure current and voltage. This experiment should also serve as a practical application of earlier series-parallel circuit concepts. As each of the multimeter circuits are presented and analyzed, specific attention is given to how series-parallel circuit theory is applied.

Current Meters

Analog current meters generally use a moving coil meter movement as shown in Figure 15–2 (p. 108). The construction consists of a coil of fine wire wrapped on a drum, which is mounted between the poles of a permanent magnet. When a direct current flows through the wire, a magnetic field is created which reacts to the permanent magnetic field. This causes the drum to rotate and deflects the meter's pointer. The level of current flowing through the wire determines the amount of pointer deflection. When the current through the moving coil stops, a restoring spring returns the pointer back to the zero position. The direction of the meter current determines the direction of pointer movement; either upscale to the right or off-scale to the left.

The two most important characteristics of this meter movement are its internal resistance, r_M, and its full-scale deflection current, I_M. The full-scale current I_M is the amount of current needed to cause the pointer to deflect to the last mark on the right side of the meter's printed scale. I_M values may range from 10 μA to over 20 mA. Typical I_M values for VOMs are 50 μA or 1 mA. A meter movement with an I_M value of 50 μA would have greater meter sensitivity than one with a 1 mA rating.

OBJECTIVES

1 To determine the characteristics of an analog meter movement

2 To calculate shunt and multiplier resistor values

3 To analyze the effects of meter loading

4 To apply series/parallel circuit concepts

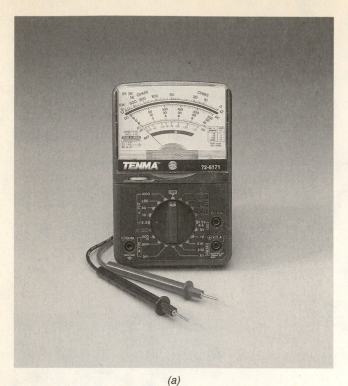

(a)

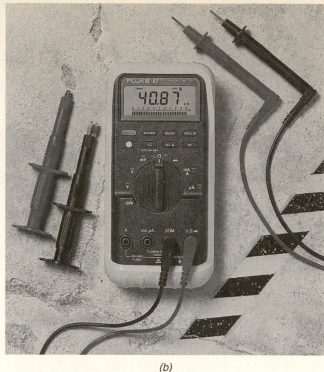

(b)

Figure 15–1. (a) Analog VOM. *Courtesy of MCM Electronics.* (b) Handheld DMM. *Courtesy Fluke Corporation.*

The internal resistance value, r_M, of the meter movement is the wire resistance of the moving coil. This r_M value will depend on the moving coil's number of turns and the wire diameter. The greater the number of turns and smaller the wire diameter, the greater the resistance will be. This resistance value generally correlates to the moving coil's I_M value. A 2 kΩ value is typical for a 50 μA movement, while 50 Ω is an approximate value for a 1 mA movement.

Multi-Range Current Meters · · · · · · · · · · ·

In most cases, the current value that is being measured will be greater than the meter's I_M value. If excessive current is allowed to flow through the meter movement, the pointer will deflect very rapidly to the extreme end of its range and will possibly be damaged when it hits the meter movement's mechanical stops. The coil itself may burn open. To safely measure these higher current values, a parallel resistor, called a meter shunt, R_{SH}, must be used. A meter shunt is a precision resistor connected across the meter movement to act as a current bypass. The shunt resistor will therefore bypass a specific fraction of the circuit's current around the meter movement, thus, extending the meter's range. This is shown in Figure 15–3. Notice that the

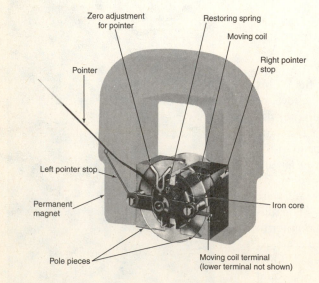

Figure 15–2. Moving coil meter movement.

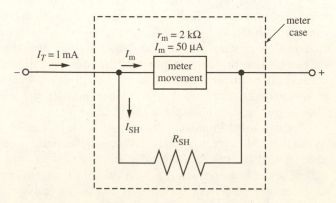

Figure 15–3. Since the shunt resistor is in parallel with the meter movement, their respective voltages will be the same.

total current, I_T, to be measured is 1 mA and the meter movement's I_M value is 50 μA.

In Figure 15–3, a total meter current of 1 mA must produce full-scale meter deflection. This basic meter is actually a simple parallel circuit where $I_T = I_{SH} + I_M$. Therefore, $I_{SH} = I_T - I_M$ or 1 mA $-$ 50 μA = 950 μa. Using parallel circuit theory to solve for the shunt resistance value, we need to know the current through the shunt and the voltage drop across it at full-scale deflection, V_M. Since the shunt resistor is in parallel with the meter movement, their respective voltages will be the same. The voltage drop across the meter movement can be found by

$$V_M = I_M \times r_M$$

$$= 50~\mu A \times 2~k\Omega$$

$$V_M = 0.1~V$$

Therefore, the voltage across R_{SH} will also equal 0.1 V at full-scale deflection.

Now the shunt resistor value can be solved for.

$$R_{SH} = V_M/I_{SH}$$

$$= 0.1~V/950~\mu A$$

$$R_{SH} = 105.3~\Omega$$

The parallel combination of the shunt resistor and the meter movement resistance results in a total meter resistance, R_M, of

$$R_M = r_M//R_{SH}$$

$$= 2~k\Omega//105.3~\Omega$$

$$R_M = 100~\Omega$$

To convert this meter to measure higher current values, R_{SH} will have to be reduced. If $I_T = 10$ mA

$$I_{SH} = I_T - I_M$$

$$= 10~mA - 50~\mu A$$

$$I_{SH} = 9.95~mA$$

$$R_{SH} = V_M/I_{SH}$$

$$= 0.1~V/9.95~mA$$

$$R_{SH} = 10.05~\Omega$$

$$R_M = 2~k\Omega/10.05~\Omega$$

$$R_M = 10~\Omega$$

Multi-Range Voltmeter

As you have seen, a moving-coil meter movement reacts to a current flowing through it. With the addition of a series resistor, called a multiplier resistor, it can also be used to measure voltage. This series combination voltmeter circuit is shown in Figure 15–4. To determine the value of resistance needed for the multiplier resistor, R_{mult}, simple series circuit concepts can be used. The total meter resistance, R_T, is equal to the sum of the series resistances as shown by the equation

$$R_T = R_{mult} + r_M$$

Since R_{mult} is in series with r_M, the full-scale current needed through the meter movement will be the same as the current through R_{mult}. If you know the full-scale voltage that you want your meter to measure, R_T can be calculated by dividing the full-scale voltage by the full-scale current.

$$R_T = \text{full-scale}~V/\text{full-scale}~I$$

Using Figure 15–4,

$$R_T = 2.5~V/50~\mu A$$

$$R_T = 50~k\Omega$$

Since $R_T = R_{mult} + r_M$, R_{mult} can be found by subtracting your known r_M value from R_T.

$$R_{mult} = R_T - r_M$$

$$= 50~k\Omega - 2~k\Omega$$

$$R_{mult} = 48~k\Omega$$

R_{mult} can also be calculated by rearranging the basic equation so that

$$R_{mult} = \frac{\text{full-scale}~V}{\text{full-scale}~I} - r_M$$

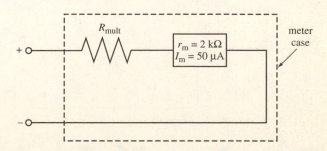

Figure 15–4. Series combination voltmeter circuit.

If a full-scale voltage range of 10 V is needed, R_{mult} can be found by

$$R_T = \text{full-scale } V/\text{full-scale } I$$

$$= 10 \text{ V}/50 \ \mu\text{A}$$

$$R_T = 200 \text{ k}\Omega$$

$$R_{mult} = R_T - r_M$$

$$= 200 \text{ k}\Omega - 2 \text{ k}\Omega$$

$$R_{mult} = 198 \text{ k}\Omega$$

Meter Loading

As the full-scale meter range of a voltmeter changes from 2.5 V to 10 V, the total meter resistance, R_T, increases from 50 kΩ to 200 kΩ. Voltmeters are often rated in ohms of resistance needed to measure 1 V of deflection. This is called its ohms per volt or Ω/V rating.

On the 2.5 V range 50 kΩ of resistance was needed. A meter's Ω/V value can be found by

$$\Omega/V = R_T/V_{range}$$

$$= 50 \text{ k}\Omega/2.5 \text{ V}$$

$$\Omega/V = 20 \text{ k}\Omega/V$$

On the 10 V range

$$\Omega/V = R_T/V_{range}$$

$$= 200 \text{ k}\Omega/10 \text{ V}$$

$$\Omega/V = 20 \text{ k}\Omega/V \text{ once again!}$$

The Ω/V rating can also be found if one knows the meter's full-scale current I_M value.

$$\Omega/V = 1/I_M$$

$$= 1/50 \ \mu\text{A}$$

$$\Omega/V = 20 \text{ k}\Omega$$

The meter's total resistance R_M value is important to know when measuring voltages. When measuring voltage, your meter is placed in parallel with the circuit or component. If the meter's resistance is too low it may seriously affect the reading, causing an inaccurate measurement. This is referred to as voltmeter loading. This is shown in Figure 15–5. The meter will measure 3.33 V, instead of the 5 V that it should measure. As illustrated here, a voltmeter with a higher R_M or Ω/V rating will reduce the meter loading. As compared to a DMM, VOM's generally have a greater loading effect because of their low Ω/V rating. A typical DMM will have an input resistance of 10 mΩ or more regardless of what voltage range it is set to.

SUMMARY

1. Safe and accurate voltage, current, and resistance measurements are critical in the job of an electronics technician.

2. Analog multimeters or VOMs are practical applications of series/parallel circuits.

3. Because of the low resistance of a current meter, it must be connected in series to prevent damage to the meter or circuit.

4. Two important characteristics of a meter movement are its full-scale deflection current, I_M, and its internal resistance, r_M.

5. A multi-range current meter uses a special precision bypass resistor called a shunt.

6. A multi-range voltmeter uses a multiplier resistor to extend its full-scale voltage range.

7. Voltmeters generally have a Ω/V rating which can be calculated if the meter's full-scale deflection current, I_M, value is known.

8. Some meters may cause meter loading of circuits, which will produce inaccurate measurements.

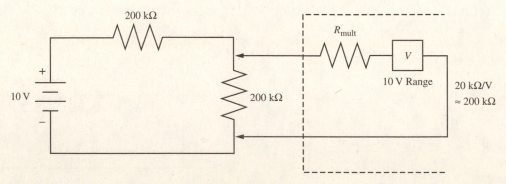

Figure 15–5. Voltmeter loading. When measuring voltage, the meter is placed in parallel with the circuit or component. If the meter's resistance is too low, it may cause an inaccurate measurement.

Check your understanding by answering the following questions:

1. An analog VOM generally uses a _____ meter movement.

2. (True/False) A meter movement with a smaller I_M value usually will have a larger r_M value.

3. A current meter must be connected in _____ with the component being measured.

4. When using a 1 mA meter movement, the shunt resistor must carry _____ mA of current when measuring 100 mA.

5. A _____ resistor is added to a basic meter movement to extend its voltage range.

6. The total resistance of a voltmeter can be determined if you know the meter's full-scale V rating and its _____ rating.

7. (True/False) The ohms-per-volt value is the same for all voltage ranges on a given analog voltmeter.

8. A voltmeter's resistance should be much _____ than the circuit being measured to prevent voltmeter loading.

MATERIALS REQUIRED

Power Supplies:
- Variable 0–15 V dc, regulated

Instruments:
- DMM

Resistors:
- 2 10 kΩ, (½-W, 5%)
- Assorted resistors for shunt combinations

Miscellaneous:
- 1 1 mA full-scale meter movement
- 1 Resistance Decade Box
- 1 SPDT Switch

PROCEDURE

A. Current Meter Characteristics

A1. With the power supply switch in the **off** position, connect the circuit shown in Figure 15–6.

A2. Set the voltage source to zero volts and turn the power switch to the **on** position.

A3. Adjust the variable dc voltage source until the test meter displays its full-scale current, I_M. Read the full-scale current value measured with the series connected DMM. Record this value in Table 15–1 (p. 113).

A4. Adjust the variable dc voltage source to obtain each of the test meter readings shown in Table 15–1.

At each level, record the corresponding DMM current value in Table 15–1.

A5. Next, you need to determine the meter's internal resistance, r_M, value. Use the circuit shown in Figure 15–7 to do this. With the power supply turned **off**, connect the circuit without the parallel resistance decade box.

A6. Turn **on** the power supply and adjust its output to produce full-scale deflection on the test meter.

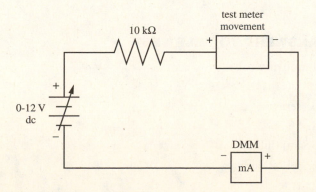

Figure 15–6. Circuit for procedure steps A1–A4.

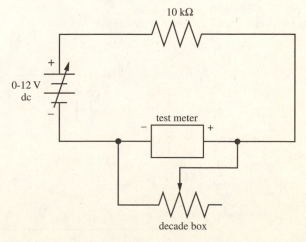

Figure 15–7. Circuit for procedure steps A5 through A9.

A7. Set the decade box to a value greater than 5 kΩ and connect it across the test meter.

A8. Now, adjust the resistance of the decade box so that the test meter deflects to ½ scale. This will occur when the decade box resistance is equal to r_M.

A9. Remove the decade box from the circuit and measure its resistance. Record this value in the provided space.

B. Multi-Range Current Meters

B1. Knowing the meter's r_M and I_M values, design the shunt resistor values needed for the multi-range current meter shown in Figure 15–8. The full-scale current ranges are 10 mA and 50 mA. Show your calculations in the provided space.

B2. Construct and test your circuit for each of the currents listed in Table 15–2. You may need to use resistor combinations to obtain the calculated shunt values. Turn **off** the power supply when changing current ranges and building your circuit.

C. Multi-Range Voltmeters

C1. Next, you will now design and construct a multi-range voltmeter as shown in Figure 15–9. This volt-

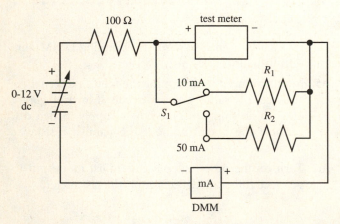

Figure 15–8. Circuit for part B of procedure.

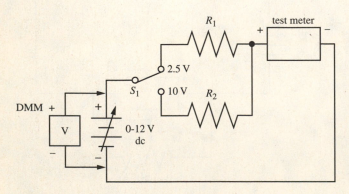

Figure 15–9. Circuit for part C of procedure.

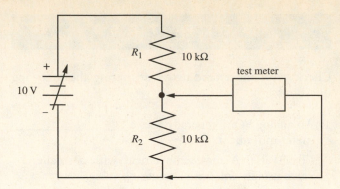

Figure 15–10. Circuit for part D of procedure.

meter will have the full-scale voltage ranges of 2.5 V and 10 V. For each of these ranges, calculate the required multiplier resistor needed. Show your calculation in the provided space.

C2. Draw a picture of the face of your test meter indicating the 2.5 V full-scale range and the corresponding deflection current value.

C3. Now, construct and test your voltmeter for each of the applied voltages shown in Table 15–3. Use a DMM to set the actual applied voltage. Use a resistor decade box for obtaining your multiplier values. Enter your results in Table 15–3.

D. Meter Loading

D1. Using the multi-range voltmeter that you constructed, determine its Ω/V rating.

D2. Next, construct the circuit shown in Figure 15–10.

D3. Calculate the voltage drop across R_2 without the voltmeter connected across it. Enter this value in Table 15–4.

D4. Using your meter's Ω/V rating, calculate the voltage that you should measure across R_2 due to meter loading on the 10 V range. Enter this value in Table 15–4.

D5. Now, use your constructed voltmeter to measure V_{R2}. Also measure this value with your DMM. Record these values in Table 15–4.

ANSWERS TO SELF TEST

1. moving coil
2. true
3. series
4. 99
5. multiplier
6. full-scale I
7. true
8. greater

Name _____   Date _____

TABLE 15–1

Test Meter Reading	DMM Reading, mA
I_M (full-scale)	
$0.8 \times I_M$	
$0.6 \times I_M$	
$0.4 \times I_M$	

$I_M =$ _____ $r_M =$ _____

Shunt Resistor Calculations:

R_1:

R_2:

TABLE 15–2

Test Current Value, mA	Actual Meter Movement Current, mA	Equivalent Scale Current Reading, mA	DMM Reading, mA
4			
6			
8			
10			
20			
30			
40			
50			

Multiplier Resistor Calculations:

2.5 V Range R_1:

10 V Range R_2:

TABLE 15–3.

Test Voltage Value, V	Actual Meter Movement Current, mA	Equivalent Scale Voltage Reading, V
0.5		
1.0		
1.5		
2.5		
3.0		
5.0		
7.0		
10		

TABLE 15–3.

Calculated V_{R2} without Loading, V	calculated V_{R2} with Loading, V	Measured V_{R2} with Voltmeter, V	Measured V_{R2} with DMM, V

QUESTIONS

1. Based on actual measured values, was your constructed current meter linear or nonlinear? Support your conclusion with lab values.

_____

_____

_____

_____

2. As you changed your meter's current range from 10 mA to 50 mA, what happened to your meter's total resistance R_M value?

_____

_____

_____

_____

3. Why was it necessary to turn off the power supply when changing ranges on your constructed current meter?

_____

_____

_____

4. What happens to a VOM's Ω/V rating as you change from the 2.5 V range to the 10 V range?

_____

_____

_____

5. How large should your meter's Ω/V rating be to prevent meter loading when measuring voltage?

_____

_____

_____

6. How does your constructed voltmeter's Ω/V rating compare to a typical DMM rating?

_____

_____

_____

_____

KIRCHHOFF'S VOLTAGE LAW (ONE SOURCE)

BASIC INFORMATION

Kirchhoff's voltage law is used to solve complex electric circuits. The law, named for Gustav Robert Kirchhoff (1824–1887), the physicist who formulated it, is the basis for modern circuit analysis.

Voltage Law

In the circuit of Figure 16–1, the series resistors R_1, R_2, R_3, and R_4 can be replaced by their total or equivalent resistance R_T, where

$$R_T = R_1 + R_2 + R_3 + R_4 \qquad (16\text{–}1)$$

Use of R_T will not affect the total current I_T. The relationship between I_T, R_T, and the voltage source V is given by Ohm's law,

$$V = I_T \times R_T \qquad (16\text{–}2)$$

Substituting formula (16–1) in (16–2) gives

$$V = I_T (R_1 + R_2 + R_3 + R_4)$$

which, upon multiplication, becomes

$$V = I_T R_1 + I_T R_2 + I_T R_3 + I_T R_4 \qquad (16\text{–}3)$$

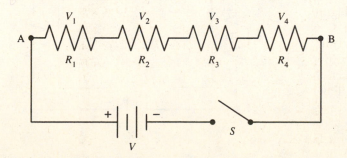

Figure 16–1. Voltages across resistors in a series circuit.

OBJECTIVES

1 To find a relationship between the sum of the voltage drops across series-connected resistors, and the applied voltage

2 To verify experimentally the relationship found in objective 1

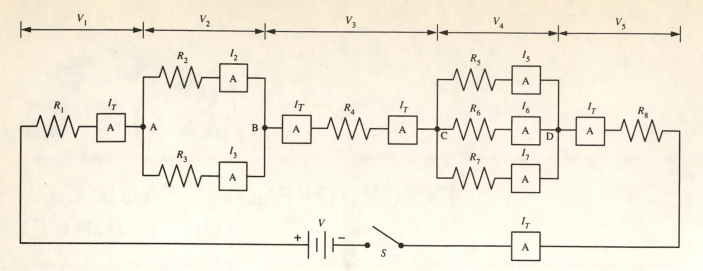

Figure 16–2. Application of Kirchhoff's law to a series-parallel circuit.

Because Ohm's law applies to any part of a circuit as well as the entire circuit, formula (16–3) shows that

$I_T R_1$ = the voltage drop across $R_1 = V_1$

$I_T R_2$ = the voltage drop across $R_2 = V_2$

$I_T R_3$ = the voltage drop across $R_3 = V_3$

$I_T R_4$ = the voltage drop across $R_4 = V_4$

Formula (16–3) can now be rewritten as

$$V = V_1 + V_2 + V_3 + V_4 \qquad \textbf{(16–4)}$$

Formula (16–4) is the mathematical expression of Kirchhoff's voltage law.

Formula (16–4) may be generalized for circuits containing one or more series-connected resistors in a closed circuit. The law also applies to series-parallel circuits (Figure 16–2). Here, $V = V_1 + V_2 + V_3 + V_4 + V_5$, where V_1, V_3, and V_5 are the voltage drops across R_1, R_4, and R_8, respectively. Voltages V_2 and V_4 are across the parallel circuits between A and B and between C and D, respectively.

Expressed in words, formula (16–4) states that in a closed circuit or loop, the applied voltage equals the sum of the voltage drops in the circuit.

The use of algebraic signs or polarity is helpful in solving electric circuit problems. The circuit of Figure 16–3 illustrates the convention used in assigning a + or a − sign to a voltage in a circuit. In the case of electron-flow current, electrons move from a negative to positive potential. The arrow in Figure 16–3 shows the direction of current, and the − and + signs indicate the following: Point A is negative with respect to B; point B is negative with respect to C; point C is negative with respect to point D; and D is negative with respect to point E. This is consistent with our assumption of electron-flow current in this circuit. With regard to the voltage source,

point E is positive with respect to point A, indicating a voltage rise.

To establish the algebraic sign for the voltages in the closed circuit, move in the direction of assumed current. Consider as positive any voltage source or voltage drop whose + (positive) terminal is reached first, and negative any voltage source or drop whose − (negative) terminal is reached first. Starting at point A in Figure 16–3 and moving in the direction of current, we have $-V_1$, $-V_2$, $-V_3$, $-V_4$, and $+V$. With this convention in mind, Kirchhoff's voltage law may now be generalized as follows.

The algebraic sum of the voltages in a closed circuit equals zero.

Applying the convention on signs and Kirchhoff's law to the closed circuit of Figure 16–3 and starting at point A, we may write the following:

$$-V_1 - V_2 - V_3 - V_4 + V = 0 \qquad \textbf{(16–5)}$$

Is this formula consistent with formula (16–4)? Yes, for by transposing the terms on the right side of formula

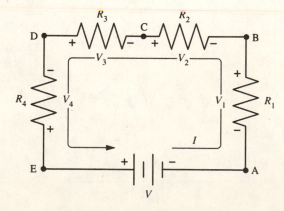

Figure 16–3. Convention for assigning polarity to voltages in a closed circuit.

(16–4) to the left side, we get $V - V_1 - V_2 - V_3 - V_4 = 0$, a result identical with that in formula (16–5).

Kirchhoff's Voltage Law (KVL) is a powerful and valuable tool when analyzing or solving many types of circuits, as well as in troubleshooting.

SUMMARY

Kirchhoff's voltage law can be expressed in two ways.

1. The sum of the voltage drops in a closed circuit equals the applied voltage.
2. The algebraic sum of the voltages in a closed circuit equals zero.

S E L F   T E S T

Check your understanding by answering the following questions:

1. In Figure 16–1 $V_1 = 3$ V, $V_2 = 5.5$ V, $V_3 = 6$ V, and $V_4 = 12$ V. The applied voltage V must then equal _____ V.

2. In Figure 16–2 $V_1 = 1.5$ V, $V_2 = 2.0$ V, $V_4 = 2.7$ V, $V_5 = 6$ V, and $V = 15$ V. The voltage $V_3 = $ _____ V.

MATERIALS REQUIRED

Power Supply:
- Variable 0–15 V dc, regulated

Instruments:
- DMM or VOM

Resistors (½-W, 5%):
- 1 330-Ω
- 1 470-Ω
- 1 820-Ω
- 1 1-kΩ
- 1 1.2-kΩ
- 1 2.2-kΩ
- 1 3.3-kΩ
- 1 4.7-kΩ

Miscellaneous:
- SPST switch

PROCEDURE

1. Measure each of the resistors supplied and record its value in Table 16–1 (p. 121).
2. With $V_{PS} = 15$ V and using the rated value of each of the resistors, calculate the individual voltage drops across $R_1(V_1)$, $R_2(V_2)$, $R_3(V_3)$, and $R_4(V_4)$ of Figure 16–4. Record the calculated values in Table 16–2. Also, record V_{PS} and the sum of the calculated voltages.

3. With power **off** and switch S_1 **open,** connect the circuit of Figure 16–4. Turn the power **on.** Adjust the power supply so that $V_{PS} = 15$ V.
4. **Close** S_1. Measure the voltage across R_1 (V_1), R_2 (V_2), R_3 (V_3), and R_4 (V_4) and record the values in Table 16–2. Calculate the sum of the voltages V_1, V_2, V_3, and V_4 and record your answer in Table 16–2. S_1 **open;** power **off.**

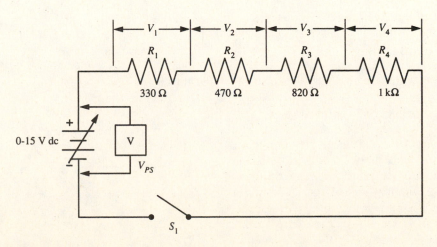

Figure 16–4. Circuit for procedure step 2.

5. With $V_{PS} = 15$ V and using Figure 16–5, calculate the voltage drops of V_1, V_2, V_3, V_4, and V_5. Record the values in Table 16–2. Also, record V_{PS} and the sum of the calculated voltages.

6. Connect the circuit in Figure 16–5. Power **on.** Adjust the power supply so that $V_{PS} = 15$ V.

7. **Close** S_1. Measure voltages V_1, V_2, V_3, V_4, and V_5, as shown in Figure 16–5. Record the values in Table 16–2.

Calculate the sum of the voltages V_1, V_2, V_3, V_4, and V_5 and record your answer in Table 16–2. S_1 **open;** power **off.**

ANSWERS TO SELF TEST

1. 26.5
2. 2.8

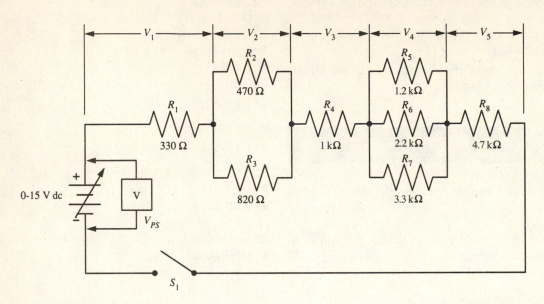

Figure 16–5. Circuit for procedure step 5.

TABLE 16–1. Color-Coded Values of Experimental Resistors

	R_1	R_2	R_3	R_4	R_5	R_6	R_7	R_8
Rated value, Ω	330	470	820	1 k	1.2 k	2.2 k	3.3 k	4.7 k
Measured value, Ω								

TABLE 16–2. Verifying Kirchhoff's Voltage Law

Step	V_{PS}, V	V_1, V	V_2, V	V_3, V	V_4, V	V_5, V	Sum of Vs, V
2							
4							
5							
7							

QUESTIONS

1. State, in your own words, the relationship between the voltage drops across series-connected resistors and the voltage applied to the entire series circuit.

——

——

——

——

——

——

——

——

——

——

——

——

——

2. Express your answer for Question 1 as a mathematical formula.

——

——

——

3. Refer to Table 16–2. Do your experimental data support your answers to Questions 1 and 2? (Refer to actual data in the table.) If not, explain the discrepancy.

_____

_____

_____

_____

_____

_____

_____

_____

4. On a separate sheet of 8½ × 11 paper, design a series-parallel circuit resembling Figure 16–2. The applied voltage is 35 V. The current supplied by the power supply is 5 mA. Use only the eight resistors listed in the Materials Required list (use the rated values shown). Draw a fully labeled circuit diagram; show all design calculations and formulas used. The design specification allows a deviation of current ±1%. The supply voltage cannot vary.

Optional Activity ·

If electronic simulation software is available, build and test your design. Include all necessary meters to measure the power supply voltage and current. Did your circuit meet the design specifications? If not, and you are able to substitute a variable resistor for any of the fixed resistors, how would you make the circuit meet the design specifications?

KIRCHHOFF'S CURRENT LAW

BASIC INFORMATION

Current Law

Experiment 11 verified that the total current I_T in a circuit containing resistors connected in parallel is equal to the sum of the currents in each of the parallel branches. This was one demonstration of Kirchhoff's current law, limited to a parallel network. The law is general, however, and applies to any circuit. Kirchhoff's current law states that

The current entering any junction of an electric circuit is equal to the current leaving that junction.

In the series-parallel circuit of Figure 17–1 (p. 124), the total current is I_T. It enters the junction at A in the direction indicated by the arrow. The currents leaving the junction at A are I_1, I_2, and I_3, as shown. The currents I_1, I_2, and I_3 then enter the junction at B, and I_T leaves the junction at B. What is the relationship between I_T, I_1, I_2, and I_3?

The voltage across the parallel circuit can be found using Ohm's law:

$$V_{AB} = I_1 \times R_1 = I_2 \times R_2 = I_3 \times R_3$$

The parallel network may be replaced by its equivalent resistance R_T, in which case Figure 17–1 is transformed into a simple series circuit and $V_{AB} = I_T \times R_T$. It follows, therefore, that

$$I_T \times R_T = I_1 \times R_1 = I_2 \times R_2 = I_3 \times R_3 \qquad \textbf{(17–1)}$$

Formula (17–1) may be rewritten as

$$I_1 = I_T \times \frac{R_T}{R_1}$$

$$I_2 = I_T \times \frac{R_T}{R_2} \qquad \textbf{(17–2)}$$

$$I_3 = I_T \times \frac{R_T}{R_3}$$

Formula (17–2) is sometimes referred to as the Current Divider Rule. Adding I_1, I_2, and I_3 gives

$$I_1 + I_2 + I_3 = I_T \times \frac{R_T}{R_1} + I_T \times \frac{R_T}{R_2} + I_T \times \frac{R_T}{R_3}$$

OBJECTIVES

1. To find a relationship between the sum of the currents entering any junction of an electric circuit and the current leaving that junction

2. To verify experimentally the relationship found in objective 1

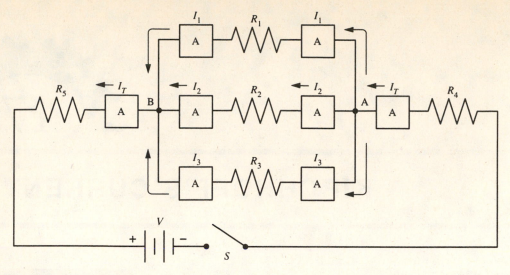

Figure 17–1. The total current through the supply is the sum of the currents in each of the branches.

$$I_1 + I_2 + I_3 = I_T \times R_T \left(\frac{1}{R_1} + \frac{1}{R_2} + \frac{1}{R_3} \right)$$

But

$$\frac{1}{R_1} + \frac{1}{R_2} + \frac{1}{R_3} = \frac{1}{R_T}$$

Therefore,

$$I_1 + I_2 + I_3 = I_T \times R_T \times \frac{1}{R_T} = I_T$$

That is,

$$I_T = I_1 + I_2 + I_3 \tag{17–3}$$

Formula (17–3) is a mathematical statement of Kirchhoff's law, applied to the circuit of Figure 17–1. In general, if I_T is the current entering a junction of an electric circuit, and $I_1, I_2, I_3, \ldots, I_n$ are the currents leaving that junction, then

$$I_T = I_1 + I_2 + I_3 + \cdots + I_n \tag{17–4}$$

This applies equally well if I_T is a current leaving a junction and $I_1, I_2, I_3, \ldots, I_n$ are the currents entering that junction.

Kirchhoff's current law is often stated in another way:

The algebraic sum of the currents entering and leaving a junction is zero.

Recall that this is similar to the formulation of Kirchhoff's voltage law: *The algebraic sum of the voltages in a closed path or loop is zero.*

Just as it was necessary to agree on a polarity convention for voltages in a loop, so it is necessary to agree on a current convention at a junction. If the current entering a junction is considered positive (+) and the current

leaving a junction is considered negative (−), then the statement that the algebraic sum of the currents entering and leaving a junction is zero can be shown to be identical with formula (17–4). Consider the circuit of Figure 17–2. The total current I_T enters the junction at A and is considered +. The currents I_1 and I_2 leave the junction at A and are designated −. Then,

$$+ I_T - I_1 - I_2 = 0 \tag{17–5}$$

and

$$I_T = I_1 + I_2 \tag{17–6}$$

Obviously, the two statements of Kirchhoff's current law lead to the same formula.

An example shows how Kirchhoff's current law may be applied to solve circuit problems. Suppose in Figure 17–3

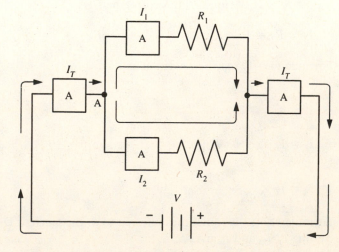

Figure 17–2. The algebraic sum of the currents entering and leaving a junction is equal to zero.

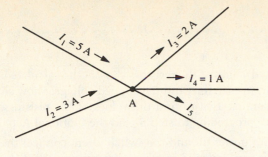

Figure 17–3. Currents entering and leaving junction A.

that I_1 and I_2 are currents entering the junction at A and are, respectively, +5 A and +3 A. Currents I_3, I_4, and I_5 are leaving A. Currents I_3 and I_4 are, respectively, 2 A and 1 A. What is the value of I_5? Applying Kirchhoff's current law,

$$I_1 + I_2 - I_3 - I_4 - I_5 = 0$$

and substituting the known values of current, we get

$$5 + 3 - 2 - 1 - I_5 = 0$$

$$5 - I_5 = 0$$

$$I_5 = 5 \text{ A}$$

SUMMARY

1. Kirchhoff's current law states that the current entering any junction of an electric circuit is equal to the current leaving that junction.

2. To use Kirchhoff's current law in solving circuit problems, polarity is assigned to current entering a junction (assume it is +) and to current leaving a junction (assume it is −).

3. Using the polarities given in 2, Kirchhoff's current law may be stated as follows: The sum of the currents entering and leaving a junction is zero. Thus, for Figure 17–1, at junction A

$$I_T - I_1 - I_2 - I_3 = 0$$

S E L F   T E S T

Check your understanding by answering the following questions:

1. In Figure 17–1 the current entering the junction at A is 0.5 A. $I_1 = 0.25$ A, $I_2 = 0.1$ A. The current I_3 must therefore equal _____ A.

2. In Figure 17–1 the current leaving the junction at B is 1.5 A. The sum of currents I_1, I_2, and I_3 must be _____ A.

3. In applying Kirchhoff's current law to junction B in Figure 17–1, the polarity given to each current by convention is as follows:
   (a) I_1 _____
   (b) I_2 _____
   (c) I_3 _____
   (d) I_T _____

4. The equation that describes the relationship among the currents at junction A in Figure 17–3 is _____.

5. In Figure 17–3, $I_2 = 4$ A, $I_3 = 4$ A, $I_4 = 3$ A, $I_5 = 1$ A. $I_1 =$ _____ A.

MATERIALS REQUIRED

Power Supply:
- Variable 0–15 V dc, regulated

Instruments:
See Note under Procedure for quantities
- DMM or VOM
- 0–10-mA milliammeter

Resistors (½-W, 5%):
- 1 330-Ω
- 1 470-Ω
- 1 820-Ω
- 1 1-kΩ
- 1 1.2-kΩ
- 1 2.2-kΩ
- 1 3.3-kΩ
- 1 4.7-kΩ

Miscellaneous:
- SPST switch

PROCEDURE

NOTE: This experiment requires numerous measurements of current in series-parallel circuits. If only one ammeter is available, it will be necessary to break the line for which a current reading must be taken. Disconnect power to the circuit by opening S_1 each time the position of the ammeter is changed.

1. Measure the resistance of each of the resistors supplied and record its value in Table 17–1.

2. With power **off** and S_1 **open,** connect the circuit of Figure 17–4. Power **on.** Adjust the power supply so that $V_{PS} = 15$ V.

3. **Close** S_1. Measure currents I_{TA}, I_2, I_3, I_{TB}, I_{TC}, I_5, I_6, I_7, I_{TD}, and I_{TE}. Record the values in Table 17–2. Calculate the sum of I_2 and I_3 and the sum of I_5, I_6, and I_7. Record your answers in Table 17–2. S_1 **open;** power **off.**

4. Design a series-parallel circuit consisting of three parallel branches and two series resistors similar to the circuit of Figure 17–1. The currents in the three parallel branches should be such that the current in the second branch is approximately twice the current in the first branch, and the current in the third branch is approximately three times the current in the first branch. (Stated another way, the currents in the three parallel branches are in the ratio 1:2:3, approximately.) Use only the resistors supplied for this experiment. The total current in the circuit is 6 mA. The maximum voltage available is 15 V. Show the position of the meters used to measure current through each resistor and I_T. Include a switch to disconnect power to your circuit. Draw a complete diagram of the circuit showing the

rated values of the resistors chosen, the calculated current in each line, and the applied voltage. Show all calculations used to find the values of the resistors. Before proceeding to step 5 read the Optional Activity.

5. With power **off** and the switch **open,** connect the circuit you designed in step 4. After approval of the circuit by your instructor, turn power **on.** Adjust the power supply to your design voltage. Read all currents in the circuit and record the values in Table 17–3. S_1 **open;** power **off.**

Optional Activity

This activity requires the use of electronic simulation software. Build a simulation of the circuit designed in step 4. Include the meters necessary to measure current in each parallel branch and in each series resistor. Set the power supply to your design voltage and record all the ammeter readings. Compare the results with the values found in step 5. Explain any discrepancies.

ANSWERS TO SELF TEST

1. 0.15
2. 1.5
3. (a) +; (b) +; (c) +; (d) −
4. $I_1 + I_2 - I_3 - I_4 - I_5 = 0$
5. 4

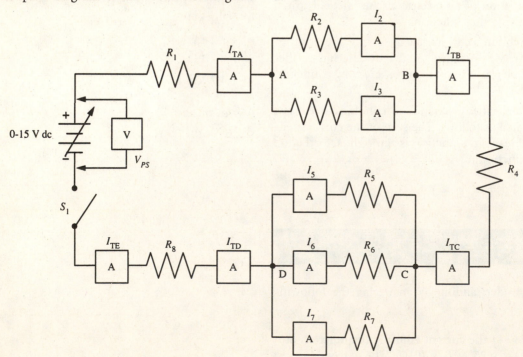

Figure 17–4. Circuit for procedure step 2.

TABLE 17–1. Measured Value of Resistors

	R_1	R_2	R_3	R_4	R_5	R_6	R_7	R_8
Rated value, Ω	330	470	820	1 k	1.2 k	2.2 k	3.3 k	4.7 k
Measured value, Ω								

TABLE 17–2. Verifying Kirchhoff's Current Law

	I_{TA}	I_2	I_3	I_{TB}	I_{TC}	I_5
current, mA						

	I_6	I_7	I_{TD}	I_{TE}	$I_2 + I_3$	$I_5 + I_6 + I_7$
current, mA						

TABLE 17–3. Design Problem Data

Calculated Value, mA				Measured Value, mA			
Branch 1 I_1	Branch 2 I_2	Branch 3 I_3	I_T	Branch 1 I_1	Branch 2 I_2	Branch 3 I_3	I_T

QUESTIONS

1. Explain, in your own words, the relationship between the currents entering and leaving a junction point in a circuit.

_____

_____

_____

_____

_____

_____

2. Write the relationship explained in Question 1 as a mathematical formula.

_____

_____

_____

_____

3. Refer to Figure 17–4. What information would you need to find I_2 and I_3 in this circuit?

_____

_____

_____

_____

_____

_____

_____

_____

4. Steps 4 and 5 in the procedure required you to design a series-parallel circuit with V_{PS} max = 15 V and I_T = 6 mA. Resistors R_1, R_2, and R_3 were in the ratio 1:2:3. Discuss your results, referring to your experimental data in Table 17–3. Your report should indicate whether you were able to meet the design specifications exactly. If not, explain the discrepancy. Also, explain how you went about choosing (or eliminating) each of the given resistors. (Submit your design circuit diagram with this report.)

_____

_____

_____

_____

_____

_____

_____

_____

_____

_____

_____

_____

_____

_____

_____

_____

_____

_____

_____

_____

_____

_____

VOLTAGE-DIVIDER CIRCUITS (LOADED)

BASIC INFORMATION

In the simple dc voltage-divider circuits studied in Experiment 10, no load current was drawn. The only current was the "bleeder" current in the divider network itself. In electronics, divider networks are frequently used as a source to supply voltage to a load, which draws current. When this is the case, the divider-voltage relationships that were obtained for no-load conditions no longer hold. The actual changes depend on the amount of current drawn and on the circuit connections. In the circuit of Figure 18–1, V is a constant-voltage source that maintains 15 V across the divider network with and without load. Without load, points A, B, and C are 5, 10, and 15 V, respectively, with respect to G. Moreover, a 5-mA bleeder current I_1 is drawn. Now if we add a load resistor R_L (as shown in Figure 18–2 p. 130 at point B) that draws 2 mA of load current I_L, the voltages at A and B will be different from the condition of no load. We can calculate the voltages in the loaded circuit from the information given.

Assume that there is a bleeder current I_1 when a load current of I_L is drawn by R_L. By Kirchhoff's law we can set up the formula

$$I_1 (R_2 + R_3) + (I_1 + I_L)(R_1) = 15 \qquad (18-1)$$

(see Figure 18–2 p. 130)

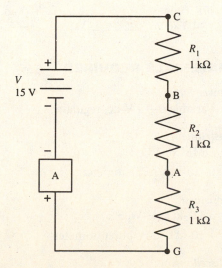

Figure 18–1. Voltage divider without load.

OBJECTIVES

1 To find what effect load has on the voltage relationships in a voltage-divider circuit

2 To verify the results of objective 1 by experiment

129

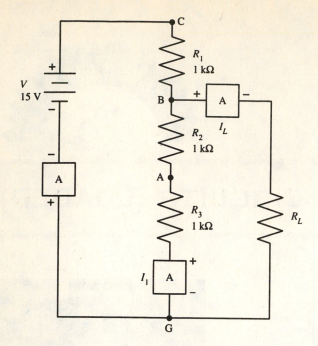

Figure 18–2. Voltage divider with fixed load R_L.

Solving for I_1 (since this is the only unknown), we get

$$I_1 (R_1 + R_2 + R_3) = 15 - (R_1)I_L \qquad \text{(18–2)}$$

$$I_1 = \frac{15 - (I_L \times R_1)}{R_1 + R_2 + R_3} \qquad \text{(18–3)}$$

Substituting 1 kΩ for each of R_1, R_2, and R_3, and $I_L = 2$ mA in formula (18–3), we get

$$I_1 = 0.00433 \text{ A, or } 4.33 \text{ mA} \qquad \text{(18–4)}$$

The voltage from A to G is

$$V_{AG} = I_1 \times R_3$$

$$V_{AG} = 4.33 \text{ mA} \times 1 \text{ k}\Omega = 4.33 \text{ V} \qquad \text{(18–5)}$$

and the voltage from B to G is

$$V_{BG} = 4.33 \text{ mA} \times 2 \text{ k}\Omega = 8.66 \text{ V} \qquad \text{(18–6)}$$

This shows that both the voltages and the bleeder current are affected when a load is added to a voltage-divider circuit.

To verify this effect experimentally, it is necessary to have a voltage source across the divider that will remain constant with and without load. In Figure 18–2 the divider network consists of R_1, R_2, and R_3 connected to a voltage source. When a load resistor (R_L) is added as shown, the voltage source is readjusted if necessary to maintain the same voltage across the network as under no-load conditions. The required measurements are made under no-load and load conditions.

SUMMARY

1. A series circuit such as in Figure 18–1 is a simple unloaded voltage divider.

2. In an unloaded voltage divider such as the circuit of Figure 18–1 the voltage V across R_1 can be found by using the formula

$$V_1 = V \times \frac{R_1}{R_T}$$

where V is the source voltage and R_T is the sum of the series resistances. The voltages at points A, B, and C are given with respect to G. They can be calculated using the preceding formula.

3. If a load is added to a voltage divider, as in Figure 18–2, the bleeder current and the voltages across the resistors in the voltage divider will change.

4. To solve for the voltages and bleeder current in a loaded voltage divider, Kirchhoff's voltage and current laws can be used.

SELF TEST

Check your understanding by answering the following questions:

1. In the circuit of Figure 18–1 let $R_1 = 1$ kΩ, $R_2 = 1.8$ kΩ, $R_3 = 2.2$ kΩ, and $V = 15$ V.
   (a) What is the voltage V_1 across R_1?
       $V_1 = $ _____ V.
   (b) What is the voltage across the combination of R_2 and R_3? $V_{BG} = $ _____ V.
   (c) What is the voltage across R_3? $V_{AG} = $ _____ V.

2. The bleeder current in question 1 is _____ A, or _____ mA.

3. In the voltage divider of Question 1, a load resistor is placed across R_2 and R_3, as shown in Figure 18–2. If the load current I_L is 5 mA, what is the bleeder current? $I_1 = $ _____ A, or _____ mA.

4. For the conditions of Question 3, what are the following voltages? $V_1 = $ _____ V, $V_{BG} = $ _____ V, and $V_{AG} = $ _____ V.

MATERIALS REQUIRED

Power Supply:
- Variable 0–15 V dc, regulated

Instruments:
- DMM or VOM
- 0–10-mA milliammeter

Resistors:
- 3 1.2-kΩ (½-W, 5%)
- 1 10-kΩ, 2-W potentiometer

Miscellaneous:
- SPST switch

PROCEDURE

1. With power **off** and S_1 **open,** connect the circuit in Figure 18–3(*a*).

2. Power **on** and S_1 **closed.** Adjust the power supply so that $V_{PS} = 10$ V. Measure I_1 (called the *bleeder current*), voltage V_{BD}, and voltage V_{CD}. Record your answers in Table 18–1 (p. 133). **Open** S_1.

3. Connect the 10-kΩ potentiometer across points B and D as shown in Figure 18–3(*b*). With the voltmeter across the power supply, adjust the potentiometer until $I_L = 2$ mA with $V_{PS} = 10$ V. Measure I_1, V_{BD}, and V_{CD}. Record the values in Table 18–1. Do not change the setting of the potentiometer. **Open** S_1.

4. Disconnect the potentiometer from the circuit and use an ohmmeter to measure the load resistance R_L across points E and F. This is the load setting for $I_L = 2$ mA. Record this value in Table 18–1.

5. Reconnect the potentiometer across BD. **Close** S_1. Adjust the potentiometer until $I_L = 4$ mA. Again adjust the power supply as necessary to maintain $V_{PS} = 10$ V. Measure I_1, V_{BD}, and V_{CD}. Record the values in Table 18–1. Do not change the setting of the potentiometer. **Open** S_1.

6. Disconnect the potentiometer from the circuit and measure the load resistance as in step 4. This is the load setting for $I_L = 4$ mA. Record this value in Table 18–1.

7. Reconnect the potentiometer across BD. **Close** S_1. Adjust the potentiometer until $I_L = 6$ mA with $V_{PS} = 10$ V. Measure I_1, V_{BD}, and V_{CD}. Record the values in Table 18–1. Do not change the setting of the potentiometer. **Open** S_1.

8. Disconnect the potentiometer from the circuit and measure the load resistance, as in step 4. This is the load setting for $I_L = 6$ mA. Record this value in Table 18–1. **Open** S_1; turn the power **off.**

9. Use the methods discussed in the BASIC INFORMATION section, calculate the bleeder current I_1, voltages V_{BD} and V_{CD}, and the load resistance R_L for each load condition (0 mA, 2 mA, 4 mA, and 6 mA) in this experiment. Record your answers in Table 18–1.

ANSWERS TO SELF TEST

1. (a) 3; (b) 12; (c) 6.6
2. 0.003; 3
3. 0.002; 2
4. 7; 8; 4.4

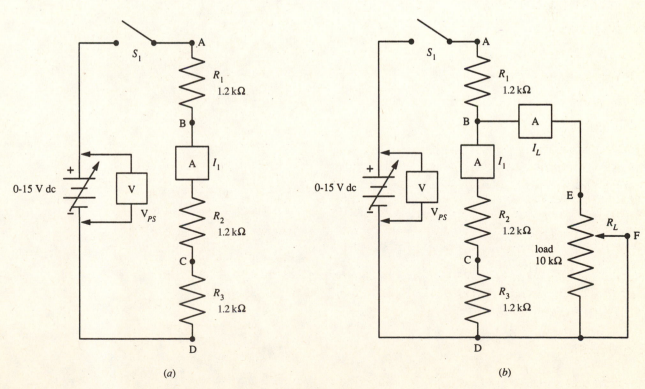

Figure 18–3. (*a*) Circuit for procedure step 1. (*b*) Circuit for procedure step 3.

TABLE 18–1. Effect of load on a Voltage Divider

			Measured Values				Calculated Values			
Steps	V	I_L (load current), mA	I_1, mA	V_{BD}, V	V_{CD}, V	R_L, Ω	I_1, mA	V_{BD}, V	V_{CD}, V	R_L, Ω
2	10	0				✕				
3, 4	10	2								
5, 6	10	4								
7	10	6								

QUESTIONS

1. Explain, in your own words, the effect load has on the voltage relationships in a voltage-divider circuit.

_____

_____

_____

_____

_____

_____

_____

_____

_____

_____

2. Refer to your data in Table 18–1. What effect does changing the load resistance have on load current? Give specific examples from your data. Explain the reason for this effect.

_____

_____

_____

_____

_____

_____

_____

_____

3. Refer to your data in Table 18–1. How is the bleeder current, I_1, affected by changes in the load current, I_L? Give specific examples from your data. Explain the reason for this effect.

_____

_____

_____

_____

_____

_____

_____

_____

4. Refer to Figure 18–3(b) and Table 18–1. What effect do changes in the load current, I_L, have on the divider tap voltages, V_{CD} and V_{BD}? Give specific examples from your data. Explain the reason for this effect.

_____

_____

_____

_____

_____

_____

_____

_____

_____

_____

5. Compare the measured values of I_1, V_{BD}, V_{CD}, and R_L with the calculated values. Explain any differences.

_____

_____

_____

_____

_____

_____

_____

_____

_____

DESIGNING VOLTAGE- AND CURRENT-DIVIDER CIRCUITS

BASIC INFORMATION

Designing a Voltage Divider for Specific Loads · · · · ·

The electronics technician may be called upon to design a voltage divider that will meet specified requirements of voltage, load current, and bleeder current. The process to follow is illustrated by a sample problem.

Problem 1. Design a voltage-divider circuit for a 30-V power supply. The loads that must be served are 50 mA at 30 V, and 40 mA at 25 V.

A "rule-of-thumb" value of bleeder current is 10 percent (approximately) of load current. The bleeder current under these conditions is 10 mA.

Solution. Draw a diagram, as in Figure 19–1. Label the known load currents and voltages. Designate the *unknown resistors* as R_1 and R_2. (The loads are shown as resistors drawing the required currents.)

Apply Kirchhoff's current law and show the currents entering and leaving the junctions A, B, and C. The total current I_T that must be sup-

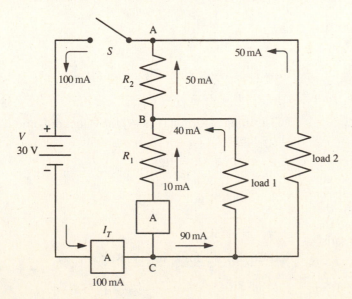

Figure 19–1. Voltage-divider circuit for problem 1.

135

plied is $I_T = 50\text{ mA} + 40\text{ mA} + 10\text{ mA} = 100\text{ mA}$. Current I_T is shown entering the junction at C and leaving the junction at A.

$$I_1 \times R_1 = 25\text{ V} \qquad (19\text{--}1)$$

or

$$R_1 = \frac{25\text{ V}}{I_1} \qquad (19\text{--}2)$$

Substituting the bleeder current value (10 mA) for I_1 gives

$$R_1 = \frac{25\text{ V}}{10\text{ mA}} = 2.5\text{ k}\Omega \qquad (19\text{--}3)$$

The current in R_2 is 50 mA, the sum of the bleeder current and load 1 current. Moreover, as is evident from Figure 19–1, the voltage from A to B is 5 V. Therefore,

$$(50\text{ mA})(R_2) = 5\text{ V} \qquad (19\text{--}4)$$

and

$$R_2 = \frac{5\text{ V}}{50\text{ mA}} = 100\ \Omega \qquad (19\text{--}5)$$

We have thus found the values of R_1 and R_2 that will deliver 40 mA at 25 V and 50 mA at 30 V with a bleeder current of 10 mA, from a 30-V power supply.

Designing Current-Divider Circuits · · · · · ·

It is sometimes necessary for the technician to design current-divider circuits. To avoid trial-and-error methods, study the conditions of the problem and apply the appropriate circuit formulas. The exact steps to follow in this procedure will depend on the nature of the design problem, but in general, the following steps should be taken to find the solution:

1. Draw a circuit diagram labeling all the known values and the unknowns.

2. Write the Ohm's and Kirchhoff's laws formulas that describe the electrical relationships involved.

3. Solve these formulas for the unknown component values.

4. Construct the circuit from the calculated values and test the circuit to see if the design requirements have been met.

Two problems illustrate this process.

Problem 2. Design the three-branch parallel circuit shown in Figure 19–2 so that a 15-V source V will deliver a total current I_T of 1.2 A. A further requirement is that the current I_T be distributed among R_1, R_2, and R_3 so that $I_1{:}I_2{:}I_3 = 2{:}4{:}6$. What values of R_1, R_2, and R_3 will satisfy these conditions?

Solution. We know V, I_T, and the ratios $I_1{:}I_2{:}I_3$. If we could find the actual values of I_1, I_2, and I_3, we would then

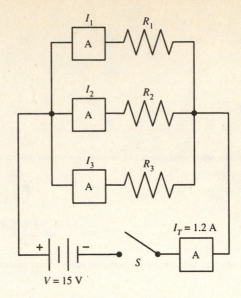

Figure 19–2. Current divider for problem 2.

know two of the Ohm's law values for each of the branches—namely, V and I. From V and I, R can be found, since

$$R = \frac{V}{I} \qquad (19\text{--}6)$$

We can set up two equations for I from the current ratios given.

$$\frac{I_1}{I_2} = \frac{2}{4}$$

$$\frac{I_1}{I_3} = \frac{2}{6} \qquad (19\text{--}7)$$

Solving for I_2 and I_3 in terms of I_1 yields

$$I_2 = \frac{4}{2}I_1 = 2I_1$$

$$I_3 = \frac{6}{2}I_1 = 3I_1 \qquad (19\text{--}8)$$

Now, applying Kirchhoff's current law,

$$I_T = I_1 + I_2 + I_3 = 1.2 \qquad (19\text{--}9)$$

Substituting in formula (19–9) the values of I_2 and I_3 from (19–8) gives

$$I_1 + 2I_1 + 3I_1 = 1.2 \qquad (19\text{--}10)$$

Solving,

$$6I_1 = 1.2$$

$$I_1 = 0.2\text{ A} \qquad (19\text{--}11)$$

Substituting 0.2 for I_1 in (19–8) gives

$$I_2 = 0.4\text{ A}$$

$$I_3 = 0.6\text{ A} \qquad (19\text{--}12)$$

Now we can find the values of R_1, R_2, and R_3.

$$R_1 = \frac{V}{I_1} = \frac{15\,V}{0.2\,A} = 75\,\Omega$$

$$R_2 = \frac{V}{I_2} = \frac{15\,V}{0.4\,A} = 37.5\,\Omega \qquad (19\text{–}13)$$

$$R_3 = \frac{V}{I_3} = \frac{15\,V}{0.6\,A} = 25\,\Omega$$

Problem 3. Design a divider network that will supply three resistive loads from a 24-V source such that loads 1, 2, and 3 receive 20 mA, 30 mA, and 50 mA, respectively. All three loads are in parallel, and the resistance of load 1 is 300 Ω.

Solution. Draw the three parallel loads R_1, R_2, and R_3. Since the current and resistance of load 1 are given, we can find the voltage across load 1 and also across each of the other two parallel loads. Thus,

$$V_1 = I_1 \times R_1 = (20\,mA)(300\,\Omega) = 6\,V \qquad (19\text{–}14)$$

We have a 24-V source. Therefore, we will have to drop 18 V before we reach the three parallel loads. A series-parallel circuit, such as that in Figure 19–3, will serve our requirements. We can now find the values of R_4 and loads R_2 and R_3.

First, solving for R_2 and R_3,

$$R_2 = \frac{V_{AB}}{I_2} = \frac{6\,V}{30\,mA} = 200\,\Omega$$

$$R_3 = \frac{V_{AB}}{I_3} = \frac{6\,V}{50\,mA} = 120\,\Omega \qquad (19\text{–}15)$$

Finally, the total current I_T in R_4 can be found by Kirchhoff's current law:

$$I_T = I_1 + I_2 + I_3 = 20\,mA + 30\,mA + 50\,mA$$

$$= 100\,mA$$

Therefore,

$$R_4 = \frac{18}{100\,mA} = 180\,\Omega \qquad (19\text{–}16)$$

SUMMARY

1. In designing either voltage- or current-divider circuits, a mathematical solution is found first. Ohm's and Kirchhoff's laws are applied in the process.

2. The general steps in solving a design problem are as follows:

   (a) Draw a diagram of the proposed circuit, labeling all the knowns and unknowns.

   (b) Set up one or more formulas that reflect the electrical relationships involved.

   (c) Solve these formulas for the values of the unknown components.

   (d) Construct the circuit from the calculated values and test the circuit to see if the design requirements have been met.

3. In designing voltage-divider circuits for specified load currents, a rule-of-thumb value for bleeder current is 10 percent (approximately) of load current.

SELF TEST

Check your understanding by answering the following questions:

1. It is required to design a voltage-divider circuit like that in Figure 19–4 (p. 138) so that a 20-V source V can supply two loads: load 1 is 100 mA at 5 V; load 2 is 50 mA at 20 V. Assume a bleeder current I_1 of 50 mA. The values of R_1 and R_2 that will accomplish this result are:

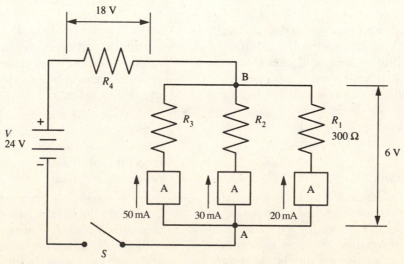

Figure 19–3. A voltage- and current-divider circuit that supplies three resistors, R_1, R_2, and R_3.

DESIGNING VOLTAGE- AND CURRENT-DIVIDER CIRCUITS **137**

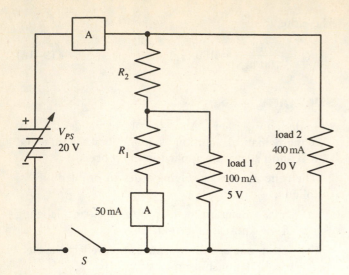

Figure 19–4. Circuit for Self-Test Questions 1 and 2.

(a) $R_1 =$ _____ Ω
(b) $R_2 =$ _____ Ω

2. For the same conditions as in Question 1, the resistances of the two loads are:
   (a) Resistance of load 1 is _____ Ω
   (b) Resistance of load 2 is _____ Ω

3. In a divider circuit similar to that in Figure 19–5 it is required to supply three equal-load resistors, drawing a total current of 150 mA, at a load voltage of 5 V. The supply source V is 50 V. The values of R_1, R_2, R_3, and R_4 that will achieve this are:
   (a) $R_1 =$ _____ Ω
   (b) $R_2 =$ _____ Ω
   (c) $R_3 =$ _____ Ω
   (d) $R_4 =$ _____ Ω

MATERIALS REQUIRED

Power Supply:
- Variable 0–15 V dc, regulated

Instruments:
- DMM or VOM
- 0–10-mA milliammeter

Resistors:
- Specify ½-W, 5% resistors as required by your design

Miscellaneous:
- SPST switch

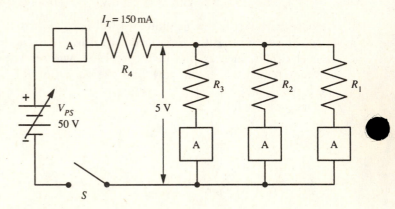

Figure 19–5. Circuit for Self-Test Question 3.

PROCEDURE

The following experiments require you to design circuits according to a number of specifications. Use commercial values of ½-W, 5% resistors. Check with your instructor as to the values available in your shop or laboratory.

If electronic simulation software is available, build and test your simulated circuit designs first. Measure and record all currents and voltages and include these values in the tables prepared in steps A3 and B3.

A. Voltage-Divider Design

A1. Design a voltage-divider circuit for a 15-V regulated power supply. The circuit must feed a 3-mA load at 9 V. Draw a circuit diagram showing the values of all components and the significant voltages and currents throughout the circuit. Your circuit should include an SPST switch for turning power to the circuit **on**

and off. Present your design in a neat organized report, listing all components required. Show all calculations leading to your design components. Have your instructor approve your design before proceeding to the next step.

A2. With power **off** and the switch **open,** connect the voltage-divider circuit you designed in step A1. If certain values of resistors are unavailable, resistors can be combined in series-parallel arrangements to approximate the design value.

A3. Prepare a table for recording your calculated and measured values of design voltages and currents. Also measure the resistance of each resistor in your design with an ohmmeter. Measure the significant voltages and currents in your circuit and record the values in the table you prepared.

B. Current-Divider Circuit

B1. Design a current-divider circuit that will feed three parallel branch loads from a 15-V power supply. The total current to be delivered by the power supply is 5 mA at 10 V. The branch currents are to divide as follows:

(a) The current in branch 2 must be 1½ times more than the current in branch 1.

(b) The current in branch 3 must be 2½ times more than the current in branch 1.

(In other words, the currents in the three branches are to divide in the ratio 1:1.5:2.5.)

Draw a circuit diagram of your design showing the values of all components and the significant voltages and currents throughout the circuit. Your circuit should include an SPST switch for turning power to the circuit on and off.

Present your design in a neat organized report, listing all components required. Show all calculations leading to your design components. Have your instructor approve your design before proceeding to the next step.

B2. With power **off,** and the switch **open,** connect the current-divider circuit you designed in step B1. If certain values of resistors are unavailable, resistors can be combined in series-parallel arrangements to approximate the design value.

B3. Prepare a table for recording your calculated and measured values of design voltages and currents. Also measure the resistance of each resistor in your design with an ohmmeter. Measure the significant voltages and currents in your circuit and record the values in the table you prepared.

ANSWERS TO SELF TEST

1. (a) 100; (b) 100
2. (a) 50; (b) 50
3. (a) 100; (b) 100; (c) 100; (d) 300

QUESTIONS

1. (a) Circuit diagram and design calculations for part A:

_____
_____
_____
_____
_____
_____
_____
_____
_____
_____
_____
_____
_____
_____
_____
_____
_____
_____
_____
_____
_____
_____

(b) Table of measured and calculated values of design voltages and currents (step A3).

_____
_____
_____
_____
_____
_____

2. (a) Circuit diagram and design calculations for part B:

_____

_____

_____

_____

_____

_____

_____

_____

_____

_____

_____

_____

_____

_____

_____

_____

_____

_____

_____

(b) Table of measured and calculated values of design voltages and currents
(step B3).

_____

_____

_____

_____

_____

_____

_____

_____

_____

_____

3. Compare your measured values of current with the design values in part A. Refer to your data in the table of Question 1(b). Explain any differences.

_____

_____

_____

_____

_____

_____

_____

_____

_____

_____

_____

_____

4. Compare your measured values of current with the design values in part B. Refer to your data in the table of Question 2(b). Explain any differences.

_____

_____

_____

_____

_____

_____

_____

_____

_____

_____

_____

_____

_____

_____

TROUBLESHOOTING ELECTRIC CIRCUITS USING VOLTAGE, CURRENT, AND RESISTANCE MEASUREMENTS

BASIC INFORMATION

Electronic technicians must be competent in many areas. They must be able to choose and use the tools and instruments required of a particular job. They must also know the physical as well as the electrical characteristics of the electronic equipment and components with which they must deal.

One of the most interesting and challenging jobs of the technician is troubleshooting and repairing defective equipment. *Troubleshooting* is the process of examining and testing a piece of defective equipment to find the source of its trouble. The technician may simply use his or her eyes and ears to locate the problem. If necessary, one or more instruments can be used in the process. In this experiment three instruments, or meters, will be used—the ammeter, the voltmeter, and the ohmmeter.

To introduce the idea of troubleshooting, this experiment uses circuits consisting of a network of resistors fed by a dc power source. Somewhere in the network a defective resistor is changing the expected behavior of the circuit.

If this circuit was part of a complex piece of equipment, the first job would be to find the circuit whose behavior was not as specified by the manufacturer or designer. Once that circuit was discovered and isolated, the next job would be to find the defective component in that circuit. By sight, touch, or even smell a suspected defective component can be located. After this component is replaced with a known good component, the circuit should operate normally, thus confirming that the defective component had been found. If sight, touch, and smell do not lead to the defective component, more systematic methods must be used. In this experiment you will apply your knowledge of dc circuits and the use of basic dc instruments to find the defective resistors in a number of different circuits.

Defects of a Circuit

Changes in the operation of a circuit can be caused by changes in the characteristics of individual parts of the circuit.

Resistors. The resistance of a resistor may change. If the circuit path through the resistor opens, the result will be equivalent to an open switch. The effect would be that of having a tremendously high resistance in the circuit. A resistor can also be short-circuited, in which case the resistance across it will be very close to zero.

Switches. Mechanical switches can introduce very high resistance in a circuit due to corrosion of their parts or poor pressure on their contacts. Defective switches may also fail to close, which causes the circuit to remain open at all times. Switches may also be short-circuited, in which case the circuit will always remain closed no matter what position the switch is in.

Circuit Conductors. Conductors may break, causing that part of the circuit to open. Faulty insulation or improper placement of components may cause adjacent conductors to touch, producing a short circuit.

Power Source. As a battery ages, its internal resistance increases. When current is drawn from the battery, the internal voltage drop will decrease the terminal voltage below its rated value.

The output voltage of an unregulated power supply will also drop as current is drawn from the supply. This is a normal condition, but it must be taken into account when testing the behavior of a circuit under load. A defective power supply will produce little or no voltage.

Troubleshooting Circuit Components

Resistors, switches, and circuit conductors can be tested with an ohmmeter when out of a circuit. Such a test is called a *static test,* or *measurement.*

When measuring the resistance of a resistor against its rated value, consideration must be given to the tolerance rating of resistor. For example, a 1.2-kΩ, 10 percent resistor is considered good if its resistance measures anywhere between 1080 Ω (1200–120) and 1320 Ω (1200 + 120).

The operation of a switch can be tested by connecting an ohmmeter across its terminals. When the switch is **open** (or **off**), the meter should have an extremely high (or infinite) reading in ohms. When the switch is **closed** (or **on**), the meter should have an extremely low (or zero) reading in ohms.

Circuit conductors—whether individual wire, multiwire cable, heavy busses, or printed circuit traces—can be given a *continuity check* using an ohmmeter. This type of check is used primarily to verify that a complete path exists between the ends of the conductors. It can also be used to discover whether a conductor is making improper contact with the equipment frame or chassis (a condition called *grounding*) or whether improper contact is being made with other conductors (a condition called *shorting*).

CAUTION: Ohmmeter tests should never be performed in live circuits—that is, circuits to which voltage has been applied.

Often, intermittent defects in circuit components can be discovered by tapping, vibrating, or flexing the part while it is connected to the ohmmeter. Any sudden changes in the reading of the meter during this test may indicate a defect that could become permanent over time or else require tightening, cleaning, or other processes to correct the intermittent behavior.

The first, and probably most important, test is that given to the power supply before the load or circuit is connected. Is the source delivering the correct no-load voltage? Is the source regulated (varying very little if at all under load) or unregulated (varying with load)? Is the source capable of delivering the current likely to be required by the load or circuit? The self-contained meter of the source or an external voltmeter can be used to measure the no-load output or terminal voltage of the power supply. The nameplate on the supply will give its current rating. This will indicate whether it has the capacity to deliver enough current to the circuit.

Troubleshooting with Dynamic Measurements

Resistors, switches, conductors, and other circuit elements may test "good" when measured outside of a circuit or when the circuit is not on. However, these same elements may behave differently when tested in a circuit with a voltage supplied. Readings taken under operating conditions are known as *dynamic measurements.* Dynamic measurements may be necessary because it is difficult or even impossible to remove a particular component from a circuit.

Comparisons of calculated values of voltage and current with the values obtained through dynamic measurements give clues to the nature of defects and help pinpoint the part that is likely to be defective.

Basic Troubleshooting Rules

The conclusions drawn from the results of dynamic measurements are based on two related principles:

1. In a circuit that is operating normally and that contains only resistors, the voltage across each resistor and the current in every part of the circuit must follow Ohm's and Kirchhoff's laws.

2. If the voltage across any resistor or the current in any part of the circuit does not follow Ohm's and Kirchhoff's laws, then the circuit is not operating normally, and it can be assumed there is a defect in the circuit.

In applying the two preceding principles it should be understood that the value of resistance used in the Ohm's law and Kirchhoff's law calculations must be the actual measurement resistance rather than the rated or color-coded value.

Behavior of DC Series Circuits · · · · · · · ·

The simple series circuit in Figure 20–1 contains a 100-V dc power source V; a control device, switch S; and three resistors: $R_1 = 2$ kΩ, $R_2 = 3$ kΩ, and $R_3 = 5$ kΩ. An ammeter is placed in the circuit to measure current. When S is open, as shown, there is no complete path for current and the meter will measure 0 A. Since there is no current, the voltage drop across each resistor is zero, as given by Ohm's law.

$$V_1 = I \times R_1 = 0 \times 2\text{ k} = 0\text{ V}$$

$$V_2 = I \times R_2 = 0 \times 3\text{ k} = 0\text{ V}$$

$$V_3 = I \times R_3 = 0 \times 5\text{ k} = 0\text{ V}$$

When S is closed, the circuit is completed and current will flow. Applying Ohm's law, with I as the circuit current, V the applied voltage, and R_T the total resistance of the series circuit, we have

$$I = \frac{V}{R_T} = \frac{100\text{ V}}{(2\text{ k} + 3\text{ k} + 5\text{ k})} = \frac{100\text{ V}}{10\text{ k}}$$

$$= 0.010\text{ A, or } 10\text{ mA}$$

Similarly, using Ohm's law we can calculate the voltage drop across each resistor:

$$V_1 = 10\text{ mA} \times 2\text{ k}\Omega = 20\text{ V}$$

$$V_2 = 10\text{ mA} \times 3\text{ k}\Omega = 30\text{ V}$$

$$V_3 = 10\text{ mA} \times 5\text{ k}\Omega = 50\text{ V}$$

If any of the circuit parts in Figure 20–1 change in value or become defective, the measured values of current and voltage will be different from those calculated. This condition will affect the operation of any load dependent on the rated (or calculated) voltage or current. For example, if one of the resistors is actually a small motor that requires a particular voltage across it to operate, a change in the circuit may reduce (or increase) the voltage across the motor, and it will not operate as intended.

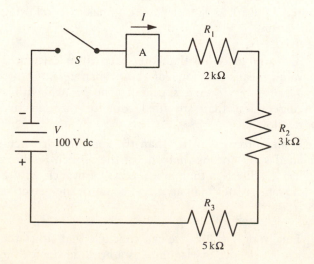

Figure 20–1. Troubleshooting a series dc circuit.

Troubleshooting a Series Circuit · · · · · · · ·

Again, the circuit in Figure 20–1 is used to demonstrate the troubleshooting process. Because the circuit is not delivering 10 mA, it is suspected of being defective. Voltage and current measurements are taken around the circuit with the following results: $V = 100$ V, $I = 4$ mA, $V_1 = 8$ V, $V_2 = 12$ V, $V_3 = 80$ V. Because 4 mA is everywhere in the circuit, the values of the three resistors can be calculated using Ohm's law and the measured values of V and I.

$$R_1 = \frac{V_1}{I} = \frac{8\text{ V}}{4\text{ mA}} = 2\text{ k}\Omega$$

$$R_2 = \frac{V_2}{I} = \frac{12\text{ V}}{4\text{ mA}} = 3\text{ k}\Omega$$

$$R_3 = \frac{V_3}{I} = \frac{80\text{ V}}{4\text{ mA}} = 20\text{ k}\Omega$$

Because the circuit specified the value of R_3 to be 5 kΩ to obtain 10 mA, replacing R_3 with a known 5 kΩ resistor will restore the circuit to its specified value of current.

The measured values can be interpreted in another way. Since current has *decreased* from 10 mA to 4 mA, the total resistance must have *increased*. It is usually correct to make an initial assumption that just one part of a circuit is defective. If we make that assumption in this problem, we can compare some measured values of voltage with the calculated values in a good circuit. Here we notice that with a decrease to 40 percent of the specified current in the circuit, the voltage across R_1 and R_2 are 40 percent of their calculated value, but the voltage across R_3 is 160 percent of its calculated value. Thus, the voltage drops across R_1 and R_2 decreased as expected, but the voltage across R_3 rose, and that behavior was not expected. Therefore, R_3 is suspected of being defective (or at least not the specified value of resistance). All the foregoing is valid only if the voltage applied to the circuit is as specified. Thus, the first voltage to verify is the voltage supplying the circuit.

Characteristics of a Parallel Circuit · · · · ·

The circuit of Figure 20–2 (p. 148) is used to explain the important characteristics of a parallel circuit. We can make the usual assumptions that the voltage supply is constant at 100 V and the circuit conductors have zero resistance. The three resistors will draw current according to Ohm's law.

$$I_1 = \frac{V}{R_1} = \frac{100\text{ V}}{1\text{ k}\Omega} = 0.100 \text{ or } 100\text{ mA}$$

$$I_2 = \frac{V}{R_2} = \frac{100\text{ V}}{3\text{ k}\Omega} = 0.033 \text{ or } 33\text{ mA}$$

$$I_3 = \frac{V}{R_3} = \frac{100\text{ V}}{6\text{ k}\Omega} = 0.017 \text{ or } 17\text{ mA}$$

By Kirchhoff's law,

$$I_T = I_1 + I_2 + I_3$$
$$= 100 \text{ mA} + 33 \text{ mA} + 17 \text{ mA}$$
$$= 150 \text{ mA}$$

The total resistance of the circuit can be found using Ohm's law

$$R_1 = \frac{V}{I_T} = \frac{100 \text{ V}}{150 \text{ mA}} = 667 \text{ } \Omega$$

The total resistance can also be found from the formula:

$$R_T = \cfrac{1}{\cfrac{1}{R_1} + \cfrac{1}{R_2} + \cfrac{1}{R_3}}$$

$$R_T = \cfrac{1}{\cfrac{1}{1 \text{ k}} + \cfrac{1}{3 \text{ k}} + \cfrac{1}{6 \text{ k}}}$$

$$= \frac{1}{10 \times 10^{-4} + 3.33 \times 10^{-4} + 1.67 \times 10^{-4}}$$

$$= \frac{1}{15 \times 10^{-4}} = 667 \text{ } \Omega$$

Troubleshooting DC Parallel Circuits · · · ·

Again, using the circuit of Figure 20–2, if the values of voltage, current, or resistance differ from the calculated values given in the previous section, we may assume the circuit has a defect.

First we measure the supply voltage. It measures 100 V, so we know the circuit is getting its proper voltage. However, the total current measures only 105 mA. It is therefore likely that one of the resistors is defective and is not drawing its rated current. (Remember, always assume in the beginning that only one part is defective.)

We could place an ammeter in each branch and measure the branch currents, or we could disconnect one lead of each resistor from the circuit and measure its resistance with an ohmmeter.

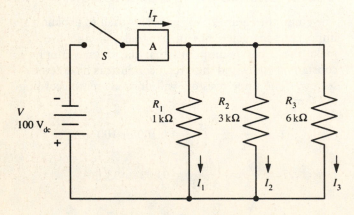

Figure 20–2. Troubleshooting a parallel dc circuit.

In the preceding example I_1 was measured and found to be 55 mA. Resistor R_1 was therefore more than 1000 Ω. Its calculated value was

$$R_1 = \frac{V}{I_1} = \frac{100 \text{ V}}{55 \text{ mA}} = 1818 \text{ } \Omega$$

An ohmmeter check was able to verify this value. Replacement with a known 1-kΩ resistor restored the circuit to its specified values.

An examination of the specified values of current would have led to another clue to the defective resistor. Note that the specified currents I_1 and I_2 combined equal 133 mA. The other combinations are $I_2 + I_3 = 117$ mA and $I_1 + I_3 = 50$ mA.

Since we know one of the resistors is drawing less than its specified current, the total current may be inconsistent with these two-resistor currents. Obviously $I_1 + I_2$ and $I_1 + I_3$ as specified (133 mA and 117 mA, respectively) exceed the actual total current of 105 mA. The combination $I_2 + I_3$ has a specified current of 50 mA, but its actual value is not known. But the combinations $I_1 + I_2$ and $I_1 + I_3$ do tell us something. If R_2 is higher than specified, R_1 and R_3 would be good, and that would result in at least 117 mA in the circuit. If R_3 is higher than specified, then R_1 and R_2 would be good, and at least 133 mA would be drawn by the circuit.

Thus, the evidence points to R_1 as being defective. The next step is to remove R_1 from the circuit and measure its resistance to verify its condition. Thus, only one measurement and knowledge of the circuit specification led directly to the suspected defective resistor. Remember, our first premise is to assume only one defect. Further investigation may in fact reveal that more than one defect is responsible for a circuit malfunction.

As another example, a value of I_T greater than the specified value would have led to the conclusion that one of the branches had a resistor whose value was less than specified. A process similar to the one just described would have led to the defective resistor.

Therefore, in a parallel circuit with the value of the source voltage V as specified, we can make the following conclusions:

1. A measured I_T *less* than the specified or calculated value means that one of the branch resistors has a value *higher* than specified. The branch with a current less than specified contains the defective resistor.

2. A measured I_T *greater* than the specified or calculated value means that one of the branch resistors has a value less than specified. The branch with a current greater than specified contains the defective resistor.

Finally, if $I_T = 0$ A, we can conclude that either the switch is defective or else there is a break in one of the main conductors feeding the three parallel branches.

Troubleshooting a DC Series-Parallel Circuit

The resistors that make up a series-parallel circuit must be tested for their individual voltage drops and current much in the manner of the series and parallel circuits previously discussed. The total current I_T must also be measured. Figure 20–3 is a series-parallel circuit with R_2 in parallel with R_3 and R_1 in series with R_4. The resistance of the parallel combination R_2 and R_3 is

$$R_{2,3} = \frac{R_2 \times R_3}{R_2 + R_3} = \frac{3 \text{ k} \times 6 \text{ k}}{3 \text{ k} + 6 \text{ k}} = 2 \text{ k}\Omega$$

For the series combination, $R_1 + R_4 = 2 \text{ k}\Omega + 5 \text{ k} = 7 \text{ k}\Omega$. Therefore $R_T = 2 \text{ k} + 7 \text{ k} = 9 \text{ k}\Omega$. The total current I_T is

$$I_T = \frac{V}{R_T} = \frac{90 \text{ V}}{9 \text{ k}\Omega} = 0.010 \text{ A, or } 10 \text{ mA}$$

Using this current, we can calculate the voltage drop across each resistor.

$$V_1 = 2 \text{ k}\Omega \times 10 \text{ mA} = 20 \text{ V}$$

$$V_4 = 5 \text{ k}\Omega \times 10 \text{ mA} = 50 \text{ V}$$

$$V_{2,3} = 2 \text{ k}\Omega \times 10 \text{ mA} = 20 \text{ V}$$

The current through R_2 and R_3 can be found using $V_{2,3}$.

$$I_2 = \frac{V_{2,3}}{R_2} = \frac{20 \text{ V}}{3 \text{ k}\Omega} = 0.0067 \text{ A, or } 6.7 \text{ mA}$$

$$I_3 = \frac{V_{2,3}}{R_3} = \frac{20 \text{ V}}{6 \text{ k}\Omega} = 0.0033 \text{ A, or } 3.3 \text{ mA}$$

Troubleshooting can begin with a measurement of I_T (after the value of V has been verified, of course). If I_T is greater than the specified or calculated value, then R_T is less than specified. Similarly, if I_T is less than specified, R_T is greater than specified. In each case, measurements of voltage drops and currents must be made.

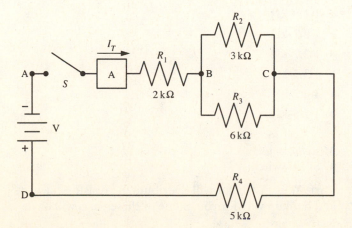

Figure 20–3. Troubleshooting a series-parallel dc circuit.

For example, in the circuit of Figure 20–3, current and voltage measurements were made indicating $I_T = 8$ mA and $V = 90$ V. Since I_T is less than the calculated value, R_T must be higher than specified. Voltage $V_{2,3}$ was then measured and found to be 34 V. Based on the specified $R_{2,3} = 2 \text{ k}\Omega$ and the measured current, $I_T = 8$ mA, the expected voltage across the parallel combination of R_2 and R_3 should have been

$$V_{2,3} = R_{2,3} \times I_T = 2 \text{ k}\Omega \times 8 \text{ mA} = 16 \text{ V}$$

Thus, either R_2 or R_3 is higher than specified. A check of the current I_2 or I_3 would reveal which is lower than specified and, consequently, which resistor is higher than specified. If R_2 and R_3 were separated and disconnected from the circuit, an ohmmeter would indicate immediately which resistor was defective (or out of specification). Remember, just to take current readings, the branches containing R_2 and R_3 would need to be broken, an ammeter inserted, and the circuit reconnected.

The method of troubleshooting a series-parallel circuit involves the following steps:

1. Knowing the specified values of resistors, current, and voltage. The specified values are those given in the schematic diagram of the circuit or calculated from known values of resistance, voltage, and current in a normally operating circuit.

2. Measuring the actual values of current and voltage around the circuit.

3. Comparing the actual with the specified values to determine any differences.

4. Examining the circuit in terms of its individual series and parallel circuits and applying the rules of troubleshooting that these types of circuits require.

SUMMARY

1. The basic elements of a dc circuit are resistors, switches, conductors, and power sources. Any of these may be defective or have intermittent troubles.

2. Each element of a circuit can be tested for resistance and/or continuity with an ohmmeter.

3. The first step in troubleshooting is measurement of the voltage source under no load and verification of its capacity to deliver the necessary current to the load.

4. Dynamic measurements of voltage and current are made with the specified voltage source connected to the circuit.

5. In troubleshooting circuits, the initial assumption is that only one element of the circuit is defective.

6. Troubleshooting generally involves comparing dynamic measurements of voltage and current with specified values of voltage and current as given by a schematic diagram or calculated using Ohm's and Kirchhoff's laws.

7. In a series circuit, a lower-than-specified current means a higher-than-specified resistance. Similarly, a higher-than-specified current means a lower-than-specified resistance.

8. In a parallel circuit, a lower-than-specified total current means that one of the branch resistors has a higher-than-specified resistance. Similarly, a higher-than-specified total current means that one of the branch resistors has a lower-than-specified resistance.

9. In troubleshooting a series-parallel circuit, the procedures for troubleshooting individual series and parallel circuits are followed.

SELF TEST
. . . .

Check your understanding by answering the following questions:

1. In the circuit of Figure 20–1, the rated value of R_1 is 2.2 kΩ, R_2 is 4.7 k Ω, and R_3 is 8.2 kΩ. The applied voltage is $V = 150$ V. When the switch is closed, the ammeter reads 4.5 mA. The voltage V_1 (measured across R_1) is 10.0 V, $V_2 = 103$ V, and $V_3 = 37.0$ V.
   (a) If the circuit were normal, the values of current and voltage would be
      (i) $I =$ _____ mA
      (ii) $V_1 =$ _____ V
      (iii) $V_2 =$ _____ V
      (iv) $V_3 =$ _____ V
   (b) The total resistance in the circuit has _____ (increased/decreased).
   (c) The defective component is _____.

2. In the circuit of Figure 20–1 the resistors are the same as in Question 1 and $V = 150$ V. When the switch is closed, an ammeter in good working condition indicates there is no current in the circuit. The measured voltages are $V = 150$ V, $V_1 = 0$ V, $V_2 = 0$ V, $V_3 = 0$ V. The component that is possibly defective (open) is _____.

3. In the circuit of Figure 20–2, the rated values of the resistors are as follows: $R_1 = 1.2$ kΩ, $R_2 = 1.8$ kΩ, $R_3 = 3$ kΩ, and $V = 300$ V. When the switch S is closed, the measured currents are as follows: $I_T = 400$ mA, $I_1 = 133$ mA, $I_2 = 167$ mA, and $I_3 = 100$ mA.
   (a) If the circuit were normal, the total current I_T would measure _____ mA.
   (b) The total resistance in the circuit has therefore _____ (increased/decreased).

(c) The normal currents should be
   (i) $I_1 =$ _____ mA
   (ii) $I_2 =$ _____ mA
   (iii) $I_3 =$ _____ mA
(d) The defective resistor is _____, whose resistance has _____ (increased/decreased).

4. In the circuit of Figure 20–2, when switch S is closed, $I_T = 0$, $I_1 = 0$, $I_2 = 0$, $I_3 = 0$, $V = 100$ V. With switch S open, an ohmmeter check from the positive terminal of V to the right-hand terminal of S shows continuity. Assuming all the wiring is good, the defective component must be (a) _____, which is (b) _____ (open/shorted).

5. In the circuit of Figure 20–3 the measured voltages are as follows: $V_{AB} = 18$ V, $V_{BC} = 27$ V, $V_{CD} = 45$ V. The defective component is (a) _____, which is (b) _____ (open/shorted).

MATERIALS REQUIRED

Power Supply:
- Variable 0–15 V dc, regulated

Instruments:
- DMM or VOM
- 0–10-mA ammeter
- 0–100-mA ammeter
   (A second multimeter with the required ammeter ranges can be substituted for the ammeters listed.)

Resistors (½-W, 5%):
   (The resistors used in this experiment will be selected from the following group.)
- 1 680-Ω
- 1 820-Ω
- 1 1-kΩ
- 1 1.2-kΩ
- 1 1.8-kΩ
- 1 2.2-kΩ
- 1 2.7-kΩ
- 1 3.3-kΩ
- 1 4.7-kΩ
- 1 5.6-kΩ

Miscellaneous:
- SPST switch

NOTE: Your instructor will distribute the resistors, switch, and hookup wire for you. Some of these parts will be defective.

PROCEDURE

A. Series Circuit

A1. Using the components assigned to you by your instructor, connect the circuit shown in Figure 20–4. Make sure switch *S* is **open** and the power supply is turned **off.** The total rated resistance of the three resistors you receive from your instructor should not exceed 3 kΩ. Check the values by color code only; if the total exceeds 3 kΩ, notify your instructor. (If acceptable, this value is needed in step A4.)

A2. Draw a diagram of your circuit. Show the rated values of all resistors on the diagram. (Use the color code to determine the rated values.)

A3. Turn on the supply and **close** *S*. Adjust the power supply until there is 15 V across its output terminals. Show this value on your circuit diagram. Maintain the output at 15 V throughout this experiment. Check the voltage periodically and adjust if necessary.

A4. Using *V* = 15 V and the rated values of the resistors, calculate the voltage drop across each resistor, and current through each resistor, and the total resistance of the circuit. Record your answers in the appropriate "Calculated" space in Table 20–1 (p.153).

A5. Troubleshoot the circuit by measuring each of the voltages and currents noted in Table 20–1.

A6. Determine which component is defective and briefly state your reasons for this choice. Record this information in Table 20–1. After troubleshooting the circuit, **open** *S* and turn **off** the supply.

B. Parallel Circuit

B1. Using the components assigned to you by your instructor, connect the circuit shown in Figure 20–5 (p. 152). Make sure switch *S* is **open** and the power supply is turned **off.** The three resistors you receive from your instructor should have a total parallel resistance of more than 500 Ω. Calculate the total resistance using the rated values (as indicated by their color code) only; if it is less than 500 Ω, notify your instructor. (If acceptable, this value is needed in step B4.)

B2. Draw a diagram of your circuit. Show the rated values of all the resistors on the diagram. (Use the color code to determine the rated values.)

B3. Turn **on** the supply and **close** *S*. Adjust the power supply until there is 15 V across its output terminals. Show this value on your circuit diagram. Maintain the output at 15 V throughout this experiment. Adjust if necessary.

B4. Using *V* = 15 V and the rated values of the resistors, calculate the voltage drop across each resistor, the current through each resistor, the total current, and the total resistance of the circuit. Record your answers in the appropriate "Calculated" space in Table 20–2 (p. 153).

B5. Troubleshoot the circuit by measuring each of the voltages and currents noted in Table 20–2. If only one ammeter is available, you will need to break and reconnect the circuit a number of times to disconnect and reconnect the ammeter as required. Before making any circuit changes, **open** *S*. Check the supply voltage after each change and after **closing** *S*; adjust to 15 V if necessary.

B6. Determine which component is defective and briefly state your reasons for this choice. Record this information in Table 20–2. After troubleshooting the circuit, **open** *S* and turn **off** the power supply.

C. Series-Parallel Circuit

C1. Using the components assigned to you by your instructor, connect the circuit shown in Figure 20–6 (p. 152).

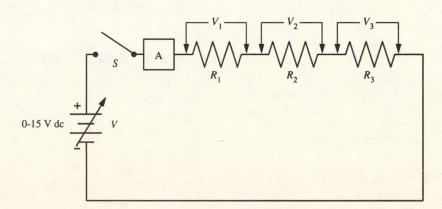

Figure 20–4. Circuit for procedure step A1. Resistors R_1, R_2, and R_3 will be assigned by the instructor.

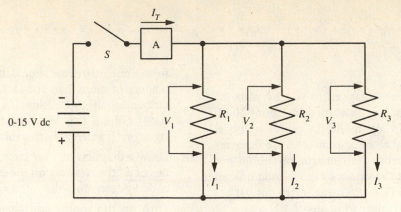

Figure 20–5. Circuit for procedure step B1. Resistors R_1, R_2, and R_3 will be assigned by the instructor.

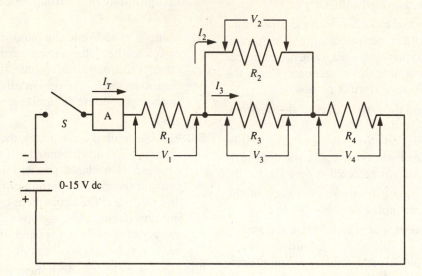

Figure 20–6. Circuit for procedure step C1. Resistors R_1, R_2, R_3, and R_4 will be assigned by the instructor.

Make sure switch S is **open** and the power supply is turned **off**.

C2. Draw a diagram of your circuit. Show the rated values of all the resistors on the diagram. (Use the color code to determine the rated values.)

C3. Turn on the supply and **close** S. Adjust the power supply until there is 15 V across its output terminals. Show this value on your circuit diagram. Maintain the output at 15 V throughout this experiment. Adjust if necessary.

C4. Using $V = 15$ V and the rated values of the resistors, calculate the voltage drop across each resistor, the current through each resistor, the total current, and the total resistance of the circuit. Record your answers in the appropriate "Calculated" space in Table 20–3.

C5. Troubleshoot the circuit by measuring each of the voltages and currents noted in Table 20–3. If only one ammeter is available, you will need to break and reconnect the circuit a number of times to disconnect and reconnect the ammeter as required. Before

making any circuit changes, **open** S. Check the supply voltage after each change and after **closing** S; adjust to 15 V if necessary.

C6. Determine which component is defective and briefly state your reasons for this choice. Record this information in Table 20–3. After troubleshooting the circuit, **open** S and turn **off** the power supply.

ANSWERS TO SELF TEST

1. (a) (i) 9.93; (ii) 21.8; (iii) 46.7; (iv) 81.5;
   (b) increased;
   (c) R_2

2. the switch

3. (a) 0.517;
   (b) increased;
   (c) (i) 250; (ii) 167; (iii) 100;
   (d) R_1; increased

4. (a) S; (b) open

5. (a) R_3; (b) open

TABLE 20–1. Troubleshooting a Series Circuit

	Rated Value of Resistor, Ω		Voltage, V			Current, mA	
			Calculated	Measured		Calculated	Measured
R_1		V_1			I_1		
R_2		V_2			I_2		
R_3		V_3			I_3		

Applied voltage (measured): $V =$ _____   Total current (measured): $I_T =$ _____

Total calculated resistance: $R_T =$ _____   Defective component: _____

Total calculated current: $I_T =$ _____   Reason: _____

TABLE 20–2. Troubleshooting a Parallel Circuit

	Rated Value of Resistor, Ω		Voltage, V			Current, mA	
			Calculated	Measured		Calculated	Measured
R_1		V_1			I_1		
R_2		V_2			I_2		
R_3		V_3			I_3		

Applied voltage (measured): $V =$ _____   Total current (measured): $I_T =$ _____

Total calculated resistance: $R_T =$ _____   Defective component: _____

Total calculated current: $I_T =$ _____   Reason: _____

TABLE 20–3. Troubleshooting a Series-Parallel Circuit

	Rated Value of Resistor, Ω		Voltage, V			Current, mA	
			Calculated	Measured		Calculated	Measured
R_1		V_1			I_1		
R_2		V_2			I_2		
R_3		V_3			I_3		
R_4		V_4			I_4		

Applied voltage (measured): $V =$ _____   Total current (measured): $I_T =$ _____

Total calculated resistance: $R_T =$ _____   Defective component: _____

Total calculated current: $I_T =$ _____   Reason: _____

QUESTIONS

1. What initial assumption is made when troubleshooting a circuit?

_____

_____

_____

2. The resistance of one resistor in a series circuit changes. It is not known whether the resistance has increased or decreased. Discuss the effect on the current in the circuit and the voltage across each resistor for each case (decreased resistance and increased resistance). Assume the source voltage remains constant.

_____

_____

_____

3. The resistance of one resistor in a parallel circuit changes. It is not known whether the resistance has increased or decreased. Discuss the effect on the total current in the circuit and the current in each branch (including the defective branch) for each case (decreased resistance and increased resistance). Assume the source voltage remains constant.

_____

_____

_____

4. Describe the troubleshooting procedures to find each of the following defective parts:
   (a) A resistor (not in a circuit)   (c) Wire conductors in a circuit
   (b) SPST switch   (d) 1.5–V dry cell

_____

_____

_____

_____

5. Describe the dynamic measurements you would use to find an open resistor in a series circuit.

_____

_____

_____

6. Describe the dynamic measurements you would use to find an open resistor in a parallel circuit.

_____

_____

_____

_____

MAXIMUM POWER TRANSFER

BASIC INFORMATION

Measuring Power in a DC Load

Power is defined as the rate of doing work. The unit of power is the *watt* (W). The relationships among the power P dissipated by a resistive load R, the voltage V across R, and the current I in R are given by the following formulas:

$$P = VI$$
$$P = I^2R \tag{21-1}$$
$$P = \frac{V^2}{R}$$

where P is given in watts, V in volts, I in amperes, and R in ohms.

Since the power consumed by a resistance is equal to the product of the voltage across the resistance and the current in the resistance ($P = VI$), power can be found by measuring V with a voltmeter and I with an ammeter. The two readings can then be multiplied to find power.

For example, if 12.5 V produces a current of 0.25 A in a resistance, the power consumed by the resistance is

$$P = VI$$
$$= 12.5 \times 0.25$$
$$= 3.125 \text{ W}$$

If the value of the resistance in the preceding example were known and either the voltage or current were also known, the other power formulas could be used.

If 12.5 V is measured across a 50-Ω resistance, the power consumed by the resistance is

$$P = \frac{V^2}{R}$$
$$= \frac{12.5 \times 12.5}{50} = \frac{156.25}{50}$$
$$= 3.125 \text{ W}$$

If a current of 0.25 A is measured in the 50-Ω resistance, the power consumed by the resistance is

$$P = I^2R$$
$$= 0.25 \times 0.25 \times 50$$
$$= 3.125 \text{ W}$$

NOTE: In the preceding examples, the values of I, V, and R satisfy Ohm's law. Therefore, it should not be surprising that the power calculated in each case is exactly the same. Each of the power formulas is derived by substituting Ohm's law for either V, I, or R.

Since

$$V = I R$$

$$P = V I = (I R)I = I^2R$$

and since

$$I = \frac{V}{R}$$

$$P = V I = V\left(\frac{V}{R}\right) = \frac{V^2}{R}$$

Power can also be measured directly with a *wattmeter*. The wattmeter is a four-terminal instrument. Two terminals are connected across a load (as a voltmeter would be connected), and two terminals are connected in series with the load (as an ammeter would be connected). Internally the analog wattmeter consists of two coils that interact to produce a single direct reading of power by the pointer of the meter. Wattmeters are often used by industrial electronics technicians and are widely used as panelboard meters and portable instruments in the electric power field.

Maximum Transfer of Power ···········

Figure 21–1 shows a circuit delivering power to a load resistance R_L from a power source V_S. The circuit itself con-tains resistance R_C, which includes the internal resistance of the power source. This simple circuit leads to a question of considerable importance in circuit design: Is there some value of circuit resistance that will allow the maximum transfer of power between the source and the load?

The power consumed by R_L is

$$P_L = I^2R_L \qquad (21-2)$$

The value of I is determined by the voltage V (assumed to be a constant value) and the total resistance of the circuit.

$$R_T = R_C + R_L$$

where R_C is the internal resistance of the source and all other resistances in the circuit delivering power to the load.

Thus, the current in formula (21–2) is

$$I = \frac{V_S}{R_C + R_L}$$

and formula (21–2) becomes

$$P_L = \left(\frac{V_S}{R_C + R_L}\right)^2 R_L$$

$$= \frac{V_S^2 R_L}{(R_C + R_L)^2}$$

The direct mathematical solution of this formula that will give the value of R_C for maximum power transfer is beyond the scope of this book. However, graphical and experimental methods can lead to the same solution.

Graphically, constant values can be assigned to V_S and R_C and the power formula (21–2) solved for a range of values for R_L. A graph can then be plotted of P_L versus R_L. If a maximum P does occur, the graph will show the value of R_L at that point.

Assume a constant voltage of 10 V and a circuit resistance of 100 Ω. The load resistance is varied from 0 to 100 kΩ and the value of P_L is calculated at each step. The

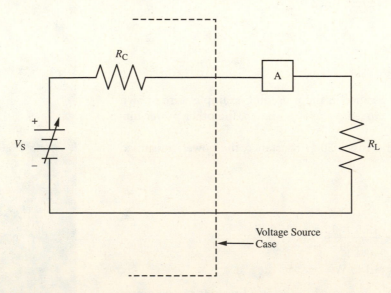

Figure 21–1. A voltage source delivers power to a load, R_L. The internal resistance of the source plus the conductors is represenrted by resistance R_c.

results are shown in Table 21–1. The values in this table are plotted in Figure 21–2.

The maximum power P_L delivered to the load R_L as seen from the table and graph is 0.25 W. This occurs when $R_L = R_C$. Further calculations with other values of R_C and V_S and a varying R_L will confirm the general rule that maximum power transfer occurs when $R_L = R_C$. The value of R_C can be interpreted two ways. If the resistance of the conductors connecting the constant voltage source to the load were negligible, then R_C would represent only the internal resistance of the voltage source, and the rule for maximum power transfer could be stated as follows:

Maximum power transfer to a load by a constant voltage source occurs when the internal resistance of the voltage source is equal to the resistance of the load.

On the other hand, if the circuit delivering power to the load were more complex, then R_C would be determined by the resistance of the circuit as viewed from the output terminals of that circuit. Remember, however, that R_C cannot be measured by simply putting an ohmmeter across the output terminals of the network connected to the load, because this network has an active voltage source. Later in this manual you will learn techniques for solving such problems. At this point we will assume R_C is known or can be easily determined. With this new definition of R_C, a more general rule for maximum power transfer can be stated:

Maximum power transfer to a load from a constant voltage source occurs when the resistance of the circuit supplying the power, as viewed from the output terminals of the circuit, is equal to the load resistance.

SUMMARY

1. The power P, in watts, dissipated by a resistor R (ohms), across which there is a dc voltage V (volts), is given by the formula $P = V^2/R$.

TABLE 21–1. Calculating Maximum Power Transfer

R_L	R_C	P_L
0	100	0
10	100	0.0826
20	100	0.139
30	100	0.178
40	100	0.204
50	100	0.222
60	100	0.234
70	100	0.242
80	100	0.247
90	100	0.249
100	100	0.250
110	100	0.249
120	100	0.248
130	100	0.246
140	100	0.243
150	100	0.240
200	100	0.222
400	100	0.160
600	100	0.122
800	100	0.0988
1,000	100	0.0826
10,000	100	0.00980
100,000	100	0.000998

2. The power P, in watts, dissipated by a resistor R (ohms) in which there is direct current I (amperes) is given by the formula $P = I^2R$.

3. If the dc voltage is V (volts) across a resistor and current in the resistor is I (amperes), the wattage dissipated by the resistor is given by the formula $P = V \times I$.

4. Power in a dc circuit can be measured directly with a wattmeter, or indirectly by measuring V and R, or V and I, or I and R, and substituting the measured values in the proper formula for power.

5. When power is delivered by a power source V_S, whose internal resistance is R_C, to a load R_L, maximum

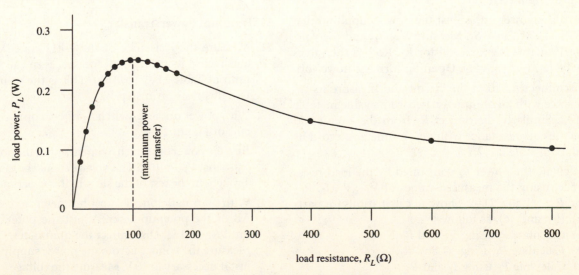

Figure 21–2. Graph of Table 2–1. Maximum power transfer occurs when the resistance of the voltage source is equal to the load resistance.

transfer of power occurs when the load resistance equals the internal resistance of the source, that is, when $R_C = R_L$.

6. If R_C represents the resistance of a network containing a constant voltage source, maximum power transfer to a load R_L occurs when $R_C = R_L$, where R_C is the resistance of the network as viewed from its output terminals.

SELF TEST

Check your understanding by answering the following questions:

1. The current in a 330-Ω resistor is 0.1 A. The power dissipated by the resistor is _____ W.

2. The voltage across a resistor is 12 V, and the current in the resistor is 50 mA. The power dissipated by the resistor is _____ W.

3. The voltage across a 220-Ω resistor is 5.5 V. The power dissipated in the resistor is _____ W.

4. A power supply with an internal resistance of 25 Ω delivers power to a 50-Ω load connected across its terminals. If the voltage delivered by the supply without load is 15 V, the power dissipated by the load is _____ W.

5. A power supply with an internal resistance of 120 Ω delivers power to a resistive load. If the no-load voltage at the output of the supply is 12 V, the maximum power would be delivered to a load whose resistance is _____ Ω.

6. The power delivered by the supply to the load in Question 5 is _____ W.

MATERIALS REQUIRED

Power Supply:
- Variable 0-15 V dc, regulated

Instruments:
- DMM or VOM
- 0–100-mA milliammeter

Resistors (½-W, 5%):
- 2 100-Ω
- 1 330-Ω
- 1 470-Ω
- 1 1-kΩ
- 1 2.2-kΩ
- 1 10-kΩ potentiometer

Miscellaneous:
- SPST switch
- DPST switch

PROCEDURE

A. Measuring Power in a DC Circuit

A1. With power **off** and switch S_1 **open,** connect the circuit of Figure 21–3.

A2. Turn the power **on.** Adjust the power supply so that $V_{PS} = 10$ V. **Close** S_1. Measure V_{PS}, I_L, and the voltage V_L across the load resistor R_L. Record the values in Table 21–2 (p. 161). **Open** S_1; turn the power **off.**

A3. Disconnect R_C from the circuit and measure its resistance with an ohmmeter. Record its value in Table 21–2. Similarly, disconnect R_L from the circuit and measure its resistance with an ohmmeter. Record its value in Table 21–2.

A4. Calculate the power P_L consumed by the load resistor R_L using the measured values of V_{PS}, V_L, I_L, R_C, and R_L, of Table 21–2. Show all calculations for parts (a), (b), and (c), as follows:
   (a) Calculate P_L using V_L and I_L.
   (b) Calculate P_L using V_L and R_L.
   (c) Calculate P_L using I_L and R_L.
   Record your answers in Table 21–2.

A5. Calculate the P_T delivered by the power supply. Show all calculations. Record your answer in Table 21–2.

B. Maximum Power Transfer

B1. Measure the resistance of the 1-kΩ (rated) resistor and record its value in all the spaces in the first column of Table 21–3 (p. 161). This is the value of R_C. Use this resistor in step B2.

B2. With power **off,** and switch S_2 **open,** connect the circuit of Figure 21–4.

B3. Turn the power **on;** switch S_2 **open.** Adjust the power supply so that $V_{PS} = 10$ V. This voltage should be maintained throughout the rest of the steps in this experiment.

B4. With S_2 **open,** connect an ohmmeter across points AB of the potentiometer. Adjust the potentiometer until $R_L = 0$ Ω. Disconnect the ohmmeter. **Close** S_2. Measure the voltage across the power supply and adjust if necessary to 10 V. Measure the voltage V_L across the load R_L. Record the value in Table 21–3.

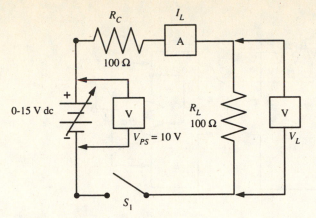

Figure 21–3. Circuit for procedure step A1.

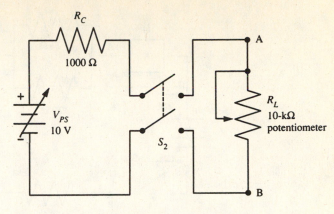

Figure 21–4. Circuit for procedure step B2.

B5. Repeat step B3 for each of the values of load given in Table 21–3 (100 Ω to 10 kΩ). Following the sequence of operations in step 3 carefully.
- (a) With S_2 **open,** adjust the potentiometer for the load resistance in Table 21–3. Measure this value with an ohmmeter and record in Table 21–3.
- (b) Without changing the potentiometer setting, **close** S_2.
- (c) Adjust V_{PS} to 10 V if necessary.
- (d) Measure V_L and record the value in Table 21–3.

B6. Using the corresponding measured values of R_C, R_L, and V_L, calculate P_L for each row of Table 21–3. Use the formula

$$P_L = \frac{V_L{}^2}{R_L}$$

Convert your answers to milliwatts and record the values in Table 21–3.

B7. Using the corresponding measured values of R_C, R_L, and V_{PS} (which should be 10 V in each case), calculate P_{PS}, the power delivered by the power supply for each row of Table 21–3. Use the formula

$$P_{PS} = \frac{V_{PS}{}^2}{R_C + R_L}$$

Convert your answers to milliwatts and record the values in Table 21–3.

C. Determining R_L for Maximum Power Transfer

C1. With S_2 **open,** connect the resistors, switch, and potentiometer, as shown in Figure 21–5. The load resistor R_L is a 10-kΩ potentiometer. Connect an ohmmeter across R_L and adjust the potentiometer so that $R_L = 0$. Do not change this setting.

C2. With S_2 **closed,** measure the resistance R_T across AB using an ohmmeter. Since this is the total resistance of the circuit feeding R_L, maximum power transfer will occur when $R_L = R_T$. Record the values of R_T and R_L in the first row of Table 21–4 (p. 162). **Open** S_2.

C3. Using an ohmmeter, adjust the resistance of the potentiometer so that $R_L = R_T$. Do not change this setting.

C4. With power **off** and S_1 **open,** connect the power supply, milliammeter, and S_1 across the circuit of Figure 21–5 as shown in Figure 21–6 (p. 160).

C5. Power **on;** S_1 **open.** Adjust V_{PS} to obtain a voltage between 10 and 15 V.

C6. **Close** S_1 and S_2. Measure I_T, V_{PS}, and V_L. Record the values in Table 21–4.

C7. Calculate the values of P_{PS} and P_L using the formulas shown in Table 21–4. Record your answers in the table.

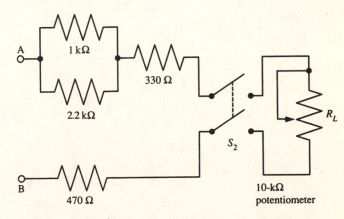

Figure 21–5. Connection of resistors, switch, and potentiometer for procedure step C1.

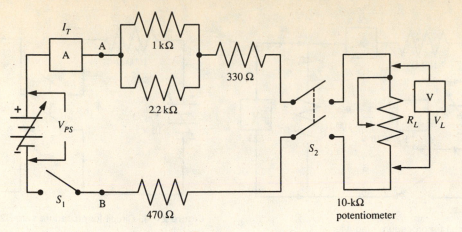

Figure 21–6. Circuit for procedure step C4.

ANSWERS TO SELF TEST

1. 3.3

2. 0.6

3. 0.138

4. 2.0

5. 120

6. 6

TABLE 21–2. Measuring Power in a DC Circuit: Part A

Steps	V_{PS}, V	V_L, V	I_L, A	R_C, Ω	R_L, Ω
2, 3					

	Power Formula	Power, W
4	(a) $P_L =$	
	(b) $P_L =$	
	(c) $P_L =$	
5	$P_T =$	

TABLE 21–3. Experimental Data to Determine Maximum Power in R_L: Part B

R_C, Ω	R_L, Ω	$R_C + R_L$, Ω	V_L, V	$P_L = \dfrac{V_L^2}{R_L}$, mW	$P_{PS} = \dfrac{V_{PS}^2}{R_C + R_L}$, mW
	0				
	100				
	200				
	400				
	600				
	800				
	850				
	900				
	950				
	1 k				
	1.1 k				
	1.2 k				
	1.5 k				
	1.7 k				
	2 k				
	4 k				
	6 k				
	8 k				
	10 k				

TABLE 21–4. Determining R_L for Maximum Power Transfer: Part C

R_T (measured), Ω	$R_L = R_T$, Ω	V_L, V	V_{PS}, V	I_T, mA	$P_L = \dfrac{V_L^{\,2}}{R_L}$	$P_{PS} = V_{PS} \times I_T$	*R_T (calculated), Ω

QUESTIONS

1. Explain, in your own words, the relationship between load resistance and the transfer of power between a dc source and a load.

_____

_____

_____

_____

2. Refer to your data in Table 21–3. For what value of R_L was there maximum power transfer? Does this value confirm the relationship discussed in Question 1? Explain any discrepancies.

_____

_____

_____

_____

3. Refer to your data in Table 21–3. Discuss the relationship between the voltage, V_L, across the load and the resistance of the load, R_L.

_____

_____

_____

_____

4. Refer to your data in Table 21–3. What is the relationship between the power delivered by the power supply P_{PS}, and the resistance of the load, R_L?

_____

_____

_____

5. How does the power delivered to the load, P_L, vary with the resistance of the load, R_L?

_____

_____

_____

6. On a separate sheet of 8½ × 11 graph paper, plot a graph of P_L versus R_L using your data from Table 21–3. Let R_L be the horizontal (or x) axis and P_L be the vertical (or y) axis. Label the graph "P_L vs. R_L."

7. On the graph paper used in Question 6, plot a graph of P_{PS} versus R_L. Use the same axes as in Question 6. Label the graph "P_{PS} vs. R_L."

SOLVING CIRCUITS USING MESH CURRENTS

BASIC INFORMATION

Linear Circuit Elements

Resistors are known as *linear devices* or *linear circuit elements*. If a circuit contains only resistors or other types of resistive elements, it is known as a *linear circuit*.

A linear element is one whose voltage and current behavior obeys Ohm's law. That is, if the voltage across a device is doubled, the current through the device is doubled; if the voltage is reduced by one-third, the current decreases by one-third. Stated another way, the ratio of voltage to current is a constant. A circuit element that behaves in this fashion is the resistor.

The meaning of the word linear can be demonstrated more clearly if a graph of the voltage-current relationship is plotted. Using the circuit of Figure 22–1, the behavior of a 1 kΩ resistor was observed. A dc voltage was varied from 5 to 25 V in steps 5 V. The current was measured at each step and recorded in Table 22–1 (p. 164). From these data the *V* and *I* points were plotted and the graph was drawn (Figure 22–2) (p. 164). The straight-line graph gives rise to the term linear as applied to the ordinary resistor. In later experiments you will use circuit elements that are not linear.

Mesh Current Method

Series-parallel circuits can be solved using Kirchhoff's voltage and current laws and Ohm's law. However, these methods are laborious and time consuming if the circuit contains more than two branches and more than

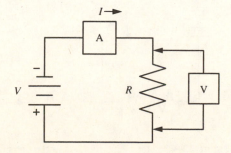

Figure 22–1. Verifying the characteristics of a linear circuit.

TABLE 22–1. Current/Voltage Relationship in a 1 k-Ω Resistor

Voltage, V	Current, mA
0	0
5	5
10	10
15	15
20	20
25	25

one voltage source. The *mesh current method* of solving circuits uses Kirchhoff's voltage law in such a way as to eliminate much of the mathematical labor. It does this by establishing voltage equations for complete circuit loops, or meshes, and solving these equations simultaneously.

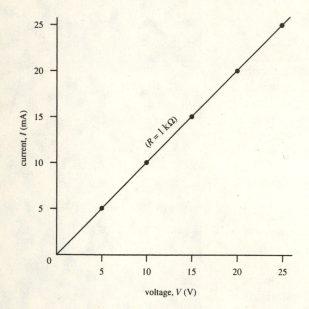

Figure 22–2. A plot of the voltage-current characteristics of the circuit of Figure 22–1.

Mesh Current Equations

The circuit of Figure 22–3 contains three branches and one voltage source. Suppose we wish to find the current in R_L. It can be found using the methods previously learned—that is, combining series and parallel resistors until we obtain the total resistance. We can then solve for the total current and using Kirchhoff's and Ohm's laws solve for I_L.

The mesh current method provides a direct procedure for finding current in any resistor using simultaneous equations. The first step in the procedure is to identify *closed paths* (also called *loops* or *meshes*) in the original circuit. The path need not contain voltage sources, but all voltage sources must be included when choosing the closed paths. We select the fewest paths possible that will include all resistors and voltage sources. Each closed path

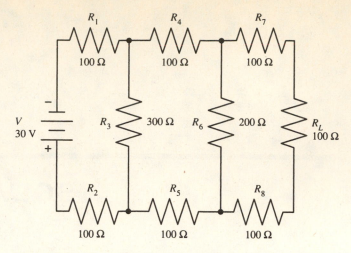

Figure 22–3. A three-branch series-parallel circuit.

is assumed to have a circulating current. It is customary to assume current direction in each case as being in the clockwise direction. Using this assumed current, called a *mesh current,* we write the Kirchhoff's voltage law equation for each path.

The loops often include resistors that are part of other loops. The voltage drops caused by the current in these other loops must be taken into account when writing the Kirchhoff's equation. An equation must be written for each loop chosen. This will result in a system of related equations that must then be solved simultaneously. The currents found by this procedure are the currents in the various resistors. If more than one current is found in a resistor, their algebraic sum is the actual current in that resistor.

A sample problem demonstrates this method.

Problem. Find the load current I_L in the load resistor R_L of the circuit in Figure 22–3.

Solution. Figure 22–3 is redrawn in Figure 22–4 showing the assumed mesh currents. In this case three meshes will be necessary to include all resistors and voltage sources.

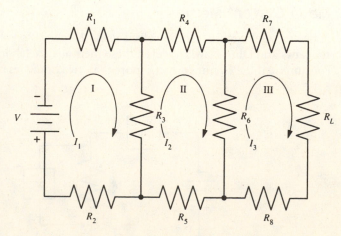

Figure 22–4. Establishing meshes and mesh currents in the circuit of Figure 22–3.

The voltages around mesh I are

$$I_1R_1 + I_1R_3 + I_1R_2 - I_2R_3 = V \qquad \textbf{(22–1)}$$

Because the voltage drop due to I_2 in R_3 (in mesh II) is opposite the voltage drop I_1R_3, I_2R_3 is subtracted from the other voltage drops. This equation can be simplified to

$$I_1(R_1 + R_2 + R_3) - I_2R_3 = V$$

Similarly, the voltages around meshes II and III can be written using I_2 and I_3. However, there are no voltage sources in meshes II and III.

For mesh II,

$$I_2R_3 + I_2R_4 + I_2R_6 + I_2R_5 - I_1R_3 - I_3R_6 = 0$$

Rearranging the terms and simplifying the equation, we have

$$- I_1R_3 + I_2 (R_3 + R_4 + R_5 + R_6) - I_3R_6 = 0 \qquad \textbf{(22–2)}$$

For mesh III,

$$I_3R_6 + I_3R_7 + I_3R_L + I_3R_8 - I_2R_6 = 0$$

Rearranging and simplifying the equation, we have

$$- I_2R_6 + I_3 (R_6 + R_7 + R_8 + R_L) = 0 \qquad \textbf{(22–3)}$$

The three mesh equations can now be rewritten as a system of simultaneous equations.

$$I_1(R_1 + R_2 + R_3) - I_2R_3 = V \qquad \textbf{(22–1)}$$
$$- I_1R_3 + I_2(R_3 + R_4 + R_5 + R_6) - I_3R_6 = 0 \qquad \textbf{(22–2)}$$
$$- I_2R_6 + I_3(R_6 + R_7 + R_8 + R_L) = 0 \qquad \textbf{(22–3)}$$

The values from Figure 22–3 can now be substituted in the equations.

$$+ I_1 (100 + 100 + 300) - I_2 (300) = 30$$
$$- I_1 (300) + I_2 (300 + 100 + 100 + 200) - I_3 (200) = 0$$
$$- I_2 (200) + I_3 (200 + 100 + 100 + 100) = 0$$

which becomes

$$500\, I_1 - 300\, I_2 = 30$$
$$- 300\, I_1 + 700\, I_2 - 200\, I_3 = 0$$
$$- 200\, I_2 + 500\, I_3 = 0$$

The mathematical methods used to solve simultaneous equations will not be covered here. If necessary, review the methods for solving simultaneous equations in any standard algebra text. Simultaneous equations can also be solved quickly using programmable and scientific calculators and computer programs.

Solution of the set of simultaneous equations (22–1), (22–2), and (22–3) results in the following currents:

$$I_1 = 84.5 \text{ mA}$$
$$I_2 = 40.9 \text{ mA}$$
$$I_3 = 16.4 \text{ mA}$$

Because I_3 is the only current through R_L, it is also I_L. Therefore, the answer to the problem is 16.4 mA. Note also that $I_1 = 84.5$ mA represents the total current delivered to the circuit by V.

Although it was not called for in this sample problem, we can find the current in each resistor and the voltage across each resistor using the values of I_1, I_2, and I_3.

The current delivered by V is $I_1 = 84.5$ mA. Since $V = 30$ V, the total resistance of the circuit is

$$R_T = \frac{V}{I_1} = \frac{30 \text{ V}}{84.5 \text{ mA}} = 355 \ \Omega$$

The current in R_1 is also the current in R_2. The voltage drops across R_1 and R_2 are

$$V_{R1} = V_{R2} = I\,R = 84.5 \text{ mA} \times 100 \ \Omega = 8.45 \text{ V}$$

The current in mesh II, I_2, is 40.9 mA. Recall that to find the voltage across R_3, we subtracted I_2 from I_1. Therefore, the actual current across R_3 is 84.5 mA − 40.9 mA = 43.6 mA. The positive sign for this current indicates that the current through R_3 is in the direction indicated by mesh current I_1.

The voltage across R_3 is $IR_3 = 43.6$ mA $\times 300 \ \Omega = 13.1$ V. This can be verified from the voltage drops across R_1 and R_2.

$$\text{Total voltage drop} = 8.45 \text{ V} + 8.45 \text{ V} = 16.9 \text{ V}$$

$$\text{Voltage across } R_3 = 30.0 \text{ V} - 16.9 \text{ V} = 13.1 \text{ V}$$

The current in R_4 and R_5 is equal to mesh current I_2, or 40.9 mA. The current in R_6 is $I_2 - I_3 = 40.9$ mA − 16.4 mA = 24.5 mA in the direction of I_2.

The voltage across R_4 is equal to the voltage drop across R_5.

$$V_{R4} = V_{R5} = 40.9 \text{ mA} \times 100 \ \Omega = 4.09 \text{ V}$$

The voltage drop across R_6 is 24.5 mA $\times 200 \ \Omega = 4.900$ V. Again, this can be verified from previous voltage drops. The voltage across R_3 is reduced by the voltage drops across R_4 and R_5.

$$13.1 \text{ V} - 2(4.09) \text{ V} = 13.1 \text{ V} - 8.18 \text{ V} = 4.920 \text{ V}$$

(The difference comes from rounding the voltage 13.1 up from 13.08.)

Finally, the current in R_7, R_8, and R_L is 16.4 mA. The voltage drops across R_7, R_8, and R_L are equal:

$$16.4 \text{ mA} \times 100 \ \Omega = 1.64 \text{ V}$$

The total voltage drop is therefore equal to the voltage drop across R_6,

$$1.64 \text{ V} \times 3 = 4.920 \text{ V}$$

which checks with the value of R_6 previously calculated.

S E L F   T E S T

Check your understanding by answering the following questions:

1. A circuit containing only resistors is known as a _____ circuit.

2. The voltage-current graph of a resistor is a _____.

3. If the voltage in Figure 22–1 were 10 V and the original resistor were replaced by one having twice the resistance, the current would (drop/rise) _____ by _____ percent.

4. (True/False) The mesh current method should be used to solve the circuit of Figure 22–1. _____

5. (True/False) Kirchhoff's current law is the basis for writing loop current equations. _____

6. If the direction of mesh current I_2 were taken as counterclockwise instead of clockwise, the actual direction of current in R_L _____ (would, would not) change.

7. The least number of mesh currents to use in solving the circuit of Figure 22–5 is _____.

8. Use the mesh current method to find the current delivered to the circuit by V in Figure 22–5. All resistors are 10 Ω; $V = 10$ V. $I = $ _____ A

MATERIALS REQUIRED

Power Supply:
- Variable 0–15 V dc, regulated

Instruments:
- DMM or VOM

Resistors:
- 7 100-Ω, ½-W, 5%

Miscellaneous:
- SPST switch

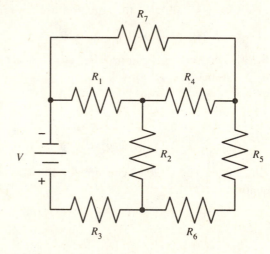

Figure 22–5. Circuit for Self-Test Questions 7 and 8.

PROCEDURE

1. Using an ohmmeter, measure the resistance of each resistor and record the value in Table 22–2 (p. 169).

2. With the power supply **off,** and switch S_1 **open,** connect the circuit of Figure 22–6.

3. Turn **on** the power supply and **close** S_1. Adjust the output of the supply to 10 V. Maintain this voltage throughout this experiment. Check the voltage from time to time and adjust it if necessary.

4. Measure the voltage across each resistor R_1 through R_6 and R_L. Record the values in Table 22–2.

5. Using Ohm's law and the measured value of resistance, calculate the current through each resistor. Record your answers in Table 22–2.

6. Using the rated value of the resistors and the three meshes shown in Figure 22–6, calculate mesh currents I_1, I_2, and I_3. Record your answers in Table 22–2. Show all calculations on a separate sheet of paper.

7. Using your answers for I_1, I_2, and I_3, calculate the current in resistors R_2 and R_4. Record your answers in Table 22–2.

ANSWERS TO SELF TEST

1. linear
2. straight line
3. drop; 50
4. false
5. false
6. would not
7. 3
8. 0.458

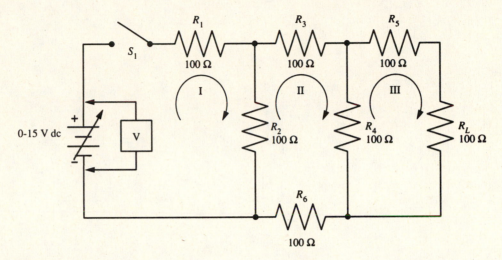

Figure 22–6. Circuit for procedure step 2.

TABLE 22–2.   Verifying Mesh Current Calculations

Resistor	Resistance		Voltage Drop Measured, V	Current Calculated, mA	Mesh Current Calculated, mA	
	Rated Ω	Measured Ω				
R_1	100				I_1	
R_2	100					
R_3	100				I_2	
R_4	100					
R_5	100				I_3	
R_6	100				I_2	
R_L	100				I_3	

$I_{R2} = I_1 - I_2 =$ _____ mA

$I_{R4} = I_2 - I_3 =$ _____ mA

QUESTIONS

1. Explain, in your own words, the characteristics of a linear circuit.

_____

_____

_____

_____

_____

_____

_____

_____

_____

_____

_____

_____

_____

_____

_____

_____

_____

_____

2. Refer to Figure 22–6. Write the three mesh current equations but assume I_2 (the current in mesh II) is counterclockwise. Solve for the currents in R_1, R_2, R_3, R_4, R_5, R_6, and R_L using the three mesh current equations. Compare your answers with the answers and measurements in Table 22–2. Discuss any discrepancies.

3. Discuss the effect that reversing the polarity of the voltage source would have on the direction and magnitude of the current in R_1, R_2, R_3, R_4, R_5, R_6, and R_L.

_____

_____

_____

_____

_____

_____

_____

_____

_____

_____

_____

_____

_____

_____

_____

_____

BALANCED-BRIDGE CIRCUIT

BASIC INFORMATION

Resistance can be measured directly using the ohms scales of a volt-ohm-milliammeter or digital multimeter. On the VOM a calibrated scale and a moving needle (pointer) indicate the value of resistance. On the DMM a digital display shows the numerical value of resistance directly.

Accurate measurement of resistance can also be obtained using a bridge circuit. A resistance bridge circuit, called a *Wheatstone bridge,* is shown in Figure 23–1. This type of bridge uses a highly sensitive galvanometer and a calibrated variable-resistance standard. The galvanometer is used to indicate that point A is at the exact same potential as point C. The voltages at A and C are varied by the variable-resistance standard. If a galvanometer with a centered zero scale is used, the pointer will indicate not only when V_A and V_C are equal but also the direction of current flow when the voltages are unequal. A highly sensitive microammeter (digital) may be used in place of the galvanometer.

In Figure 23–1 the Wheatstone bridge is connected to measure the resistance of unknown resistor R_X. Resistors R_1 and R_2 are precision fixed resistors. They are called the *ratio arms* of the bridge. A precision variable resistance R_3 is called the *standard arm.* When the voltage difference between A and C is zero, no current will flow through the galvanometer,

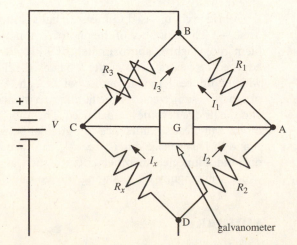

Figure 23–1. Wheatstone bridge circuit used for making accurate measurements of resistance.

and the pointer will be centered at zero. In this condition, the bridge is said to be *balanced*. Power to the circuit is supplied by a regulated dc power supply.

The Balanced Bridge . . . .

The circuit of Figure 23–1 shows the currents in each of the arms of the Wheatstone bridge. The voltage drops in each arm can therefore be found using Ohm's law:

$$V_{AB} = I_1 R_1$$

$$V_{DA} = I_2 R_2 \qquad\qquad (23\text{–}1)$$

$$V_{CB} = I_3 R_3$$

$$V_{DC} = I_X R_X$$

For the bridge to be balanced—that is, $V_{AC} = V_{CA} = 0$—the voltage at A with respect to B, V_{AB}, must be equal to the voltage at C with respect to B, V_{CB}, or

$$V_{AB} = V_{CB}$$

Substituting the *IR* values for these voltages from formula (22–1), we have

$$I_1 R_1 = I_3 R_3$$

Dividing both sides by $I_3 R_1$ leads to the ratio

$$\frac{I_1}{I_3} = \frac{R_3}{R_1} \qquad\qquad (23\text{–}2)$$

Similarly,

$$V_{DA} = V_{DC}$$

and

$$I_2 R_2 = I_X R_X$$

from which

$$\frac{I_2}{I_X} = \frac{R_X}{R_2} \qquad\qquad (23\text{–}3)$$

At balance there is no current in line AC. From Kirchhoff's current law,

$$I_1 = I_2 \qquad\qquad (23\text{–}4)$$

$$I_3 = I_X$$

From formula (23–4), I_X is substituted for I_3 in formula (23–2).

$$\frac{I_1}{I_X} = \frac{R_3}{R_1} \qquad\qquad (23\text{–}5)$$

From formula (23–4), I_1 is substituted for I_2 in formula (23–3).

$$\frac{I_1}{I_X} = \frac{R_X}{R_2} \qquad\qquad (23\text{–}6)$$

Combining formulas (23–5) and (23–6), we obtain

$$\frac{I_1}{I_X} = \frac{R_3}{R_1} = \frac{R_X}{R_2}$$

or

$$\frac{R_3}{R_1} = \frac{R_X}{R_2} \qquad\qquad (23\text{–}7)$$

Solving for R_X yields

$$R_X = \frac{R_2 \times R_3}{R_1} = \frac{R_2}{R_1}(R_3) \qquad\qquad (23\text{–}8)$$

Formula (23–8) can be used to find the value of an unknown resistor in a bridge circuit if the values of the other three arms of the bridge are known. The formula states that the unknown resistor is equal to the ratio of R_2 to R_1 multiplied by the value of the variable resistor R_3.

Since R_2 and R_1 are fixed precision resistors, the value of R_2/R_1 can be set as required. For maximum accuracy and sensitivity in adjusting R_3, R_2 should be equal to R_1. In this case $R_2/R_1 = 1$ and formula (23–8) becomes

$$R_X = R_3$$

The limit of R_X measurements is therefore the maximum value of R_3. It can be seen that the ratio R_2/R_1 determines the maximum value of R_X that can be measured. For example, if $R_2/R_1 = 3$, then $R_X = 3R_3$, or the maximum value of R_X can be three times the maximum value of R_3. Such a ratio will also decrease the sensitivity of the measurement, because any slight adjustment of R_3 will not be easily detected by the galvanometer, thus decreasing the accuracy of the resistance value obtained from reading R_3. It is also obvious that when the ratio R_2/R_1 is less than 1, the range of resistance values that can be measured will be less than the maximum value of R_3. The Wheatstone bridge can also be used for purposes other than measuring resistance values. The bridge circuit is often used in instrumentation circuits to produce highly sensitive voltage values used to indicate changes in temperature, light, humidity, weight, or some other measurable quantity.

A bridge circuit used to measure light is shown in Figure 23–2. The resistance of the photocell at normal or ambient room light is approximately 100 kΩ. Resistor R_3 is adjusted or calibrated to create a balanced bridge circuit, which produces zero volts across points A and B. When the photocell is subjected to more light, its resistance decreases and the bridge becomes unbalanced. The unbalanced condition will create a difference in potential between points A and B. This voltage difference, V_{AB}, can then be brought out and connected to other circuits, such as amplifiers, which will magnify the changing voltage value.

SUMMARY

1. A Wheatstone bridge is used for measuring resistance values with great accuracy.

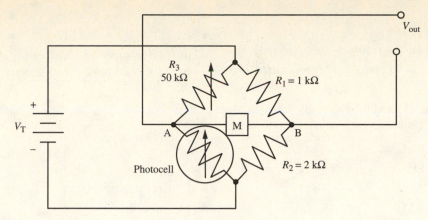

Figure 23–2. Wheatstone bridge circuit used to react to changes in light intensity.

2. The ratio arms R_1 and R_2 in Figure 23–1 are precision resistors. Their ratio, R_2/R_1, determines the range of resistance measurements of the bridge.

3. The standard arm R_3 is a variable calibrated precision resistance by means of which the bridge is balanced when measuring an unknown resistance R_X.

4. A highly sensitive zero-center galvanometer or digital microammeter is used as the indicator. A dc voltage source powers the bridge.

5. When the voltage across the indicating meter (V_{AC} in Figure 23–1) is zero, there is no current between A and C. The meter will then read zero.

6. At balance in a Wheatstone bridge, the product of the resistors of opposite arms is equal; that is, in Figure 23–1,

$$R_1R_X = R_2R_3$$

and the value of R_X is thus

$$R_X = \frac{R_2}{R_1}(R_3) \qquad \text{(23–8)}$$

Formula (23–8) states that at balance the value of the unknown resistor R_X is equal to the resistance of the calibrated resistor R_3 multiplied by the ratio R_2/R_1.

S E L F   T E S T

Check your understanding by answering the following questions.

1. The calibrated variable-resistance of the Wheatstone bridge is called the _____.

2. At balance the current through the galvanometer is _____.

3. (True/False) The direction of current through a zero-centered galvanometer can be determined by the direction the pointer moves. _____

4. The galvanometer in Figure 23–3 reads zero when R_3 is set at 3.75 kΩ. The value of R_X is _____ Ω.

5. In Figure 23–1 $R_2/R_1 = 10$. If the unknown resistor $R_X = 5$ kΩ, what setting of R_3 (in ohms) will produce a zero reading of the galvanometer? _____.

6. A Wheatstone bridge is known to be in good operating order and can measure resistors in the range 0.1 to 100 kΩ. In measuring a resistor whose value is about 50 kΩ, the bridge cannot be balanced with the present values of R_1 and R_2. To balance the bridge it is necessary to change the setting of the _____ arms.

MATERIALS REQUIRED

Power Supply:
- Variable 0–15 V dc, regulated

Instruments:
- VOM or DMM
- Zero-center galvanometer or digital microammeter

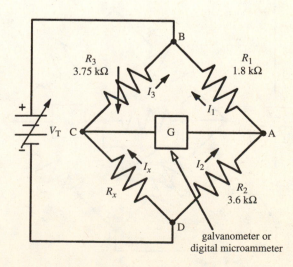

Figure 23–3. Circuit for Self Test Question 4.

Resistors (½-W, 5%):
- 1 5.6 kΩ
- 2 5.1 kΩ (If available, 1% resistors are preferred)
- 1 10 kΩ potentiometer or decade resistance box (1 Ω–10,000 Ω)

The following resistors should be available from an assortment in the shop or laboratory (at least one of each value):

- 560-Ω
- 620-Ω
- 750-Ω
- 1 kΩ
- 1.8 kΩ
- 2 kΩ
- 2.2 kΩ
- 2.4 kΩ
- 3.6 kΩ
- 3.9 kΩ
- 4.7 kΩ
- 6.8 kΩ

- 1.1 kΩ
- 1.2 kΩ
- 3 kΩ
- 3.3 kΩ
- 8.2 kΩ
- 10 kΩ
- Instructor-supplied resistors for steps 8–13 (these resistors should have values greater than 10 kΩ but less than 100 kΩ)

Miscellaneous:
- 2 SPST switches

Optional:
- 1 50 kΩ potentiometer
- 1 photocell

PROCEDURE

The purpose of this experiment is to verify the bridge formulas and simulate the use of the Wheatstone bridge in measuring resistance. It is not intended to duplicate the accuracy of laboratory quality or commercial Wheatstone bridge instruments. In this experiment a galvanometer, microammeter (analog or digital), or DMM can be used. In Figure 23–4 the label M indicates any of these instruments. The term *meter* in the procedures that follow is applicable to whatever instrument is being used.

1. With power **off** and S_1 and S_2 **open,** connect the circuit of Figure 23–4. The potentiometer or resistance box should be set at its maximum value (10 kΩ) initially. The 5.6 kΩ resistor in the meter circuit will reduce the sensitivity of the meter during the initial stages of adjusting the standard arm resistance. The unknown resistor R_X should be chosen from the assortment noted in the Materials Required list.

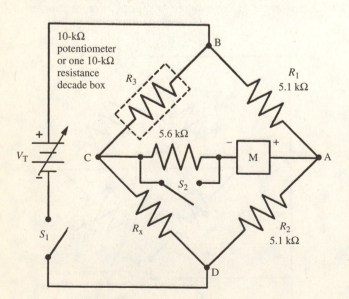

Figure 23–4. Circuit for procedure step 1.

2. Turn the Power **on.** Adjust V_T to 6 V. **Close** S_1 (S_2 should remain **open**). The meter should indicate a positive current value. (If it does not, check to see that $R_3 = 10$ kΩ.)

3. Gradually reduce the resistance of R_3. The meter current value should move down scale, approaching zero. When the current is at or very near zero, **close** S_2, shorting out the 5.6 kΩ resistor. Continue to adjust R_3 until the meter current is zero. Power **off,** S_1 and S_2 **open.**

4. If R_3 is a decade resistance box, read the setting of the dials and record this measured value in Table 23–1 (p. 177) for resistor 1, measured value. If R_3 is a potentiometer, remove it from the circuit without changing its setting. Measure the resistance of the potentiometer using an ohmmeter and record the value in Table 23–1 as noted before.

5. Remove resistor R_x, from the circuit. Record its rated resistance and tolerance values (based on the color code) in Table 23–1 under resistor 1.

6. Using the Wheatstone bridge formula

$$R_X = \frac{R_2}{R_1}(R_3)$$

calculate the resistance of resistor 1R. Record your answer in Table 23–1.

7. Repeat steps 1 through 6 for five additional resistors chosen from the assortment available. For each unknown resistor, record the value of R_3, the rated value of resistance and tolerance for R_X, and the calculated value of R_X in Table 23–1.

8. Using the basic bridge circuit of Figure 23–3, with the standard arm used in steps 1 through 7, choose new resistors for the ratio arms R_1 and R_2 such that the bridge can then be used to measure resistances up to 30 kΩ. Measure R_1 and R_2 with an ohmmeter and record their values in Table 23–2 (p. 177).

9. With power **off** and S_1 and S_2 **open,** connect the new circuit using the chosen resistors for R_1 and R_2. Standard arm R_3 should be at its maximum setting (10 kΩ). Your instructor will give you a resistor for R_X.

10. Turn the power **on.** Adjust V_T to 6 V. **Close** S_1 (S_2 should remain **open**). Gradually reduce the resistance of R_3 until the meter current is near zero. **Close** S_2 and further adjust R_3 until the meter current is zero. Power **off;** S_1 and S_2 **open.**

11. Determine the value of R_3 as in step 4. Record this value in Table 23–2.

12. Calculate R_X using the Wheatstone bridge formula as in step 6. Record your answer in Table 23–2.

13. Repeat steps 8 through 12 for values of R_1 and R_2 that would extend the range of resistance measurements to 100 kΩ. Record all data in Table 23–2. After recording your data, **open** S_1 and S_2; power **off.**

Optional Activity ·

This activity requires a 50 kΩ potentiometer or 500 kΩ dark to 500 Ω light equivalent photocell such as EG&G Optoelectronics #VT43MT. Construct the light measuring circuit of Figure 23–2. Add a SPST switch in the supply voltage circuit. Meter M can be any of the meters used in the previous Procedure. With the switch open, have your instructor approve the circuit before proceeding to the next step.

Close the SPST switch. Increase V_T slowly to 10 V. The photocell should be exposed to normal room illumination. Adjust R_3 to obtain a balanced bridge ($V_{AB} = 0$ V).

Using a flashlight or lamp, increase the amount of light on the photocell. Observe the effect on V_{AB}. Move the light source away from the photocell slowly and observe the effect on V_{AB}.

Remove the light source and cover the photocell with your hand. Observe V_{AB}. Again, slowly move your hand away from the photocell and observe the effect on the voltage.

Write a report on your observations and explain the effect of different light intensities on the output voltage.

ANSWERS TO SELF TEST

1. standard arm
2. zero
3. true
4. 7.5 kΩ
5. 500 Ω
6. ratio

EXPERIMENT 23

TABLE 23–1. Bridge Measurement of R_X: $R_2/R_1 = 1$

Resistor Number	1	2	3	4	5	6
R_3, Ω						
Rated (color code) value, R_X, Ω						
Percent tolerance, R_X, %						
Calculated value, R_X, Ω						

TABLE 23–2. Bridge Multipliers

Steps	Maximum Measurable Resistance, $k\Omega$	Measured Value R_1, Ω	Measured Value R_2, Ω	Ratio $\dfrac{R_2}{R_1}$	R_3, Ω	R_X Rated Value Ω	R_X Rated Tolerance, %	R_X Calculated Value, Ω
8, 9, 10, 11, 12	30							
13	100							

QUESTIONS

1. Explain, in your own words, the relationship among the resistors in a balanced-bridge circuit.

_____

_____

_____

_____

_____

2. Give four factors that determine the accuracy of resistance measurements made with a balanced-bridge circuit.

_____

_____

_____

_____

_____

_____

3. Why is it important to use a very sensitive galvanometer or other meter in a balanced-bridge circuit?

_____

_____

_____

_____

_____

_____

_____

4. Refer to Table 23–1. Were the resistance measurements within the tolerance ratings of the resistors? Explain any discrepancies.

_____

_____

_____

_____

_____

_____

_____

_____

5. Is the 6-V power supply critical in this experiment? Would a higher or lower voltage have affected your results? What would happen if the supply voltage were zero? Discuss fully.

_____

_____

_____

_____

_____

_____

_____

_____

_____

_____

SUPERPOSITION THEOREM

BASIC INFORMATION

We have applied Ohm's and Kirchhoff's laws and mesh methods to study simple resistive circuits—that is, circuits containing series, parallel, or series-parallel combinations of resistors. However, there are more complex circuits in electricity. Solving such circuits may become difficult using the methods we have learned thus far. For such complex circuits, more-powerful methods of solution are helpful.

Superposition Theorem

The superposition theorem states that

> *In a linear circuit containing more than one voltage source, the current in any one element of the circuit is the algebraic sum of the currents produced by each voltage source acting alone. Further, the voltage across any element is the algebraic sum of the voltages produced by each voltage source acting alone.*

To apply this theorem to the solution of a problem, we must understand what is meant by "each source acting alone." Suppose a network, say Figure 24–1, has two voltage sources V_1 and V_2 and we wish to find the effect on the circuit of each source acting alone. To determine the effect of V_1 we must replace V_2 by its internal resistance and solve the modified circuit. If any of the voltage sources are considered ideal (i.e., having no internal resistance) or if the internal resistance is very low compared with other circuit elements, the voltage source can be replaced with a short circuit. In Figure 24–2 (p. 180) we have replaced V_2 with a

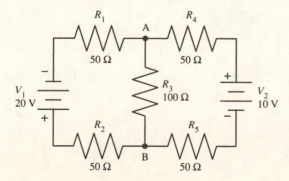

Figure 24–1. Resistor circuit with two voltage sources.

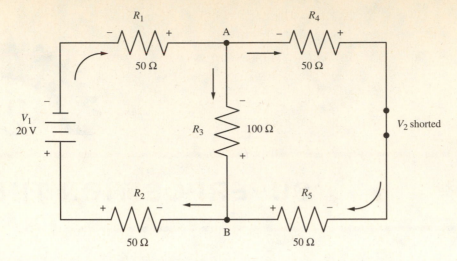

Figure 24–2. The first step when using superposition is to replace one of the voltage sources with its internal resistance. In this circuit, V_2 is an ideal source (no internal resistance), so that V_2 is replaced by a short circuit.

short circuit. The circuit of Figure 24–2 is a series-parallel circuit with one voltage source V_1. We can solve for the currents in each of the resistors R_1 through R_5 and the current supplied by V_1 using methods learned in previous experiments. We can also find the voltage across each resistor of the circuit. To determine the effect of V_2, we replace V_1 with its internal resistance, again a short circuit, as in Figure 24–3 and solve the circuit of Figure 24–3. Again, we will find the currents in R_1 through R_5 and the current supplied by V_2. Similarly, we can find the voltage across each resistor. The final step is to add algebraically the two currents in each of the resistors to find the total current in the resistor. The voltage across each resistor will also be the algebraic sum of two voltages. The current supplied by each voltage source will be the algebraic sum of the currents in the short circuits that replaced the voltage sources and the current supplied by the voltage source itself.

A sample problem demonstrates this procedure.

Problem. Using the values shown in Figure 24–1, find the current in each resistor, the voltage across each resistor, and the current supplied by each voltage source.

Solution. The first step in the solution is to replace V_2 with a short circuit and solve the new circuit shown in Figure 24–2.

Figure 24–2 shows the effect of V_1 acting on the circuit alone, when V_2 is shorted. Current directions and voltage polarities have been placed on the circuit to show V_1's effect.

Solving for R_T, we find

$$R_T = [(R_4 + R_5)\|R_3] + (R_1 + R_2)$$
$$R_T = [(50\ \Omega + 50\ \Omega)\|100\ \Omega] + (50\ \Omega + 50\ \Omega)$$
$$R_T = 150\ \Omega$$

Where the symbol $\|$ means "in parallel" we can now solve for the total current, I_T, as a result of V_1.

$$I_T = V_1/R_T$$
$$I_T = 20\ \text{V}/150\ \Omega = 133\ \text{mA}$$

Therefore,

$$I_{R1} = 133\ \text{mA} \qquad I_{R3} = 66.7\ \text{mA}$$
$$I_{R2} = 133\ \text{mA} \qquad I_{R4} = 66.7\ \text{mA}$$
$$I_{R5} = 66.7\ \text{mA}$$

The individual resistor voltage drops, as a result of V_1 will be

$$V_{R1} = R_1 \times I_{R1} = 50\ \Omega \times 133\ \text{mA} = 6.67\ \text{V}$$
$$V_{R2} = R_2 \times I_{R2} = 50\ \Omega \times 133\ \text{mA} = 6.67\ \text{V}$$
$$V_{R3} = R_3 \times I_{R3} = 100\ \Omega \times 66.7\ \text{mA} = 6.67\ \text{V}$$
$$V_{R4} = R_4 \times I_{R4} = 50\ \Omega \times 66.7\ \text{mA} = 3.33\ \text{V}$$
$$V_{R5} = R_5 \times I_{R5} = 50\ \Omega \times 66.7\ \text{mA} = 3.33\ \text{V}$$

Next, we replace V_1 with a short circuit and solve for the circuit values due to V_2 alone. This is shown in Figure 24–3. Notice, in Figure 24–3, that the current directions as the result of V_2 are in the same direction as earlier for R_1, R_2, R_4, and R_5 but in the opposite direction for R_3. This is important when we algebraically sum the currents from both sources.

Solving for R_T', we find

$$R_T' = [(R_1 + R_2)\|R_3] + (R_4 + R_5)$$
$$R_T' = [(50\ \Omega + 50\ \Omega)\|100\ \Omega] + (50\ \Omega + 50\ \Omega)$$
$$R_T' = 150\ \Omega$$

The total current, as a result of V_2 is

$$I_T' = V_2/R_2$$
$$I_T' = 10\ \text{V}/150\ \Omega = 66.7\ \text{mA}$$

Therefore,

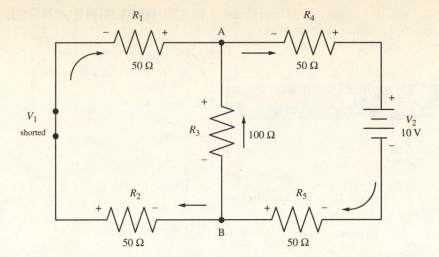

Figure 24–3. After the circuit of Figure 24–2 is solved, V_1 is replaced by a short circuit and the new circuit is solved.

$$I_{R1}' = 33.3 \text{ mA} \qquad I_{R4}' = 66.7 \text{ mA}$$

$$I_{R2}' = 33.3 \text{ mA} \qquad I_{R5}' = 66.7 \text{ mA}$$

$I_{R3}' = -33.3$ mA   (Since I_{R3}' is in the opposite direction of I_{R3} found when V_1 acted alone, a minus sign is used.

The individual resistor voltage drops, as a result of V_2 will be

$$V_{R1}' = R_1 \times I_{R1}' = 50 \text{ } \Omega \times 33.3 \text{ mA} = 1.67 \text{ V}$$

$$V_{R2}' = R_1 \times I_{R2}' = 50 \text{ } \Omega \times 33.3 \text{ mA} = 1.67 \text{ V}$$

$$V_{R3}' = R_3 \times I_{R3}' = 100 \text{ } \Omega \times -33.3 \text{ mA} = -3.33 \text{ V}$$

$$V_{R4}' = R_4 \times I_{R4}' = 50 \text{ } \Omega \times 66.7 \text{ mA} = 3.33 \text{ V}$$

$$V_{R5}' = R_5 \times I_{R5}' = 50 \text{ } \Omega \times 66.7 \text{ mA} = 3.33 \text{ V}$$

By combining each of the currents, as stated by the superposition theorem, we can find the actual current due to both voltage sources:

$$I_{R1} = 133 \text{ mA} + 33.3 \text{ mA} = 166.3 \text{ mA}$$

$$I_{R2} = 133 \text{ mA} + 33.3 \text{ mA} = 166.3 \text{ mA}$$

$$I_{R3} = 66.7 \text{ mA} + (-33.3 \text{ mA}) = 33.4 \text{ mA}$$

$$I_{R4} = 66.7 \text{ mA} + 66.7 \text{ mA} = 133.4 \text{ mA}$$

$$I_{R5} = 66.7 \text{ mA} + 66.7 \text{ mA} = 133.4 \text{ mA}$$

The voltages across each resistor can now be found using Ohm's law.

$$V_1 = 166.3 \text{ mA} \times 50 \text{ } \Omega = 8.32 \text{ V}$$

$$V_2 = 166.3 \text{ mA} \times 50 \text{ } \Omega = 8.32 \text{ V}$$

$$V_3 = 33.4 \text{ mA} \times 100 \text{ } \Omega = 3.34 \text{ V}$$

$$V_4 = 133.4 \text{ mA} \times 50 \text{ } \Omega = 6.67 \text{ V}$$

$$V_5 = 133.4 \text{ mA} \times 50 \text{ } \Omega = 6.67 \text{ V}$$

The currents and voltages are shown in Figure 24–4. You should verify these values using Kirchhoff's voltage and current laws.

NOTE: Because answers were rounded to three significant figures, verification of current and voltage values may not agree in the third significant figure.

SUMMARY

1. The superposition theorem is most useful when applied to linear circuits that have two or more voltage sources.

2. To find the current in or voltage across an element in a linear circuit, all the voltage sources except one are removed and replaced by their internal resistances (short circuits in the case of regulated or ideal voltage sources), and the effect of the one remaining source on the circuit is determined. This process is repeated

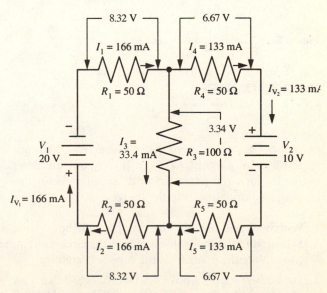

Figure 24–4. The voltages and currents of the circuit in the sample problem.

for each source. The actual currents and voltages produced by all sources is the algebraic sum of the individual currents and voltages.

SELF TEST

Check your understanding by answering the following questions:

1. (True/False) The superposition theorem may be applied to the circuit of Figure 24–5. _____

2. In Figure 24–2 the graph of voltage across R_4 versus current in R_4 is _____.

3. The 10-V source in Figure 24–1 causes _____ (more/less) current in R_3 than would be the case if V_1 were the only voltage source and V_2 were replaced by a short circuit.

4. In applying the superposition theorem to the solution of the circuit in Figure 24–1, V_2 is replaced by a _____.

5. In the circuit of Figure 24–1 the voltage across R_5 is _____ V.

6. The current through R_1 in Figure 24–6 is _____ A.

MATERIALS REQUIRED

Power Supply:
- 2 variable 0–15 V dc, regulated

Instruments:
- DMM or VOM
- 0–100-mA milliammeter

Resistors (½-W, 5%):
- 1 820-Ω
- 1 1.2 kΩ
- 1 2.2 kΩ

Miscellaneous:
- 2 SPDT switches

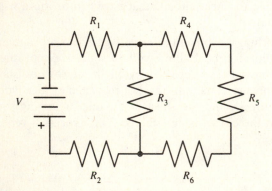

Figure 24–5. Circuit for Self-Test Question 1.

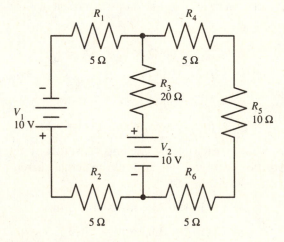

Figure 24–6. Circuit for Self-Test Question 6.

PROCEDURE

CAUTION: This experiment requires current readings in three different parts of a circuit. If only one ammeter is available, turn off power in both supplies before connecting and disconnecting the meter.

1. With power **off** in both power supplies and both switches S_1 and S_2 in the B position, connect the circuit of Figure 24–7. Note the polarity of the supplies carefully.

2. Turn power supply 1 **on.** Adjust its voltage so that $V_{PS1} =$ 15 V. Set switch S_1 to position A and S_2 to position B. This will apply power to R_1, R_2, and R_3. Measure I_1, I_2, and I_3,

voltage V_1 across R_1, voltage V_2 across R_2, and voltage V_3 across R_3. Record the values in Table 24–1 (p. 185). It is important to note the direction of current (by using + and − signs) as well as its value. Also note the direction of the voltage drops V_1, V_2, and V_3. (The voltage sign should always be opposite that of the current sign. Thus if V_1 is −, I_1 will be +.) Turn V_{PS1} **off.**

3. Set S_1 to position B. Adjust power supply 2 so that V_{PS2} = 10 V. Set S_2 to position A. This will apply power to R_1, R_2, and R_3 from V_{PS2}. Measure I_1, I_2, I_3, voltage V_1 across R_1, voltage V_2 across R_2, and voltage V_3 across R_3.

Record the values in Table 24–2 (p. 185). Note carefully the polarity of each measurement, as in step 2.

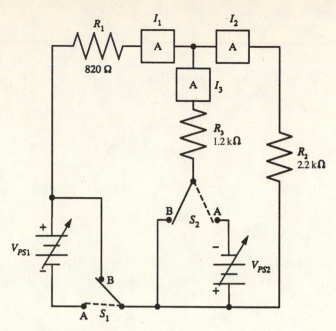

Figure 24–7. Circuit for procedure step 1.

4. With V_{PS1} = 15 V and V_{PS2} = 10 V, set S_1 to position A (S_2 should already be in position B). Both power supplies are now feeding R_1, R_2, and R_3. Measure I_1, I_2, I_3, V_1, V_2, and V_3 as in the previous two steps. Record the values in Table 24–3 (p. 185), noting carefully the polarity of each value. Turn the power **off.**

5. Using the measured values of R_1, R_2, and R_3 and V_{PS1} = 15 V and V_{PS2} = 10 V, calculate I_1, I_2, and I_3 supplied by two sources using the superposition theorem. Show all calculations and diagrams. Record your calculated values in Table 24–3.

ANSWERS TO SELF TEST

1. false
2. a straight line
3. less
4. short circuit
5. 6.65
6. 0.75

TABLE 24–1. Effect of V_{PS1} Alone

Current, mA	Voltage, V
I_1:	V_1:
I_2:	V_2:
I_3:	V_3:

TABLE 24–2. Effect of V_{PS2} Alone

Current, mA	Voltage, V
I_1:	V_1:
I_2:	V_2:
I_3:	V_3:

TABLE 24–3. Effect of V_{PS1} and V_{PS2} Acting Together

Measured Values		Calculated Values					
		V_{PS1} Only		V_{PS2} Only		V_{PS1} and V_{PS2} Together	
Current, mA	Voltage, V	Current, mA	Voltage, V	Current, mA	Voltage, V	Current, mA	Voltage, V
I_1	V_1	I_1	V_1	I_1	V_1	I_1	V_1
I_2	V_2	I_2	V_2	I_2	V_2	I_2	V_2
I_3	V_3	I_3	V_3	I_3	V_3	I_3	V_3

QUESTIONS

1. Explain, in your own words, how the superposition theorem is used to find the currents in a circuit supplied by more than one voltage source.

——

——

——

——

——

——

——

——

——

——

——

——

——

——

2. Refer to your data in Tables 24–1, 24–2, 24–3. Do the results of your experiment confirm the superposition theorem? Cite specific data from the tables.

_____

_____

_____

_____

_____

_____

_____

_____

3. Why is it important to include the polarity sign when recording the value of current in step 2?

_____

_____

_____

_____

_____

_____

_____

_____

4. If the polarity of both power supplies in Figure 24–7 were reversed, how would the current in R_2 be affected?

_____

_____

_____

_____

_____

_____

_____

_____

_____

_____

_____

THEVENIN'S THEOREM

BASIC INFORMATION

Thevenin's theorem is another mathematical tool that is very helpful in solving complex linear circuit problems. The theorem makes it possible to determine the voltage or current in any portion of a circuit. The technique employed involves reducing the complex circuit to a simple equivalent circuit.

Thevenin's Theorem

Thevenin's theorem states that any linear two-terminal network may be replaced by a simple equivalent circuit consisting of a Thevenin voltage source V_{TH} in series with an internal resistance R_{TH} causing current to flow through the load. Thus, the Thevenin equivalent for the circuit of Figure 25–1 (p. 188)(a) feeding the load R_L is the circuit of Figure 25–1(d). If we knew how to calculate the values of V_{TH} and R_{TH}, the process of finding the current I_L in R_L would be a simple application of Ohm's law.

The rules for determining V_{TH} and R_{TH} are as follows:

1. The voltage V_{TH} is the voltage "seen" across the load terminals in the original network, with the load resistance removed (open-circuit voltage); that is, it is the voltage we would measure if we placed a voltmeter across AB in Figure 25–1(a) with the load resistance removed.

2. The resistance R_{TH} is the resistance seen from the terminals of the open load, looking into the original network when the voltage sources in the circuit are short-circuited and replaced by their internal resistance.

The development of the Thevenin equivalent circuit of Figure 25–1(a) is as follows:

1. [Figure 25–1(b)] The load resistance R_L has been removed, and the voltage across AB has been calculated. In this case the voltage drop across R_3 is one-half the source voltage V, because R_3 is one-half the total resistance in the series circuit consisting of V, R_1 (assumed to be the internal resistance of the voltage source V), R_2, and R_3. The Thevenin equivalent voltage $V_{TH} = 6$ V.

2. [Figure 25–1(c)] The voltage source V has been shorted and only its internal resistance remains in the circuit. The equivalent resistance of the parallel circuit across AB is now calculated.

OBJECTIVES

1 To determine the Thevenin equivalent voltage (V_{TH}) and resistance (R_{TH}) of a dc circuit with a single voltage source

2 To verify experimentally, the values of V_{TH} and R_{TH} in solving a series-parallel circuit

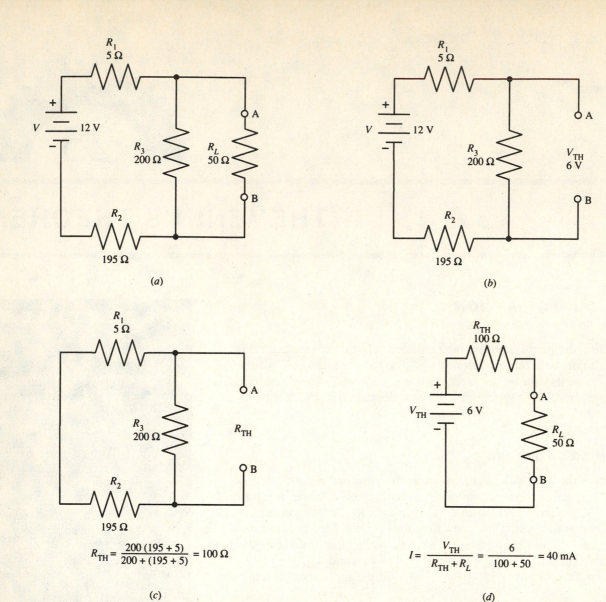

Figure 25–1. Solution of a series-parallel circuit using Thevenin's theorem.

$$R_{TH} = \frac{(R_1 + R_2) \times R_3}{(R_1 + R_2) + R_3} = \frac{(5\ \Omega + 195\ \Omega) \times 200\ \Omega}{(5\ \Omega + 195\ \Omega) + 200\ \Omega}$$

$$= \frac{40\ k\Omega}{400} = 100\ \Omega$$

3. [Figure 25–1(d)] The Thevenin equivalent voltage and resistance are connected in series with the load resistance R_L to form a simple series circuit.

The load current I_L can now be found by using Ohm's law.

$$I_L = \frac{V_{TH}}{R_L + R_{TH}} = \frac{6\ V}{50\ \Omega + 100\ \Omega} = \frac{6\ V}{150\ \Omega}$$

$$I_L = 40\ mA$$

It might seem that the Thevenin method unnecessarily adds work to solving a circuit and that Ohm's and Kirchhoff's laws could solve the problem more quickly and easily. Of course, the sample problem was intentionally made simple to illustrate the method more clearly, but even a simple circuit can prove the worth of the method. Suppose it was necessary to find I_L for a range of 10 values of R_L while the rest of the circuit remained unchanged. It would be quite laborious to apply Ohm's and Kirchhoff's laws 10 times to calculate each value of I_L. With just a single calculation of the Thevenin equivalent circuit, the current I_L for any value of R_L can be calculated quickly through a single application of Ohm's law.

Solving an Unbalanced-Bridge Circuit by Thevenin's Theorem · · · · · · · · · · · · · · ·

Figure 25–2(a) is an unbalanced-bridge circuit. It is required to find the current I in R_5. Thevenin's theorem lends itself readily to a solution of this problem.

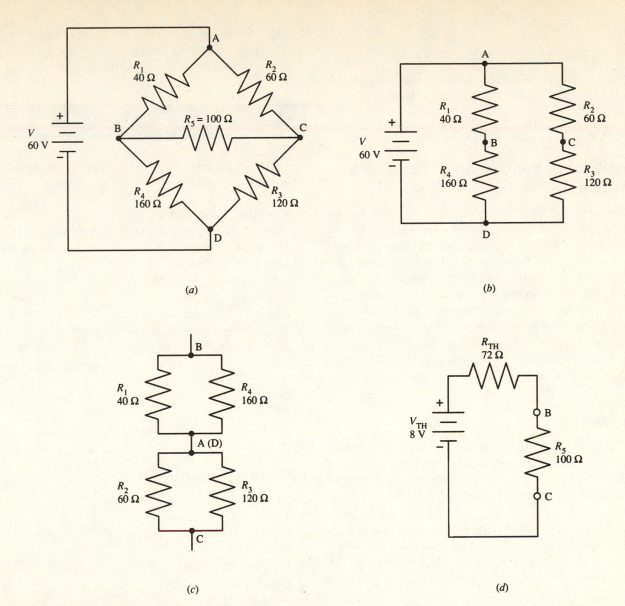

Figure 25–2. Solving an unbalanced bridge circuit using Thevenin's theorem.

For this exercise, consider R_5 the load. The problem then is to transform the circuit into its Thevenin equivalent supplying current to R_5.

The Thevenin voltage V_{TH} is found by removing R_5 from the circuit and solving for V_{BC} in Figure 25–2(b). The difference in voltage between BD and CD will be V_{BC}. Voltages V_{BD} and V_{CD} can be found directly using resistance ratios.

$$V_{BD} = \frac{R_4}{(R_1 + R_4)} \times V$$

$$= \frac{160\ \Omega}{200\ \Omega} \times 60\ V = 48\ V$$

$$V_{CD} = \frac{R_3}{(R_2 + R_3)} \times V$$

$$= \frac{120\ \Omega}{180\ \Omega} \times 60\ V = 40\ V$$

$$V_{BD} - V_{CD} = V_{BC} = 48\ V - 40\ V = 8\ V = V_{TH}$$

The Thevenin resistance R_{TH} is found by shorting the voltage source and replacing it with its internal resistance. In this case it is assumed V is an ideal voltage source, and its internal resistance is therefore zero. Thus, AD is, in effect, shorted. The resistance across BC (the Thevenin equivalent resistance) can be visualized more easily if the circuit of Figure 25–2(b) is redrawn with AD shorted. Figure 25–2(c) shows BC more clearly and thus makes it easier to find R_{BC}.

Resistor R_1 in parallel with R_4 becomes

$$\frac{40\ \Omega \times 160\ \Omega}{40\ \Omega + 160\ \Omega} = \frac{6.4\ k\Omega}{200} = 32\ \Omega$$

Resistor R_2 in parallel with R_3 becomes

$$\frac{60\ \Omega \times 120\ \Omega}{60\ \Omega + 120\ \Omega} = \frac{7.2\ k\Omega}{180} = 40\ \Omega$$

Thus,

$$R_{BC} = 32 \ \Omega + 40 \ \Omega = 72 \ \Omega = R_{TH}$$

Substituting these values in the Thevenin equivalent circuit [Figure 25–2(d)] and solving for I, we find

$$I = \frac{V_{TH}}{R_{TH} + R_5} = \frac{8 \ V}{172 \ \Omega} = 46.5 \ mA$$

Experimental Verification of Thevenin's Theorem

It is possible to determine the values of V_{TH} and R_{TH} for a load R_L in a specific network by measurement. Then we can experimentally set the output of a voltage-regulated power supply to V_{TH} by connecting a resistor whose value is R_{TH} in series with V_{TH} and with R_L. We can measure I in this equivalent circuit. If the measured value of I_L in R_L in the original network is the same as the I measured in the Thevenin equivalent, we have one verification of Thevenin's theorem. For a more complete verification, this process would have to be repeated many times with random circuits.

SUMMARY

1. Thevenin's theorem states that any linear two-terminal network may be replaced by a simple equivalent circuit that acts like the original circuit across the load connected to the two terminals.

2. The equivalent circuit consists of a Thevenin voltage source (V_{TH}) in series with the Thevenin internal resistance (R_{TH}) in series with the two terminals of the load. Thus, for the complex circuit of Figure 25–2(a), the Thevenin equivalent at the terminals BC is Figure 25–2(d).

3. To determine the Thevenin voltage V_{TH}, open the load at the two terminals in the original network and calculate the voltage at these two terminals. This *open-load* voltage is V_{TH}.

4. To determine the Thevenin resistance R_{TH}, keep the load open at the two terminals in the original network and short the original power source and replace it with its own internal resistance. Then calculate the resistance at the open-load terminals, looking back into the original circuit.

5. Thevenin's theorem can be used with a circuit having one or more power sources.

6. Once the complex network has been replaced by the Thevenin equivalent circuit, the current in the load may be found by applying Ohm's law.

SELF TEST

Check your understanding by answering the following questions:

1. Consider the circuit of Figure 25–1(a). V = 24 V, $R_1 = 30 \ \Omega$, $R_2 = 270 \ \Omega$, $R_3 = 500 \ \Omega$, and $R_L = 560 \ \Omega$. Assume the internal resistance of the supply is zero. Find each value.
   (a) V_{TH} = _____ V
   (b) R_{TH} = _____ Ω
   (c) I_L = _____ A

2. Consider the circuit of Figure 25–2(a). V = 12 V, $R_1 = 200 \ \Omega$, $R_2 = 500 \ \Omega$, $R_3 = 300 \ \Omega$, $R_4 = 600 \ \Omega$, and $R_5 = 100 \ \Omega$. Assume a voltage-regulated power supply. Find each value.
   (a) V_{TH} = _____ V
   (b) R_{TH} = _____ Ω
   (c) I_5 = _____ A (the current in R_5)

MATERIALS REQUIRED

Power Supply:
- Variable, 0–15 V dc, regulated

Instruments:
- DMM or VOM
- 0-5–mA milliammeter

Resistors (½-W, 5%):
- 1 330-Ω
- 1 390-Ω
- 1 470-Ω
- 1 1 kΩ
- 1 1.2 kΩ
- 2 3.3 kΩ
- 1 5-kΩ, 2–W potentiometer

Miscellaneous:
- 2 SPST switches

PROCEDURE

1. Using an ohmmeter, measure the resistance of each of the 7 resistors supplied. Record the values in Table 25–1 (p. 193).

2. With power **off** and both S_1 and S_2 **open,** connect the circuit of Figure 25–3 with $R_L = 330 \ \Omega$. Turn the power **on; close** S_1. Adjust V_{PS} to 15 V. **Close** S_2 and

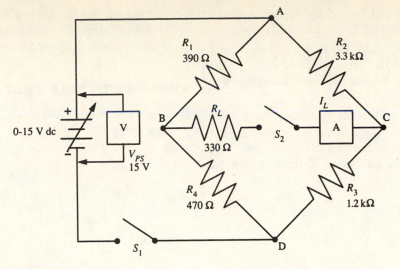

Figure 25–3. Circuit for procedure step 2.

measure I_L, the current through the load resistor R_L. Record this value in Table 25–2 (p. 193) in the 330-Ω row under "Original Circuit." **Open** S_2. S_1 should remain **closed.**

3. With S_1 **closed** and S_2 **open,** measure the voltage across BC (Figure 25–3). This is voltage V_{TH}; record the value in Table 25–2 in the 330-Ω row under the "V_{TH} Measured" column. **Open** S_1; turn **off** the power supply.

4. Remove the power supply from the circuit by disconnecting it across AD. Short AD by connecting a wire across the two points.

5. With S_2 still open, connect an ohmmeter across BC to measure the resistance across points B and C. This is R_{TH}. Record the value in Table 25–2 in the 330-Ω row under "R_{TH} Measured."

6. Adjust the power supply so that $V_{PS} = V_{TH}$. Connect the ohmmeter across the potentiometer, and adjust the resistance so that the resistance across the potentiometer is R_{TH}.

7. Disconnect the 330-Ω load resistor, S_2, and the milliammeter from the circuit of Figure 25–3 and connect them as shown in Figure 25–4. With S_2 **open,** and power **on,** check to see that $V_{PS} = V_{TH}$.

8. **Close** S_2. Measure I_L and record the value in Table 24–2 in the 330-Ω row under "Thevenin Equivalent Circuit, Measured." **Open** S_2; turn the power **off.**

9. Using the measured values of V_{PS}, R_1, R_2, R_3, and R_4 (Table 25–1), calculate V_{TH} for the circuit of Figure 25–3. Record your answer in Table 25–2 in the 330-Ω row under "V_{TH}, Calculated."

10. Calculate R_{TH} in Figure 25–3 using the measured values for R_1, R_2, R_3, and R_4. (Voltage-regulated power supplies normally have negligible internal resistance.) Record your answer in Table 25–2 in the 330-Ω row under "R_{TH}, Calculated."

11. Using the calculated values of V_{TH} and R_{TH} from steps 9 and 10 as recorded in Table 25–2, calculate I_L. Record your answer in Table 25–2 in the 330-Ω row under "I_L, Calculated."

12. Substitute a 1 kΩ resistor for R_L in the circuit of Figure 25–3. Turn the power **on** and adjust V_{PS} to 15 V; **close** S_1 and S_2. Measure I_L and record the value in Table 25–2 in the 1 kΩ row under "I_L, Measured, Original Circuit." **Open** S_2.

13. Remove the 1 kΩ load resistor R_L and connect the 3.3 kΩ load resistor. Adjust V_{PS} to 15 V, if necessary. **Close** S_2. Measure I_L and record the value in the 3.3 kΩ row under "I_L Measured, Original Circuit." **Open** S_1 and S_2; turn the power **off.**

14. Connect the Thevenin equivalent circuit as shown in Figure 25–4 using the 1 kΩ load resistor in place of the

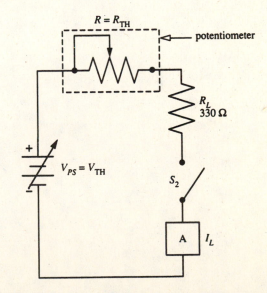

Figure 25–4. Thevenin's equivalent circuit for procedure step 7. The values of V_{TH} and R_{TH} are determined in steps 3 and 5.

THEVENIN'S THEOREM **191**

330-Ω resistor. V_{TH} and R_{TH} should be the measured values recorded in Table 25–2 in the 330-Ω row.

15. Turn the power **on;** adjust V_{PS} to V_{TH}. **Close** S_2 and measure I_L. Record the value in Table 25–2 in the 1 kΩ row under "I_L, Measured, Thevenin Equivalent Circuit." **Open** S_2.

16. Remove the 1 kΩ resistor and connect the 3.3 kΩ load resistor. **Close** S_2 and measure I_L. Record the value in Table 25–2 in the "3.3 kΩ" row, under "I_L, Measured, Thevenin Equivalent Circuit." **Open** S_2; power **off.**

17. Calculate I_L for $R_L = 1$ kΩ and $R_L = 3.3$ kΩ for the circuit of Figure 25–3, using the measured values for R_1, R_2, R_3, R_4, and R_L. Record your answers in Table 25–2.

ANSWERS TO SELF TEST

1. (a) 15; (b) 188; (c) 0.02
2. (a) 4.45; (b) 338; (c) 0.01

TABLE 25–1. Measured Resistor Values

Resistor	Rated Value, Ω	Measured Value, Ω
R_1	390	
R_2	3.3 k	
R_3	1.2 k	
R_4	470	
R_L	330	
R_L	1 k	
R_L	3.3 k	

TABLE 25–2. Measurements to Verify Thevenin's Theorem

R_L, Ω	V_{TH}, V		R_{TH}, Ω		I_L, mA		
					Measured		
	Measured	Calculated	Measured	Calculated	Original Circuit	Thevenin Equivalent Circuit	Calculated
330							
1 k							
3.3 k							

QUESTIONS

1. Explain, in your own words, how Thevenin's theorem is used to convert any linear
   two-terminal network into a simple equivalent circuit consisting of a resistance
   in series with a voltage source.

———

———

———

———

———

———

———

———

———

———

2. Refer to your data in Table 25–2. How do the values of I_L measured in the original circuit (Figure 25–3) compare with those measured in the Thevenin equivalent circuit (Figure 25–4)? Should the comparable measurements be the same? Explain why.

_____

_____

_____

_____

_____

_____

_____

_____

_____

3. Refer to Table 25–2. Compare calculated and measured values of R_{TH}. Are the results as you would have expected? Explain. Make the same comparison with the two values of V_{TH}.

_____

_____

_____

_____

_____

_____

_____

_____

_____

4. Explain an advantage of using Thevenin's theorem when finding load currents in a dc circuit

_____

_____

_____

_____

_____

_____

_____

_____

_____

_____

NORTON'S THEOREM

BASIC INFORMATION

Norton's Theorem

Thevenin's theorem simplifies the analysis of complex networks by reducing the original circuit to a simple equivalent circuit containing a constant-voltage source V_{TH} in series with an internal resistance R_{TH}. Norton's theorem utilizes a similar technique of simplification. The Norton source, however, delivers a constant current.

Norton's theorem states that any two-terminal linear network may be replaced by a simple equivalent circuit consisting of a constant-current source I_N in parallel with an internal resistance R_N. Figure 26–1(a) (p. 196) shows an actual network terminated by a load resistance R_L. Figure 26–1(b) shows the Norton equivalent circuit. The Norton current I_N is distributed between resistance R_N and the load R_L.

With reference to Figure 26–1(a), the rules for determining the constants in the Norton equivalent circuit are as follows:

1. The constant current I_N is the current that would flow in AB if the load resistance between A and B were replaced by a short circuit.

2. The Norton resistance R_N is the resistance seen from terminals AB with the load removed and the voltage sources shorted and replaced with their internal resistance. Thus, R_N is defined in exactly the same way as the Thevenin resistance R_{TH}. Therefore, $R_N = R_{TH}$.

Applications

Consider the circuit of Figure 26–2(a) (p. 197). We wish to find the current I_L in R_L using Norton's theorem. (Of course, this circuit can also be solved using Ohm's and Kirchhoff's laws as well as mesh and Thevenin's methods. Such solutions are recommended as student exercises.)

The development of the Norton equivalent circuit for Figure 26–2(a) may be followed from Figure 26–2(b), (c), and (d).

1. [Figure 26–2(b)] Load resistor R_L is short-circuited, thus shorting R_3. The current produced by V_T is I_N.

$$I_N = \frac{V}{R_1 + R_2} = \frac{20\ V}{5\ \Omega + 195\ \Omega} = \frac{20\ V}{200\ \Omega}$$

$$I_N = 100\ mA$$

OBJECTIVES

1. To determine the values of Norton's constant-current source I_N and Norton's current-source resistance R_N in a dc circuit containing one or two voltage sources

2. To verify experimentally the values of I_N and R_N in the solution of complex dc networks containing two voltage sources

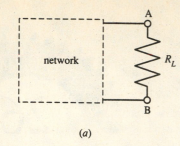

(a)

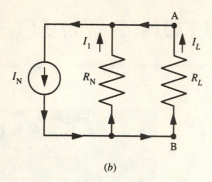

(b)

Figure 26–1. Norton's equivalent circuit consists of a constant current source, I_N, and a shunt resistance, R_N.

2. [Figure 26–2(c)] The voltage source V is shorted and replaced by its internal resistance. With R_L removed, the resistance across AB is calculated. The resistance is R_N.

$$R_N = \frac{(R_1 + R_2) \times (R_3)}{R_1 + R_2 + R_3} = \frac{(5\ \Omega + 195\ \Omega) \times (200\ \Omega)}{5\ \Omega + 195\ \Omega + 200\ \Omega}$$

$$R_N = \frac{40\ k\Omega}{400} = 100\ \Omega$$

3. [Figure 26–2(d)] The original circuit is replaced by the Norton constant-current source $I_N = 100$ mA in parallel with the Norton resistance $R_N = 100\ \Omega$. The load resistance R_L is connected across the Norton equivalent circuit.

From the current divider rule, the value of I_L can now be calculated.

$$I_L = \frac{I_N R_N}{R_L + R_N}$$

$$= \frac{(100\ \text{mA})(100\ \Omega)}{(350\ \Omega) + (100\ \Omega)} = \frac{10\ \text{V}}{450\ \Omega}$$

$$I_L = 22\ \text{mA}$$

As in the case of Thevenin's theorem, Norton's theorem is useful in applications in which it is necessary to calculate the load current as the load resistance varies over a wide range of values.

Solution of a DC Network with Two Voltage Sources

The solution of complex dc networks containing two or more voltage sources is possible using any of the methods discussed in this and previous experiments.

Norton's theorem is used to solve the following problem.

Problem. Develop a formula for finding the load current in the circuit of Figure 26–3(a) (p. 198) over a range of different load resistors. Using this formula find I_L for $R_L = 100\ \Omega, 500\ \Omega, 1\ k\Omega$. Assume that V_1 and V_2 are constant-voltage sources.

Solution. The first step is to find the Norton constant-current source I_N. This is done by replacing R_L with a short across FG and solving for the current through FG. Using the mesh currents I_1 and I_N [Figure 26–3(b)], we have

$$I_1(R_1 + R_2) - I_N R_2 = -V_1 - V_2$$

$$-I_1 R_2 + I_N R_2 = V_2$$

$$320 I_1 - 220 I_N = -30$$

$$-220 I_1 + 220 I_N = 20$$

Solving for I_N, we obtain

$$I_N = 0.009\ \text{A} = 9\ \text{mA}$$

While I_N is negative, we are concerned with its value, not its direction.

The Norton resistance R_N is the resistance measured across FG in Figure 26–3(c). This is the same as the Thevenin resistance R_{TH}. It is found by shorting all voltage sources and replacing them with their internal resistance. In this problem we assume that V_1 and V_2 are ideal voltage sources (that is, constant voltage sources), so that their internal resistance is zero. The resistance across FG in this case is R_1 in parallel with R_2

$$R_N = \frac{100\ \Omega \times 220\ \Omega}{100\ \Omega + 220\ \Omega} = \frac{22\ k\Omega}{320\ \Omega}$$

$$R_N = 68.75\ \Omega$$

To find I_L we can use the formula from the previous example:

$$I_L = \frac{I_N R_N}{R_N + R_L}$$

$$I_L = \frac{9\ \text{mA} \times 68.75\ \Omega}{68.75\ \Omega + R_L} = \frac{0.619}{68.75 + R_L}$$

We can now calculate the values of I_L for each of the values of R_L.

For $R_L = 100\ \Omega$,

$$I_L = \frac{0.619}{68.75 + 100} = \frac{0.619}{168.75} = 4\ \text{mA}$$

For $R_L = 500\ \Omega$,

$$I_L = \frac{0.619}{68.75 + 500} = \frac{0.619}{568.75} = 1\ \text{mA}$$

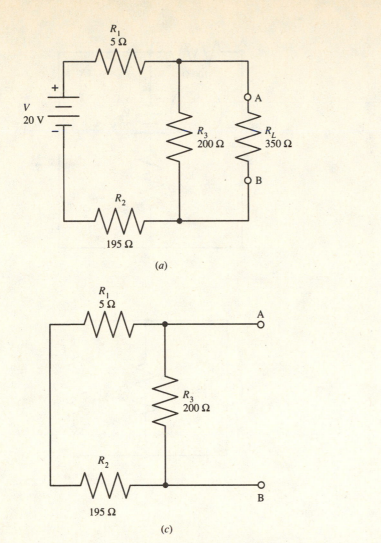

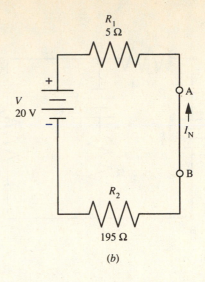

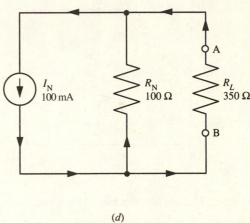

Figure 26–2. Applying Norton's theorem to the solution of a dc network.

For $R_L = 1k\Omega$,

$$I_L = \frac{0.619}{68.75 + 1000} = \frac{0.619}{1068.75} = 0.6 \text{ mA}$$

SUMMARY

1. Norton's theorem offers another method of solving complex linear circuits. Norton's theorem allows a complex two-terminal network to be replaced by a simple equivalent circuit that acts like the original circuit at the load connected to the two terminals.

2. Norton's theorem applies to a linear circuit with one or more power sources.

3. The equivalent circuit is a circuit consisting of a constant-current source I_N in parallel with the internal resistance of the source R_N, across which the load R_L is connected. The current I_N is then divided between R_N and R_L. Figure 26–1(b) shows the Norton current source and its load.

4. To determine the Norton current I_N, short-circuit the load and calculate the current through the short in the original circuit. This short-circuit current is I_N. To calculate I_N, it may be necessary to use Ohm's and Kirchhoff's laws.

5. To determine the Norton resistance R_N, which acts in parallel with the Norton current, the same technique is applied as in finding the Thevenin resistance in the preceding experiment. Proceed as follows: Open the load at the two terminals in question in the original network. Short all the voltage sources and replace them with their internal resistances. Then calculate R_N, the resistance at the open-load terminals, looking back into the circuit.

6. Once the complex network has been replaced by the Norton equivalent circuit, the current I_L through the load may be found by applying the formula

$$I_L = \frac{I_N \times R_N}{R_N + R_L}$$

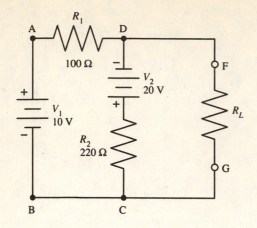

(a)

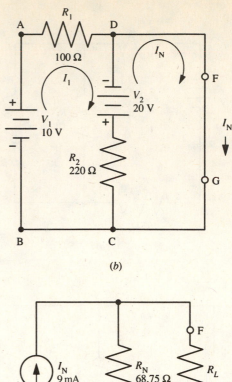

(b)

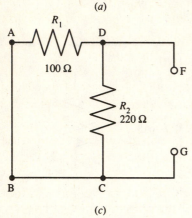

(c)

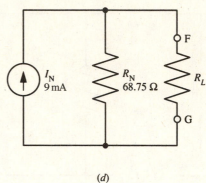

(d)

Figure 26–3. Applying Norton's theorem to a circuit with two voltage sources.

Check your understanding by answering the following questions:

1. Consider the circuit of Figure 26–2(a). $V = 12$ V, $R_1 = 1\ \Omega$, $R_2 = 39\ \Omega$, $R_3 = 60\ \Omega$, and $R_L = 27\ \Omega$. Assume the internal resistance of the voltage source V is zero. Find each value in the equivalent Norton circuit.
   (a) $I_N = $ _____ A
   (b) $R_N = $ _____ Ω
   (c) $I_L = $ _____ A

2. Consider the circuit of Figure 26–3(a). $V_1 = 30$ V, $V_2 = 30$ V. Assume the internal resistance of these voltage sources is zero. $R_1 = 45\ \Omega$, $R_2 = 150\ \Omega$, and R_L (the load) $= 47\ \Omega$. Find each value in the Norton equivalent circuit.
   (a) $I_N = $ _____ A
   (b) $R_N = $ _____ Ω
   (c) $I_L = $ _____ A

MATERIALS REQUIRED

Power Supply:
- 2 variable 0–15 V dc, regulated

Instruments:
- 2 DMM or VOM

Resistors (½-W, 5%):
- 1 390-Ω
- 1 560-Ω
- 1 680-Ω
- 1.2 kΩ
- 1 1.8 kΩ
- 1 2.7 kΩ
- 1 10-kΩ, 2-W potentiometer

Miscellaneous:
- 2 SPST switches
- 3 SPDT switches

PROCEDURE

A. Determining I_N and R_N

A1. With power **off** in both supplies; S_4 and S_5 **open;** and switches S_1, S_2, and S_3 in position Ⓐ, connect the circuit of Figure 26–4.

A2. Power **on** V_{PS1} and V_{PS2}. Adjust the supply voltages so that $V_{PS1} = 12$ V and $V_{PS2} = 6$ V. (Note carefully the correct polarity of the connections.) Maintain these voltages throughout the experiment. **Close** S_4 and S_5. Measure I_L through R_L and record the results in Table 26–1 (p. 201) in the 1.2 kΩ row under "I_L, Measured, Original Circuit."

A3. Replace R_L in turn with a 390-Ω, 560-Ω, and 1.8 kΩ resistor. In each case, measure I_L and record the values in the "I_L, Measured, Original Circuit" column.

A4. Move S_3 to position Ⓑ. This, in effect, replaces R_L with a short. The current measured by the meter is the short-circuit current of the Norton equivalent generator I_N. Record the value in Table 26–1 in the 1.2 kΩ row under "I_N, Measured."

A5. Turn the power **off.** Move S_1, S_2, and S_3 to position Ⓑ and open S_5. This, in effect, replaces the voltage sources with short circuits and opens the load circuit between D and E. (Regulated power supplies are considered to have negligible resistance.) S_4 is still **closed.**

A6. With an ohmmeter, measure the resistance across CF. This is the resistance across the Norton equivalent generator R_N. Record this value in Table 26–1 in the 1.2 kΩ row under "R_N, Measured."

A7. From the circuit of Figure 26–4, calculate the value of the Norton current I_N and record it in Table 26–1 in the 1.2 kΩ row under "I_N, Calculated."

A8. From the circuit of Figure 26–4, calculate the value of the Norton shunt resistance R_N and record it in

Table 26–1 in the 1.2 kΩ row under "R_N, Calculated."

A9. Using the values of I_N and R_N from steps A7 and A8, calculate the load current I_L for the 1.2 kΩ, 390-Ω, 560-Ω, and 1.8 kΩ load resistors in Figure 26–4. Record these values in Table 26–1 under "I_L, Calculated."

B. Using the Norton Equivalent Circuit

B1. With power **off** and S_1 **open,** connect the circuit of Figure 26–5 with $R_L = 1.2$ kΩ. Meter A1 will measure the Norton current I_N, whereas meter A2 will measure the load current I_L. The potentiometer will serve as R_N. Use an ohmmeter while adjusting the potentiometer until its resistance measures that of R_N found in step A6.

B2. Adjust the power supply to its lowest output voltage. Turn power **on** and **close** S_1. Slowly increase the output of the power supply until the *current* measured by ammeter A2 is equal to the value of I_N you found in step A4 and recorded in Table 26–1.

B3. With meter A1 measuring I_N, record the load current I_L measured by meter A2 in Table 26–1 in the 1.2 kΩ row under "I_L, Measured, Norton Equivalent Circuit." S_1 **open;** power **off.**

B4. For each of the other load resistors in Table 26–1, connect the Norton equivalent circuit (Figure 26–5) and measure I_L for each value of R_L. Record your values in Table 26–1 under "I_L, Measured, Norton Equivalent Circuit." **Open** S_1; power **off.**

ANSWERS TO SELF-TEST

1. (a) 0.3; (b) 24; (c) 0.14
2. (a) 0.47; (b) 34.6; (c) 0.199

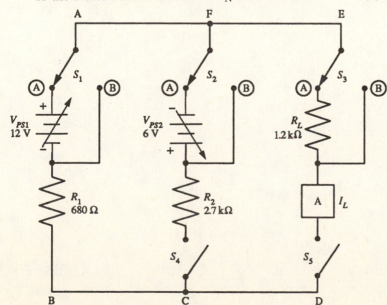

Figure 26–4. Circuit for procedure step A1.

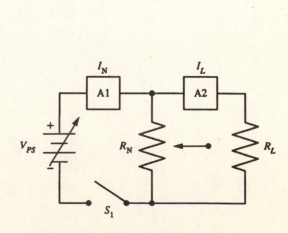

Figure 26–5. Circuit for procedure step B1.

TABLE 26–1. Measurements to Verify Norton's Theorem

R_L, Ω	I_N, mA		R_N, Ω		I_L, mA		
					Measured		
	Measured	Calculated	Measured	Calculated	Original Circuit	Norton Equivalent Circuit	Calculated
1.2 k							
390							
560							
1.8 k							

QUESTIONS

1. Explain, in your own words, how Norton's theorem is used to convert any two-terminal linear network into a simple circuit consisting of a constant current source in parallel with a resistance.

_____

_____

_____

_____

_____

_____

_____

_____

2. Refer to your data in Table 26–1. How do the values of I_L measured in the original circuit (Figure 26–4) compare with those measured in the Norton equivalent circuit (Figure 26–5)? Should the comparable measurements be the same? Explain why.

_____

_____

_____

_____

_____

3. Refer to Table 26–1. Compare measured and calculated values for I_N. Are the results as you would have expected? Explain. Make the same comparison with the two values of R_N.

_____

_____

_____

_____

_____

_____

_____

_____

_____

_____

_____

_____

_____

_____

_____

4. Explain an advantage of using Norton's theorem when finding load currents in a dc circuit.

_____

_____

_____

_____

_____

_____

_____

_____

_____

_____

_____

_____

_____

_____

MILLMAN'S THEOREM

BASIC INFORMATION

If is often necessary to find the voltage across two points to which a number of parallel branches and sources are connected. In previous experiments various techniques involving Kirchhoff's laws and superposition were used. Millman's theorem provides another method for solving such circuits. In many cases it provides a more direct and quicker method than either Kirchhoff's laws or superposition.

Millman's Theorem

If a circuit can be redrawn or viewed as having two common lines (for example, a "hot" line and a ground), Millman's theorem can be used to find the voltage between the two lines. If the circuit contains only one voltage source, the usual methods used to solve parallel circuits are best used. However, if a number of the parallel branches contain voltage sources, the usual methods become cumbersome and time consuming. Millman's theorem provides a shortcut method in those cases.

Figure 27–1 (p. 204) is a circuit containing two sources. If the voltage from X to ground must be found, any of the techniques covered in earlier experiments can be used. However, by redrawing the circuit as in Figure 27–2 (p. 204), Millman's theorem can be readily applied.

Millman's theorem is in the form of a formula:

$$V_{XG} = \frac{\dfrac{V_1}{R_1} + \dfrac{V_2}{R_2} + \dfrac{V_3}{R_3}}{\dfrac{1}{R_1} + \dfrac{1}{R_2} + \dfrac{1}{R_3}} \qquad (27\text{–}1)$$

where

V_{XG} is the voltage across the common lines

V_1 is the total voltage sources in the first branch

R_1 is the total resistance in the first branch

V_2 is the total voltage sources in the second branch

R_2 is the total resistance in the second branch

V_3 is the total voltage sources in the third branch

R_3 is the total resistance in the third branch

If any branch does not contain a voltage source, the voltage is equal to zero. Formula (27–1) can be extended to any number of branches by

OBJECTIVES

1 To verify Millman's theorem experimentally

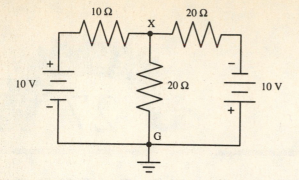

Figure 27–1. Millman's theorem can be used to solve circuits with more than one voltage source.

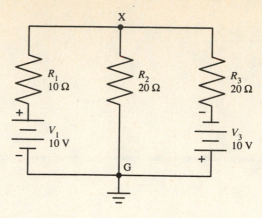

Figure 27–2. The circuit of Figure 27–1 redrawn to show its applicability to Millman's theorem.

simply adding the V_n/R_n term to the numerator and the $1/R_n$ term to the denominator.

To illustrate the application of Millman's theorem, the circuit of Figure 27–2 will be solved using formula (27–1).

$$V_{XG} = \frac{\dfrac{10}{10} + \dfrac{0}{20} - \dfrac{10}{20}}{\dfrac{1}{10} + \dfrac{1}{20} + \dfrac{1}{20}}$$

The polarity of V_3 is negative, since it would make point X negative with respect to ground. (Of course, if ground was considered positive, V_3 would be positive, and V_1 would be written as negative.)

Simplifying, we have

$$V_{XG} = \frac{1 + 0 - 0.5}{0.1 + 0.05 + 0.05} = \frac{0.5}{0.2}$$

$$V_{XG} = 2.5 \text{ V}$$

To use the Millman formula, the branches must all be in parallel. Thus, a series-parallel circuit cannot be solved directly by the Millman formula. Sometimes a series-parallel circuit lends itself to simplification, so that a two-step calculation can be used. A second example illustrates this process.

Figure 27–3 is a series-parallel circuit with two voltage sources. The current through the load R_L will be found using the Millman formula. Because the circuit as shown cannot be represented by a pure parallel circuit, it is necessary to combine some components. By combining R_2 and R_3 into a single resistance and R_5 and R_L into a single resistance, the circuit can be solved by Millman's theorem.

Since R_2 and R_3 are equal, their equivalent resistance $R_{2,3}$ is 10/2 or 5 Ω. Similarly, combining R_5 and R_L results in 20/2, or 10 Ω. Now the parallel circuit is complete, as shown in Figure 27–4. Applying the Millman formula, we obtain

$$V_{XG} = \frac{\dfrac{10}{10} + \dfrac{5}{5} + \dfrac{0}{20}}{\dfrac{1}{10} + \dfrac{1}{5} + \dfrac{1}{20}} = \frac{1 + 1 + 0}{0.1 + 0.2 + 0.05} = \frac{2}{0.35}$$

$$V_{XG} = 5.71 \text{ V}$$

This is the voltage across XE in the original circuit of Figure 27–3. The current in this part of the circuit is

$$I = \frac{5.71}{20} = 286 \text{ mA}$$

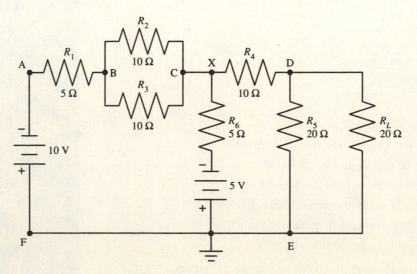

Figure 27–3. A series-parallel circuit with two voltage sources.

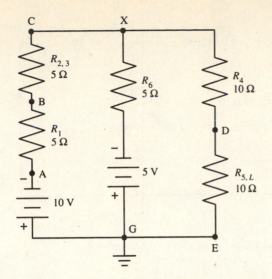

Figure 27–4. Figure 26–3 redrawn to show its three parallel branches.

This is the current in XD, through resistor R_4. At point D the current divides in half, one-half to R_5, the other half through R_L:

$$I_{RL} = \frac{286\text{mA}}{2} = 143 \text{ mA}$$

SUMMARY

1. The Millman theorem formula is

$$V_{AB} = \frac{\dfrac{V_1}{R_1} + \dfrac{V_2}{R_2} + \dfrac{V_3}{R_3} + \cdots + \dfrac{V_n}{R_n}}{\dfrac{1}{R_1} + \dfrac{1}{R_2} + \dfrac{1}{R_3} + \cdots + \dfrac{1}{R_n}}$$

where the V's are the total voltage sources in each branch and the R's are the total resistance in each branch.

2. Millman's formula can be used only in pure parallel circuits.

3. Millman's formula is most useful in parallel circuits in which more than one branch has a voltage source.

4. If polarity is assigned to the two parallel lines of the circuit, any voltage sources connected to the lines with reverse polarity must be represented in the Millman formula with a minus sign.

5. Millman's formula can be used as a shortcut in solving circuit problems requiring more than one step.

SELF TEST

Check your understanding by answering the following questions.

1. The Millman theorem can be used to solve circuit problems that can also be solved using _____ laws and _____.

2. Millman's theorem can be used to solve only pure _____ circuit problems.

3. The answer to Millman's formula is given in _____ units.

4. (Refer to Figure 27–2.) If $R_1 = R_2 = R_3 = 10 \ \Omega$, $V_1 = 5$ V, and $V_3 = 10$ V, find the current in R_2 using Millman's formula. $I_2 = $ _____ mA. The direction of current is _____ (downward from X to G/upward from G to X).

MATERIALS REQUIRED

Power Supplies:
- 2 variable 0–15 V dc, regulated

Instruments:
- 2 DMM
- 1 VOM
- 0–100-mA dc ammeter (The VOM DMM can be substituted for the ammeter if they have the necessary dc current range.)

Resistors (½-W, 5%):
- 4 68-Ω
- 1 100-Ω

Miscellaneous:
- 2 SPST switches

PROCEDURE

1. Using an ohmmeter, measure the resistance of the five resistors in this experiment. Label the resistors R_1 through R_4 and R_L, and connect them in the circuits that follow, according to their labels. Record the measured resistance values in Table 27–1 (p. 207).

2. With power supply V_1 **off** and S_1 **open,** connect the circuit of Figure 27–5 (p. 206). Use the resistors labeled R_1 and R_2. This is circuit 1.

3. Turn **on** the power supply and **close** S_1. Increase the output voltage of V_1 to 15 V. Measure the current in the circuit and record the value in Table 27–2 (p. 207). **Open** S_1 and turn **off** the power supply.

4. With power supply V_2 **off** and S_2 **open,** connect the circuit of Figure 27–6 (p. 206). Use resistor labeled R_3 in this circuit. This is circuit 2.

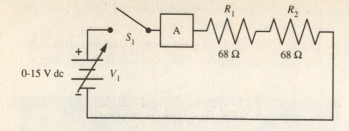

Figure 27–5. Circuit 1 for procedure step 2.

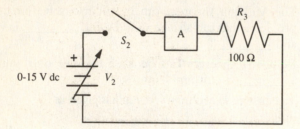

Figure 27–6. Circuit 2 for procedure step 4.

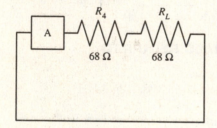

Figure 27–7. Circuit 3 for procedure step 6.

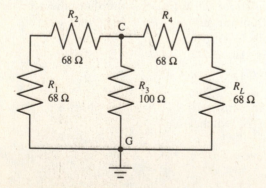

Figure 27–8. Circuit 4 for procedure step 7.

5. Turn **on** the power supply and **close** S_2. Increase the output voltage of V_2 to 10 V. Measure the current in the circuit and record the value in Table 27–2. **Open** S_2 and turn **off** the power supply.

6. Connect the circuit of Figure 27–7. Use the resistors labeled R_4 and R_L. Record the reading of the ammeter in Table 27–2. This is circuit 3.

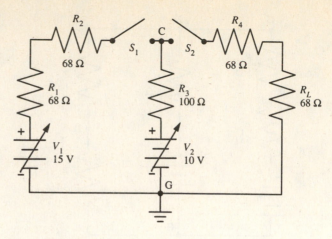

Figure 27–9. Circuit 5 for procedure step 9.

7. Reconnect the circuits of Figures 27–5, 6, and 7 without the switches and power supplies, as shown in Figure 27–8. In reconnecting the circuits be sure to observe the locations of each of the labeled resistors. This is circuit 4.

8. Using an ohmmeter, measure the resistance of the circuit across CG. Record this value in Table 27–2.

9. With the power supplies **off** and switches S_1 and S_2 **open,** connect the circuit of Figure 27–9. This is circuit 5. Note that this involves disconnecting portions of the circuit of step 7 and reconnecting the power supplies and S_1 and S_2. Both switches are **open**.

10. Turn **on** V_1 and adjust its output voltage to 15 V.

11. Turn **on** V_2 and adjust its output voltage to 10 V.

12. **Close** S_2, then **close** S_1. If necessary, readjust V_1 to 15 V and V_2 to 10 V.

13. Measure the voltage across CG and record the value in Table 27–2. After you have completed the necessary measurement, **open** S_1 and S_2 and turn **off** the power supplies.

14. Using the total measured current of circuits 1, 2, and 3 and the total resistance of circuit 4, calculate the voltage across CG.

    V_{CG} = total measured current × total resistance

    Record your answer in Table 27–2.

15. Using the rated value of the resistors, calculate the voltage across CG using Millman's theorem formula. Record your answer in Table 27–2.

ANSWERS TO SELF-TEST

1. Kirchhoff's; superposition

2. parallel

3. voltage

4. − 167; downward from X to G

TABLE 27–1. Resistance Measurements

Resistor	R_1	R_2	R_3	R_4	R_L
Rated value, Ω	68	68	100	68	68
Measured value, Ω					

TABLE 27–2. Verification of Millman's Theorem

Circuit	Voltage, V	*Measured Current, mA	Measured Total Resistance, Ω	Measured Voltage Across CG, V	Calculated V_{CG} Using Measured Current and Resistance, V	Calculated V_{CG} Using Millman's Theorem and Rated Values, V
1	15					
2	10					
3	0					
4	0					
5						

*Caution: Resistors will be hot. Power dissipation exceeds power rating of resistors. Do these steps quickly, and turn off power immediately.

QUESTIONS

1. Explain, in your own words, how Millman's theorem is used to solve a dc circuit.
   Note any limitations and restrictions on its application.

——
——
——
——
——
——
——
——
——
——
——
——

2. Compare the three values of V_{CG} in Table 27–2 for circuit 5 (Figure 27–9). Should they be equal? Explain why?

_____

_____

_____

_____

_____

_____

_____

_____

_____

_____

_____

_____

_____

_____

_____

_____

_____

3. Explain an advantage of using Millman's theorem to solve a dc circuit.

_____

_____

_____

_____

_____

_____

_____

_____

_____

_____

_____

_____

_____

_____

_____

_____

MAGNETIC FIELD ASSOCIATED WITH CURRENT IN A WIRE

BASIC INFORMATION

Current Producing a Magnetic Field

The physicist Hans Christian Oersted observed that a compass needle was deflected when placed in the vicinity of a wire carrying an electric current. Moreover, he observed that the direction in which the compass needle pointed depended on its position relative to the wire and on the direction of the current.

These observations are depicted in Figure 28–1(a) (p. 210). A current-carrying wire is shown coming out of the plane of the page. Its cross section is represented by a circle W. When a compass is placed in position 1, its needle points in the direction shown. As the compass is moved from position 1 to 2, 3, and 4, the needle also changes direction, as shown.

If the wire is not carrying current, as in Figure 28–1(b), the compass will point to north whether it is in position 1, 2, 3, or 4.

If the direction of current in the wire is reversed from that in Figure 28–1(a), then the compass will point in a direction opposite to the one in Figure 28–1(a), as shown in Figure 28–1(c).

These observations led to the following conclusions:

1. A magnetic field exists around a wire carrying current.
2. The direction of the magnetic field depends on the direction of current in the wire.
3. The magnetic field appears to be circular around the wire.

Later research confirmed that the magnetic field lies in a plane perpendicular to the current-carrying wire, that the magnetic field is circular around the wire, and that the magnetic field is strongest close to the wire. Research also showed that a magnetic field surrounds any moving electric charge whether it is being carried by a wire or not. For example, the electron beam in a cathode-ray tube produces the same type of circular magnetic field as a current-carrying wire.

Direction of Magnetic Field

Observations such as those illustrated in Figure 28–1(a) and (c) lead to a rule for predicting the direction of the magnetic field around a current-

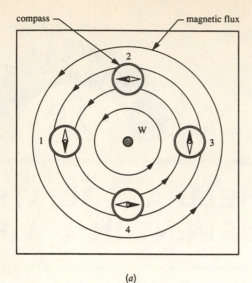

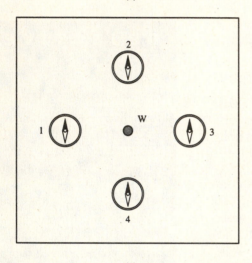

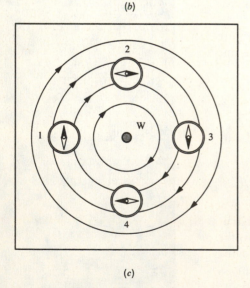

Figure 28–1. The magnetic field surrounding a current-carrying conductor.

carrying wire. The rule, which was developed when current was considered to flow from + to − (conventional current flow), is referred to in many textbooks as the *right-hand rule*. Because this text is based on current flow from − to + (electron-flow current), the rule is modified to use the left hand. The *left-hand rule* states:

> *If the current-carrying wire is grasped by the fingers of the left hand with the thumb extended and pointing in the direction of electron-flow current, the fingers encircle the wire in the direction of the magnetic lines of force* (Figure 28–2).

The right-hand rule is identical except that the right hand is used and the thumb points in the direction of conventional current flow. No matter which rule is used, the direction of the magnetic lines of force will be the same.

Magnetic Field Produced by a Coil · · · · · ·

When two magnetic fields are brought near each other, they interact and the lines of force are distorted. Further, the lines of force of two magnetic fields *aid each other* when their lines of force are in the *same direction,* and *oppose (cancel)* each other when their lines of force are in *opposite directions.*

If a current-carrying wire is wrapped in the form of a coil, as in Figure 28–3, the principle of magnetic fields aiding and opposing will apply. The circular lines of force around the conductors will combine, forming the pattern of lines of force around the coil shown in Figure 28–3. The lines of force entering and leaving the coil in effect form a south and a north pole. Again, a left-hand rule can be used to determine the direction of the lines of force and therefore the south- and north-pole ends of the coil. The left-hand rule for coils is as follows:

> *With the thumb extended, grasp the coil with the fingers of the left hand, in the direction of electron-flow current. The extended thumb will point in the direction of the magnetic lines of force. Therefore, the end of the coil to which the thumb is pointing may be considered a north pole and the other end of the coil, the south pole* (Figure 28–4).

The wires forming the coil may be wrapped around any type of material, but generally the form around which the wire is wrapped is an insulating, nonmagnetic circu-

Figure 28–2. Left-hand rule used to determine the direction of magnetic lines of force around a current-carrying conductor.

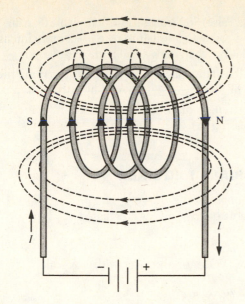

Figure 28–3. Magnetic field around a current-carrying coil.

lar, square, or rectangular tube. The tube itself may be hollow, in which case the coil is said to have an *air core*. In many applications a soft-iron bar is placed in the tube. Because the soft iron is a magnetic material, it provides a good path for the lines of force in the core of the coil, thus increasing flux density and forming a strong magnet. However, the soft iron will develop its magnetic strength only when the coil is energized. Soft iron does not retain its magnetic strength when it is not in a magnetic field, as is the case when the coil is deenergized. The arrangement of coil and soft-iron core is therefore used to provide a temporary magnet, or an *electromagnet*.

The strength of the magnet formed by a coil depends on the number of turns of wire in the coil and the amount of current in the coil. The strength of the magnet is given in terms of magnetomotive force, mmf, which is analogous to voltage in an electric circuit. The formula for magnetomotive force (mmf) is

$$\text{mmf} = IN \qquad \qquad (28\text{--}1)$$

where mmf is magnetomotive force in ampere-turns

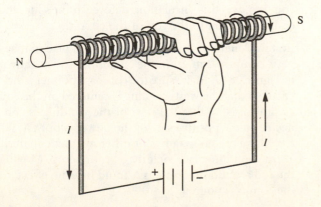

Figure 28–4. Left-hand rule for determining the poles formed by a currrent-carrying coil.

I is the current in the coil in amperes

N is the number of turns in the coil

In the magnetic circuit, magnetomotive force is dependent upon the lines of force produced by the coil ø and the reluctance of the material through which the lines of force must pass $\Re$. Thus, mmf can be expressed with the following formula:

$$\text{mmf} = \text{ø}\Re \qquad \qquad (28\text{--}2)$$

If we think of lines of force ø as analogous to current, and reluctance $\Re$ analogous to resistance, formula (28–2) can be considered the "Ohm's law" of magnetic circuits.

Coils with soft-iron cores are widely used in industry for such devices as relays, motor-control contactors, alarms, and similar devices. Coils with movable soft-iron cores are used as operating devices to open and close doors, move material, actuate other devices, and the like. Iron-core coils are frequently referred to as *solenoids*.

SUMMARY

1. A magnetic field is developed by a moving electric charge.

2. Circular magnetic lines of force appear around a wire carrying current. The magnetic field is at right angles to the wire and surrounds it. The magnetic field extends the entire length of the wire.

3. The direction of the lines of force depends on the direction of current. If the wire is grasped in the left hand with the thumb pointing in the direction of electron-flow current, the fingers point in the direction of the circular magnetic lines of force (Figure 28–2).

4. If the wire is wound in the form of a coil and current is flowing in the coil, the coil will produce a magnetic field (Figure 28–3).

5. The polarity of the poles formed by a coil may be determined by grasping the coil with the left hand, with the fingers pointing in the direction of current in the windings. The extended thumb then points to the north pole of the magnet (Figure 28–4).

6. The poles of a coil may be reversed by (a) reversing the current or (b) reversing the direction the wire is wound.

7. Coils are frequently wound on soft-iron cores. When current is turned on, the iron core acts like a temporary bar magnet, but it loses its magnetism when current is turned off.

8. The strength of the magnetic field of a coil may be increased by increasing the current in the coil or the number of turns of wire or both.

9. A magnetic circuit may be compared to an electric circuit, where (a) mmf (magnetomotive force) is the equivalent of V (electromotive force); (b) ø, the resulting lines of magnetic flux, is the equivalent of I, and (c) $\Re$, the opposition to ø, is the equivalent of R.

10. In a magnetic circuit

$$mmf = \emptyset\mathfrak{R}$$

and

$$mmf = IN$$

where mmf is magnetomotive force in ampere-turns

 $\emptyset$ is magnetic flux

 $\mathfrak{R}$ is magnetic reluctance

 I is current in amperes

 N is the number of turns of wire

11. The strength of an electromagnet is directly proportional to I, the current in the coil; that is, mmf increases as I increases.

SELF TEST

Check your understanding by answering the following questions:

1. (True/False) An electric current creates a magnetic field. _____

2. The magnetic field at any one point about a wire carrying current lies in a plane _____ to the wire at that point.

3. If a wire were perpendicular to this page and electron-flow current were in a direction toward you, the magnetic field around the wire would be in the _____ (clockwise/counterclockwise) direction.

4. (True/False) Soft iron is more permeable than air. Therefore, a coil with a soft-iron core has a more intense field in the core than does a coil with an air core. _____

5. The left-hand rule can be used to determine the location of the _____ formed by current in a coil.

6. The strength of an electromagnet can be increased by increasing (a) _____ or (b) _____.

MATERIALS REQUIRED

Power Supply:
- Variable 0–15 V dc 1 A, regulated

Instruments:
- DMM, VOM, or ammeter with 1 A scale

Resistors (½-W, 5%):
- 1 15-Ω, 25-W

Miscellaneous:
- 1 SPST switch
- 1 magnetic compass
- Iron filings
- 16 ft of no. 18 copper magnet wire
- 1 round hollow cardboard or plastic tube, about 2 in. long and 0.5 in. inside diameter (to be used as a coil form)
- 1 round soft-iron bar about 2 in. long, with an outside diameter such that it will fit snugly into the hollow tube
- 1 sheet of thin cardboard or stiff plastic about 8 1/2 × 11 in.
- Electrical tape

PROCEDURE

Solenoid with Air Core

1. Construct a solenoid coil by wrapping the no. 18 wire tightly around the hollow tube. By wrapping the turns tightly against one another, you will probably need two or three layers of wire to produce a 100-turn coil. Leave about 8 in. of wire free before starting to wrap the wire around the hollow tube. After wrapping the 100 turns around the tube, bring out at least an 8-in. length of wire. Wrap electrical tape around the coil to prevent it from unraveling (Figure 28–5). Draw an arrow on the outside of the tape to indicate the direction of the winding. Label the ends of the coil as "start" and "end" of winding.

2. With power **off** and S_1 **open,** connect the coil constructed in step 1 in the circuit of Figure 28–6. The coil is shown in simplified form to indicate the di-

rection of the winding. Magnet wire is usually insulated with a thin coat of plastic or varnish. Make sure you remove this insulation before connecting the ends of the coil to the rest of the circuit.

3. Rest the length of the coil on a table and place a magnetic compass 2 in. from the end of the coil as shown in Figure 28–7. Make sure the compass pointer is oriented so that the west-east line is centered on the coil as shown. Note that the start of the winding is connected to the positive side of the power supply. Make sure there are no magnets or other magnetic materials in the immediate vicinity of your setup. Moving the coil around the compass should have no effect on the compass pointer.

4. Adjust the power supply control to minimum output voltage. Turn power **on.**

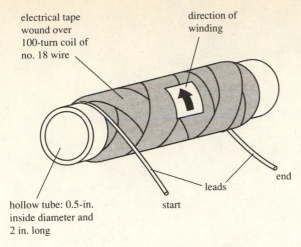

Figure 28–5. Construction of solenoid coil for procedure step 1.

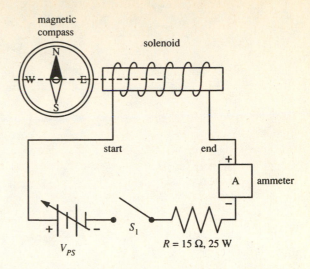

Figure 28–7. Position of compass and solenoid for procedure step 3.

5. Increase the output voltage until the ammeter reads 0.75 A. Note the new position of the compass pointer. Record in Table 28–1 (p. 215) the direction indicated by the north-seeking end of the pointer (indicate the direction as N, S, E, or W).

6. Move the coil to within 0.5 in. of the compass. Record the position of the compass pointer in Table 28–1. Do not leave the coil in this position for more than a few seconds. Go immediately to the next step.

7. With the coil still 0.5 in. from the compass, decrease the voltage until the ammeter reads 0.10 A. Record the position of the compass pointer in Table 28–1.

8. **Open** S_1. The current in the coil is now zero. Record the position of the pointer in Table 28–1.

Solenoid with Soft-Iron Core

9. Insert a 2-in. soft-iron core inside the solenoid coil. Arrange the coil with iron core and magnetic compass as in step 3 (see Figure 28–7).

10. With the power supply adjusted to its lowest voltage, **close** S_1. Increase the supply voltage slowly until the ammeter reads 0.75 A. Observe the compass pointer and record its direction (N, S, E, or W) in Table 28–1.

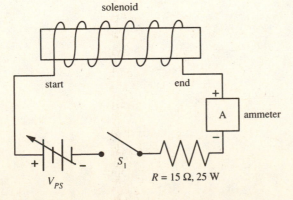

Figure 28–6. Circuit with solenoid for procedure step 2.

11. Move the solenoid to within 0.5 in. of the compass as in step 6. Record the position of the compass pointer in Table 28–1. Do not hold this position for more than a few seconds. Proceed immediately to step 12.

12. Decrease the voltage of the supply until the ammeter reads 0.10 A. Record the position of the compass pointer in Table 28–1.

13. **Open** S_1. The current in the coil is now zero. Record the position of the pointer in Table 28–1.

14. Reverse the direction of current in the solenoid by reversing the connections to the start and end leads of the coil. **Close** S_1 and increase the voltage until the ammeter reads 0.75 A. Record the position of the pointer in Table 28–1. Proceed quickly to step 15.

15. **Open** S_1; power off. Record the position of the pointer in Table 28–1.

Magnetic Field of a Solenoid

16. With the solenoid in the same position as in step 15 but without the compass, place a thin cardboard or plastic sheet directly over the coil without touching it. Sprinkle a very thin layer of iron filings evenly on the cardboard.

17. Turn **on** the power and **close** S_1. Adjust the voltage until the ammeter measures 0.75 A. Notice the movement of the iron filings when S_1 is closed. Gently tap the top of the cardboard until the iron filings align themselves in a steady pattern. Sketch the pattern on a separate piece of paper. **Open** S_1; power **off.**

ANSWERS TO SELF TEST

1. true
2. perpendicular
3. clockwise
4. true
5. poles
6. (a) number of turns;
   (b) amount of current

TABLE 28–1. **Magnetic Field About a Solenoid**

Step	Current in Solenoid, A	Solenoid Distance from E, in.	Current Polarity		Direction of Pointer
			Start of Winding	End of Winding	
5	0.75	2	+	−	
6	0.75	0.5	+	−	
7	0.10	0.5	+	−	
8	0	0.5	+	−	
10	0.75	2	+	−	
11	0.75	0.5	+	−	
12	0.10	0.5	+	−	
13	0	0.5	+	−	
14	0.75	0.5	−	+	
15	0	0.5	−	+	

QUESTIONS

1. Explain, in your own words, the relationship between current in a wire and a magnetic field. Cite specific experimental data that confirm these relationships.

———————————————————————————————————————
———————————————————————————————————————
———————————————————————————————————————
———————————————————————————————————————

2. Is it possible to determine the north and south poles at the ends of the solenoid in Figure 28–7 even before S_1 is closed? If not, what additional information must be known? If it is possible, what is the pole of the solenoid end closest to the compass?

———————————————————————————————————————
———————————————————————————————————————
———————————————————————————————————————
———————————————————————————————————————

3. Did step 14 in the procedure affect your answer to Question 2? If so, how?

———————————————————————————————————————
———————————————————————————————————————
———————————————————————————————————————

4. Explain the left-hand rule for determining the poles of a solenoid. What would be the effect of using the right hand for determining the poles?

_____

_____

_____

_____

_____

_____

_____

_____

_____

_____

_____

_____

_____

_____

_____

_____

5. Explain how each of the following affects the strength of an electromagnet:
   (a) The amount of current in the coil
   (b) The number of turns of wire in the coil
   (c) The type of material used in the core

_____

_____

_____

_____

_____

_____

_____

_____

_____

_____

_____

_____

_____

_____

INDUCING VOLTAGE IN A COIL

BASIC INFORMATION

Electromagnetic Induction

If a zero-center galvanometer or a low-reading digital ammeter is connected to the ends of a straight piece of copper conductor and the conductor is placed a few inches away from the poles of a horseshoe magnet, as in Figure 29–1(*a*) (p. 218), there will be no reading on the meter. If the conductor is then moved down quickly between the poles of the magnet, as in Figure 29–1(*b*), the meter will show a reading and then return to zero when the conductor stops moving. If the conductor is then moved up through the magnetic field of the horseshoe magnet, the meter will again have a reading but this time with the opposite polarity.

The same reaction of the meter will occur if the conductor is stationary and the horseshoe magnet is moved across the conductor.

Finally, if the conductor is allowed to remain stationary in the center of the stationary magnetic field, there will be no reading at all on the meter.

From the preceding discussion, we can conclude that when a magnetic field is cut by a conductor, a voltage is induced in the conductor, and if the conductor is part of a complete circuit, a current will flow in the conductor. This process is true whether the conductor moves across the magnetic field or the magnetic field is moved across a conductor.

The point to remember is that the conductor must *cut across* the magnetic field. If the conductor is held parallel to the direction of the field and moves through it in that fashion, the lines of force will not be cut and no voltage will be induced.

Polarity of Induced Voltage

The fact that the polarity of the meter reading changes depending on the direction of movement of the conductor is evidence that the polarity of the voltage induced in the conductor depends on the *direction* of cutting of the lines of force. The polarity of the induced voltage can be established by *Lenz' law*.

Lenz' law states that the polarity of the induced voltage must be such that the direction of the resulting current in the conductor produces a magnetic field around the conductor that will oppose the *motion* of the inducing field. An example illustrates the meaning of Lenz' law.

The ends of the air-core coil in Figure 29–2 (p. 218) are connected to a sensitive ammeter capable of indicating current direction. The north

OBJECTIVES

1 To verify experimentally that a voltage is induced in a coil when the lines of force of a magnet cut across its windings

2 To verify experimentally that the polarity of the induced voltage depends on the direction in which magnetic lines of force cut the coil windings

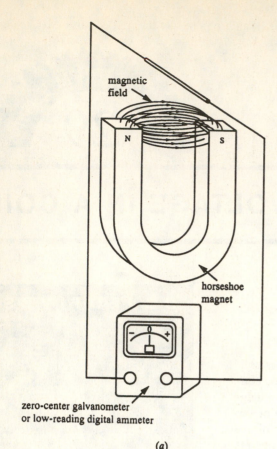

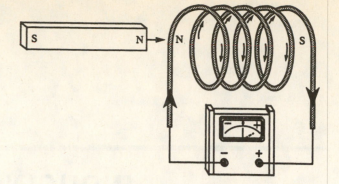

Figure 29–2. The bar magnet pushed into the center of the coil in the direction shown generates a current that produces magnetic poles; these poles tend to repel the motion of the bar magnet. This verifies Lenz' law.

tally, the reading of the ammeter will verify the polarity of the induced voltage and the direction of current. Using the left-hand rule for coils will establish the poles at the ends of the coil.

If the bar magnet is now pulled to the left out of the core of the coil, the polarity of the voltage induced in the coil will be reversed, and the current will flow in the opposite direction. This will reverse the poles at the ends of the coil. A south pole will appear at the left end of the coil. As the bar magnet is pulled from the core, the south pole of the coil will attract the north pole of the magnet back into the core, thus opposing the pulling motion of the bar magnet out of the core. Again this condition satisfies Lenz' law.

Counter Electromotive Force · · · · · · · · · · ·

To this point we have discussed that a voltage is induced in a conductor when there is relative motion between the conductor and the magnetic field of a permanent magnet—that is, when the conductor cuts the lines of force of the magnet. In fact, the effect of induction in a conductor is achieved when there is relative motion between the conductor and the field of *any* magnet, including an electromagnet.

In a previous experiment it was demonstrated that a magnetic field exists about a current-carrying coil. This coil is in fact an electromagnet and may be substituted for the bar magnet in the preceding example.

The term *relative motion* is not limited merely to the physical movement of a magnet near a conductor or of a conductor in a magnetic field. Relative motion may exist without any physical movement whatsoever. Consider a coil through which an increasing or decreasing (not a steady-state) current is flowing. In the case of an increasing current, the magnetic field about the coil is increasing or expanding; that is, it is a *moving field*. When the current in the coil is decreasing, the magnetic field about the coil is decreasing or collapsing. Again, it is a moving magnetic field. A conductor held stationary in this moving magnetic field

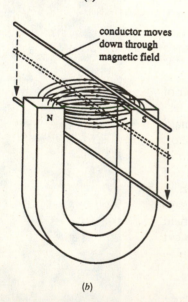

Figure 29–1. The meter indicates current in the circuit when the conductor is moved quickly through the magnetic field.

pole of a bar magnet is pushed quickly into the center of the coil from the left, inducing a voltage in the winding with the polarity shown. The resulting current sets up a magnetic field in the coil, so that a north pole appears on the left end of the coil. The north pole thus produced opposes the north pole of the bar magnet being pushed into the left side of the coil. Therefore, the original motion of the bar magnet produced a magnetic field that now opposes that motion. This satisfies Lenz' law. Experimen-

218   EXPERIMENT 29

is in effect being *cut* by the magnetic lines of force. Therefore, a voltage will be induced in the conductor. Expanding and collapsing magnetic fields will cause voltages of opposite polarity to be induced in the conductor.

The expanded or collapsing (moving) magnetic lines of force about the coil with an increasing or decreasing current *cut the windings of the coil itself.* Accordingly, a voltage is induced in the coil, which by Lenz' law will oppose the inducing force. That is, the voltage induced in the coil will have a polarity opposite that of the voltage that initially caused current to flow in the coil. The voltage induced in the coil may therefore be called a *counter electromotive force,* or simply a *counter emf.*

Magnitude of Induced Voltage · · · · · · · · ·

The value of the voltage induced in a coil depends directly on (1) the *number* of turns N of wire in the coil and (2) the *rate* at which the lines of force are cut by the windings of the coil. Thus, the more turns of wire, the higher will be the induced voltage across the ends of the coil. Similarly, the faster the magnetic flux is cut by the windings, the higher will be the voltage produced across the coil.

SUMMARY

1. A voltage is induced in a conductor when the conductor cuts the lines of force in a magnetic field. Cutting the lines of force can be achieved by moving either the conductor or the magnetic field.

2. The polarity of the voltage induced in the conductor is determined by the direction in which the lines of force are cut. Thus, if a positive voltage is induced, say, by a conductor cutting the lines of force in a *downward* direction, a negative voltage will be induced by an *upward* cutting of these same lines of force.

3. The polarity of the induced voltage can be predicted by Lenz' law, which states: The polarity of the voltage induced in a conductor must be such that the *magnetic field set up by the resulting current* in the conductor will oppose the motion of the magnetic field that originally produced it.

4. The value of the voltage induced in a coil when its windings cut magnetic lines of force depends directly on the number of turns in the winding and on the rate at which the magnetic flux lines are cut by the windings.

Check your understanding by answering the following questions:

1. A moving magnetic field will _____ a _____ in a conductor if the magnetic lines of force _____ the conductor.

2. The polarity of voltage induced in a conductor will depend on the _____ of cutting of the lines of force by the conductor.

3. A conductor moving parallel to the lines of force in a magnetic field _____ (will/will not) have a voltage induced in it.

4. (True/False) The more turns there are in a coil whose windings cut the lines of force in a magnetic field, the greater will be the voltage induced in it, all other things being equal. _____

5. _____ law can be used to predict the polarity of voltage induced in a coil if the polarity of the field inducing the voltage is known.

MATERIALS REQUIRED

Instruments:
- Zero-center galvanometer (a zero-center sensitive microammeter can be used)

Miscellaneous:
- Solenoid coil consisting of 100 turns of no. 18 magnet wire wound on a hollow cardboard or plastic cylindrical form 3 in. long and 1 in. inside diameter.
- Bar magnet, about 4 in. long and narrow enough to slide easily inside the coil form.

PROCEDURE
■ ■

1. Connect the coil and galvanometer or DMM as shown in Figure 29–3 (p. 220). Note the direction of the winding around the core form. Place the bar magnet lengthwise as shown with its north pole facing the coil and about 2 in. away from the end of the coil.

2. With the magnet and coil stationary, observe the reading of the galvanometer. Record this value and the polarity of the reading in Table 29–1 (p. 221).

3. Turn the coil on its end and rest it on the lab table. The start of the winding should be at the bottom of the coil.

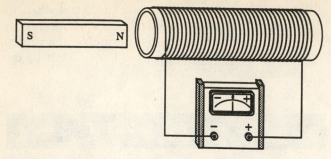

Figure 29–3. The coil and meter are connected in series for procedure step 1.

While observing the galvanometer, insert the north pole of the bar magnet down quickly into the core of the coil. Record the highest reading of the galvanometer and the polarity of the reading in Table 29–1.

4. With the magnet resting in the core of the coil, observe the galvanometer, and record the maximum reading and polarity in Table 29–1.

5. While observing the galvanometer, withdraw the magnet quickly from the core of the coil. Record the highest reading and its polarity in Table 29–1.

6. Reverse the bar magnet. While observing the galvanometer, insert the south pole down quickly into the core of the coil. Record the highest reading of the galvanometer and its polarity.

7. With the magnet stationary within the core of the coil, record the reading of the galvanometer and its polarity in Table 29–1.

8. While observing the galvanometer, withdraw the magnet quickly from the core. Record the highest reading and polarity in Table 29–1.

9. Repeat step 6, but insert the magnet much more quickly than before. Record the highest reading and polarity in Table 29–1.

10. Repeat step 6, but insert the magnet much more slowly than before. Record the highest reading and polarity in Table 29–1.

11. Stand the bar magnet vertically on end with the north pole up. With the galvanometer still connected to the coil, draw the coil quickly down over the magnet. The start of the coil winding should be at the upper end of the coil. Be careful not to move or shake the galvanometer. Record the highest reading and polarity in Table 29–1.

12. While observing the galvanometer, quickly lift the coil over and away from the magnet. Record the highest reading and polarity in Table 29–1.

ANSWERS TO SELF TEST

1. induce; voltage; cut
2. direction
3. will not
4. true
5. Lenz'

Name _____ Date _____

TABLE 29–1. Inducing Voltage in a Coil

Step	Condition	Voltage, Polarity	Current, Highest Reading, μA
2	Magnet stationary		
3	North pole of magnet inserted into solenoid		
4	Magnet stationary within the solenoid core		
5	Magnet withdrawn from solenoid		
6	South pole of magnet inserted into solenoid		
7	Magnet stationary within the solenoid core		
8	Magnet withdrawn from solenoid		
9	Same as step 6 but more rapidly		
10	Same as step 6 but more slowly		
11	Solenoid plunged down over magnet		
12	Solenoid pulled up, away from magnet		

QUESTIONS

1. Explain fully, in your own words, the results of lines of force cutting across the conductors in a winding. Describe the specific conditions and results of the process.

_____

_____

_____

_____

_____

_____

_____

_____

_____

2. Explain Lenz' law in your own words.

_____

_____

_____

_____

_____

_____

3. Does the result of your experiment verify Lenz' law? If not, what steps would need to be performed in order to verify the law? If your experimental results did verify the law, discuss these results.

_____

_____

_____

_____

_____

_____

_____

_____

_____

4. Refer to your data in Table 29–1. What results verify that it is relative motion between the magnetic field and the windings of the solenoid that produces an induced voltage?

_____

_____

_____

_____

_____

_____

_____

_____

5. Faraday's law states that the magnitude of the voltage induced in a winding depends directly on the number of turns of wire in the winding and the rate of speed at which the magnetic lines of force cut the winding. Do your results verify this law? If so, discuss your results. If not, what additional steps are needed to verify the law?

_____

_____

_____

_____

_____

_____

_____

_____

_____

_____

APPLICATIONS OF THE DC RELAY

BASIC INFORMATION

Applications of Magnetism

One of the earliest applications of magnetism was the magnetic compass. It can be said that the compass ushered in the age of modern marine navigation and exploration, for mariners were then no longer dependent on sighting the stars. With the compass they could find their way at any time of day or night and in any weather.

The next dramatic development in the application of magnetism was in the generation of electricity. The generator was designed to take advantage of the emf created when a conductor cuts magnetic lines of force. In the early dc generators used by the power companies, the armature of the generator contained many windings of copper wire rotating in the magnetic field produced by field windings and acted much like an electromagnet. In the modern generator used by utilities that arrangement is reversed: The armature windings are embedded in the stationary frame of the generator, whereas the field windings that produce the magnetic lines of force rotate. In both cases, conductors are being cut by magnetic flux, thus producing a voltage.

The electric motor also works on the principle of the interaction between the magnetic fields associated with its armature and field windings. There are electric motors in every home: the motor that circulates the furnace heat, the air-conditioner motor, the washing machine and dryer motors, the electric clock motor, and the like.

Electronics, too, depends in many cases on magnetic fields. The cathode-ray tube, for example, requires magnetic fields of various types to operate effectively. Circuits of all types use inductors and transformers, two devices whose operation is a function of electromagnetism.

The Relay

The *relay*, another electromagnetic device, performs countless tasks in industry and in the home. Relays are used for switching, for indicating, for transmitting, and for protecting circuits. Although modern solid-state electronic devices have replaced electromagnetic mechanical relays for many applications, it is not likely that the thousands of other applications of these relays could be easily transferred economically or physically to solid-state devices in the foreseeable future.

OBJECTIVES

1 To study the characteristics and operation of a dc relay

2 To explore the uses of dc relays in electric circuits

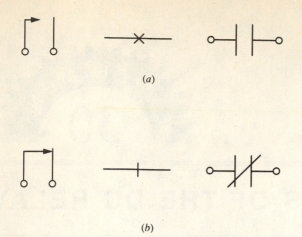

(a)

(b)

Figure 30–1. Relay symbols. The arrowhead indicates the stationary contact. (*a*) Normally open (NO) contacts; when the relay is energized, they *make* contact. (*b*) Normally closed (NC) contacts; when the relay is energized they *break* contact.

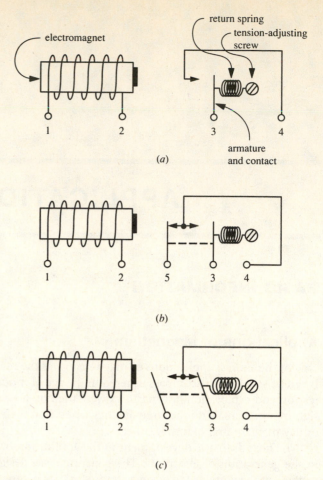

(a)

(b)

(c)

Figure 30–2. Pictorial representation of a relay. (*a*) SPST relay. (*b*) SPDT relay. (*c*) SPDT relay in energized state.

Relays are electromagnetically operated, remotely controlled switches, with one or more sets of contacts. When energized, the relay operates, opening or closing its contacts or opening some contacts and closing others. Contacts that are open when the relay is not energized are called *normally open* (NO) contacts. Contacts that are closed when the relay is not energized are called *normally closed* (NC) contacts. Figure 30–1 shows three types of schematic symbols used for open and closed contacts.

There are certain terms usually associated with relays that the technician should understand. A relay is said to "pick up" when it is energized. The *pickup value* is the smallest value of actuating current required to close a normally open contact or open a normally closed contact. When a relay is deenergized, it is said to "drop out." Relay contacts are held in their normal position either by springs or by some gravity-actuated mechanism.

A typical relay consists of an electromagnet, a small movable plate called an armature, and a set of contacts attached to the armature. When the electromagnet is energized, the armature is attracted to the electromagnet, which causes the contacts to open or close depending on whether their deenergized state is closed or open.

Figure 30–2(*a*) shows the major elements of a relay. The spring is used to return the armature to its normal deenergized position. The tension-adjusting screw can provide fine-tuning for the magnetic force, and thus the coil current, that is necessary to move the armature when the coil is energized. When the coil in Figure 30–2(*a*) is in its normal deenergized state, the circuit between terminals 3 and 4 resembles an open switch. When the electromagnet is energized, the armature is attracted to the coil, and the movable contact connected to terminal 3 closes on the stationary contact (the arrowhead) connected to terminal 4. The circuit between 3 and 4 now resembles a closed switch. The entire action is similar to that of a single-pole, single-throw (SPST) switch.

In Figure 30–2(*b*) two mechanically connected, movable contacts are attached to the armature. The figure shows the deenergized state of the relay. In this condition the contact between 4 and 5 is normally closed and the contact between 3 and 4 is normally open. When the coil is energized the contacts move to the position shown in Figure 30–2(*c*). In this condition the circuit between 3 and 4 is closed and the circuit between 4 and 5 is open. The action of the relay in Figure 30–2(*b*) and (*c*) is that of a single-pole, double-throw spot switch.

Remote-Control Operations ·············

One of the major advantages of relays is their ability to be operated from remote locations at relatively low voltages. The electromagnet coil of the relay can be energized with low current and voltage from a source some distance away from the relay itself. The circuit connected to the contacts, on the other hand, can be rated for much higher voltages and current. The circuit of Figure 30–3 uses a 12-V dc source to control a 277-V lighting circuit. The relay control switch is located on the wall, whereas the 277-V circuit and relay are located in the ceiling of the room.

Relays can be used to open circuits as well as close them. In Figure 30–4 a relay coil is connected to a heat

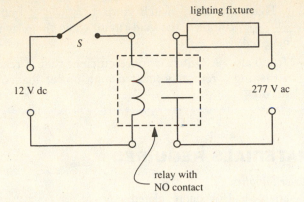

Figure 30–3. Control of high-voltage ac lighting circuit with a low-voltage dc relay.

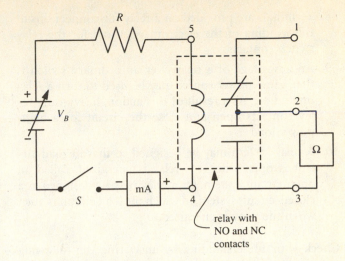

Figure 30–5. Circuit for determining the pickup value of a dc relay.

sensor. The main contact of the relay is in series with the line feeding an electric heater. When the sensor detects an overheating condition, it produces enough current to energize the relay coil. This opens the NC contacts of the relay, thereby opening the circuit and turning off the heater.

Relay Specifications

Relay manufacturers supply a specification sheet with each of their relays. This "spec" sheet contains relay ratings, designates whether the relay is dc or ac, and specifies the location and the ratings of the contacts. For example, a relay may be specified as a dc SPDT relay with a relay coil resistance of 400 Ω. The contacts of the relay may be rated at 5 A.

The relay coil and contact terminals can usually be located by inspection. If a relay is enclosed in a sealed unit, this may not be possible. An ohmmeter may then be used to identify the terminals. The ohmmeter is used to measure the resistance between any two terminals. There are only a limited number of two-terminal combinations possible. The resistance of the coil is measured between the two terminals that are connected to the relay coil. Relay coils will vary from very low to very high values of resistance. Normally closed contacts will measure zero resis-

tance. Normally open contacts will read infinite resistance. For example, in the relay in Figure 30–5, the coil resistance can be measured with an ohmmeter between terminals 4 and 5. There will be zero resistance between terminals 2 and 1 and infinite resistance between terminals 2 and 3. The resistance between terminals 1 and 3 will also be infinite. If you cannot determine by inspection whether contact 2 or 1 is the movable contact, the relay should be tripped. A resistance check will then show, as in the case of Figure 30–5, that there is now zero resistance contacts 2 and 3 and infinite resistance between contacts 2 and 1 and contacts 3 and 1. It is therefore evident that contact 2 is the movable one.

CAUTION: In making resistance checks of relay contacts, be certain the power (to the load) is **off**; otherwise you may damage the ohmmeter.

SUMMARY

1. Relays are electromagnetic switches used as protective devices, as indicating devices, and as transmitting devices.

2. Protective relays protect good components from the effects of circuit components that have failed.

3. Transmission relays are used in communication systems.

4. Indicating relays may be used to identify a component that has failed, or they may be used with attention-getting devices such as bells or buzzers.

5. Relays may be used as SPST switches, or they may have complex switching arrangements.

6. The switch contacts of relays may be normally open (NO) or normally closed (NC). The contacts are held in their normal positions by springs or by some gravity-actuated mechanism.

7. Relays are either ac- or dc-operated and generally consist of an electromagnet, a movable armature, and contacts.

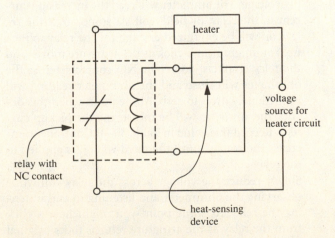

Figure 30–4. Relay with an NC contact used to prevent an electric heater from overheating.

8. Terminals are provided on a relay for connections to the winding of the electromagnet and for the relay switch contacts.

9. An advantage of a relay over an ordinary switch is that a low-power source may be used to turn a relay on and off. As a result of this action, heavy-duty relay contacts open and close the circuit for a high-power load.

10. A relay system may be designed so that its load circuit is open when the relay is energized, called an open-circuit system. Or a relay may be designed as a closed-circuit system, in which the load circuit is open when the relay is energized.

Check your understanding by answering the following questions:

1. A dc relay is one in which _____ current in the relay coil actuates the relay mechanism.

2. The movable arm of a relay is called the _____.

3. A relay that "picks up" at 10 mA is one that is turned _____ (on/off) when 10 mA flows in the _____.

4. Some important electrical specifications of a relay are the
   (a) _____ and _____ of the coil;
   (b) pickup _____ of the coil;
   (c) current-handling capacity of the _____.

5. In Figure 30–2(b), switch contacts _____ and _____ complete the load circuit when the relay is energized.

6. (True/False) A switch is more efficient than a relay in turning a remotely located high-power load on and off.

MATERIALS REQUIRED

Power Supplies:
- Variable 0–15 V dc, regulated
- 120-V, 60-Hz line voltage*

Instrument:
- DMM or VOM

Resistors (½-W, 5%):
- 1 560-Ω
- 1 1.8 kΩ

Relay:
- 1 dc, SPDT, 12-V coil voltage, 300–400-Ω field, 120-V ac, 0.5-A contacts.

Miscellaneous:
- Test lamp set consisting of 60-W incandescent lamp, on-off switch, 1-A fuse, line cord, and polarized plug

*Note: Your instructor may assign a lower voltage test lamp set.

PROCEDURE

1. You will be given a SPDT relay for this experiment. Inspect and test the relay to determine the following (the terminals have been numbered for reference purposes):
   (a) Relay (field) coil terminals
   (b) NO contact terminals
   (c) NC contact terminals
   (d) Nameplate data (if any)
   (e) Relay coil resistance, measured
   Record all information in Table 30–1 (p. 229).

2. With power **off** and S_1 **open,** connect the circuit of Figure 30–6. Set the ohmmeter range for a low-resistance reading. Adjust the power supply to its lowest voltage.

3. Turn power **on** and **close** S_1. While observing the ohmmeter and listening and watching for any changes in the relay, slowly increase the power supply voltage. The ammeter will read the increasing current in the relay, or field, coil. At some point the relay coil will be energized enough for the relay to pick up. You might hear a click as the armature moves and the relay contacts close. The NO contacts across the ohmmeter will close and the ohmmeter reading will immediately drop to zero. At the point when this occurs, the ammeter will be measuring the pickup current. Record this value in Table 30–1. Continue to increase the voltage until the rated voltage is applied to the relay coil.

4. Slowly reduce the voltage across the relay coil while observing the ohmmeter and listening to and watching the relay. At some point you might hear a click from the relay as the armature returns to its original deenergized condition and the NC contacts close again. The NO contacts will open and the ohmmeter

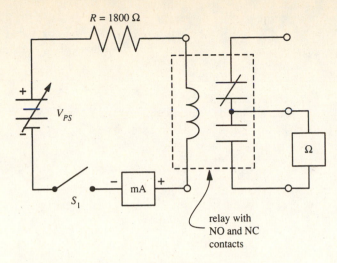

Figure 30–6. Circuit with relay for procedure step 2.

will measure infinite resistance. At this point the relay has dropped out (often referred to as being "tripped"). The ammeter will be measuring the dropout current. Record this value in Table 30–1.

5. Verify your original values of pickup and dropout currents by repeating steps 3 and 4. If care was taken to note the exact points at which the ohmmeter readings changed, the pickup and dropout current values should not vary significantly from your original readings. **Open** S_1; turn the power **off.**

6. With power **off,** S_1 **open,** and the lamp circuit unplugged, with its switch **off,** connect the circuit of Figure 30–7. If you are using a low voltage lamp circuit, replace the plug and 120 V power source with the low voltage supply. The fuse can be omitted in a low voltage system. All other parts of the circuit should be as shown in Figure 30–7. The lamp circuit is connected across the NO relay terminals A and B, as shown on the diagram. (Your relay may not be marked with these letters.)

7. **CAUTION:** This part of the experiment involves potentially lethal voltage. Use extreme care when operating this circuit. Do not touch any parts of the circuitry with both hands at the same time.

   Insert the plug into the 120-V outlet or turn on the low voltage lamp circuit power supply. Record the status of the lamp (on or off) by checking the appropriate box in Table 30–2 (p. 229).

8. **Close** S_1. Observe the state of the lamp and record whether it is on or off in Table 30–2.

9. **Open** the lamp circuit switch. Unplug the lamp circuit from the outlet or turn **off** the low voltage lamp supply. Reconnect the lamp circuit across the NC relay terminals marked BC in Figure 30–7. Your relay may not have its terminals marked in this way.

10. Insert the plug or turn on the low voltage lamp supply. **Close** the lamp circuit switch. Record the lamp status in Table 30–2.

11. **Open** the relay circuit switch S_1. Record the lamp status in Table 30–2. **Open** the lamp circuit switch, unplug the lamp circuit or turn off the **low** voltage lamp supply. Turn **off** the relay power supply.

ANSWERS TO SELF TEST

1. direct
2. armature
3. on; relay coil
4. (a) resistance, voltage (b) current (c) contacts
5. 3, 4
6. false

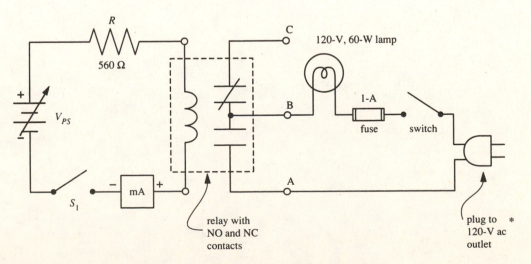

Figure 30–7. Lamp circuit connected to the relay circuit of Figure 29–6 for procedure step 6.

*A low voltage lamp circuit may be used in place of the 120-Vac source.

Name _____   Date _____

TABLE 30–1.   Relay Characteristics

Function	Terminal Connection	Relay Characteristics	
Relay coil		Relay coil resistance, Ω	
Normally open contacts		Pickup current, mA	
Normally closed contacts		Dropout current, mA	

Nameplate data: _____

TABLE 30–2.   Operation of SPDT Relay

Step	S_1	Circuit Connection to Relay	Status	
			On	Off
7	Open	Across A-B		
8	Closed	Across A-B		
9	Closed	Across B-C		
10	Open	Across B-C		

QUESTIONS

1. Explain, in your own words, what the pickup value of a relay is.

_____

_____

_____

_____

_____

_____

_____

_____

_____

_____

_____

_____

_____

_____

_____

_____

2. Is it possible to change the pickup value of the relay you used in this experiment? If so how can this be done? If not, why?

_____

_____

_____

_____

_____

_____

_____

3. When a relay's contacts return to their deenergized condition, the relay is said to be reset, or tripped. What is another term for the reset current of your relay? How was this value obtained experimentally?

_____

_____

_____

_____

_____

_____

_____

4. Could the relay in this experiment be used to control a 10 A heater circuit? Would the relay operate as specified if the relay coil was connected to a 5 V supply? Explain.

_____

_____

_____

_____

_____

_____

_____

Optional Activity ·

Note and list the equipment and appliances in the shop and laboratory and in your home that depend on the operation of relays. Note and list also those devices that operate using magnetics other than in relays. In order to better identify these devices you can use any suitable search engine on the internet.

OSCILLOSCOPE OPERATION

BASIC INFORMATION

The cathode-ray oscilloscope (CRO), or "scope," as it is familiarly known, is one of the most versatile instruments in electronics. Oscilloscopes are used in a variety of applications including consumer electronics repair, digital systems troubleshooting, control system design, and physics laboratories. Oscilloscopes have the ability to measure time and voltage levels of a signal, determine the frequency of an oscillator, view rapidly changing waveforms, and determine if an output signal is distorted. The technician must therefore be able to operate this instrument and understand how and where is it used.

Oscilloscopes can be classified as analog or digital types. Analog oscilloscopes directly apply a voltage being measured to an electron beam moving across the oscilloscope screen. This voltage deflects the beam up, down, and across tracing the waveform on the screen. Digital oscilloscopes sample the input waveform and then use an analog-to-digital converter (ADC) to change the voltage being measured into digital information. The digital information is then used to reconstruct the waveform to be displayed on the screen. Analog and digital type oscilloscopes are shown in Figure 31–1 (p. 232).

A digital or analog oscilloscope may be used for many of the same applications. Each oscilloscope type possesses unique characteristics and abilities. The analog oscilloscope can display high frequency varying signals in "real time." Whereas digital oscilloscopes allow you to capture and store information which can be accessed at a later time or interfaced to a computer. In the following experiments, we will be concentrating on the analog oscilloscope.

What an Oscilloscope Does

An oscilloscope displays the instantaneous amplitude of an ac voltage waveform versus time on the screen of a cathode-ray tube (CRT). Basically, the oscilloscope is a graph-displaying device. It has the ability to show how signals change over time. As shown in Figure 31–2 (p. 232), the vertical axis (Y) represents voltage and the horizontal axis (X) represents time. The (Z) axis or intensity is sometimes used in special measurement applications. Inside the cathode-ray tube is an electron gun assembly, vertical and horizontal deflection plates, and a phosphorescent screen. The electron gun emits a high-velocity, low-inertia beam of electrons that strike the chemical coating on the inside face of the CRT, causing it to

OBJECTIVES

1. To identify the operating controls of an oscilloscope

2. To set up the oscilloscope and adjust the controls properly to observe an ac voltage waveform

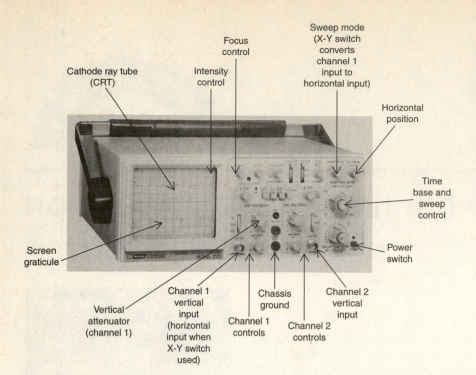

Figure 31–1. Oscilloscope types. (a) Analog oscilloscope. *Courtesy BK Precision.* (b) Digital oscilloscope. *Courtesy of Tektronix.*

emit light. The brightness (called intensity) can be varied by a control located on the oscilloscope front panel. The motion of the beam over the CRT screen is controlled by the deflection voltages generated in the oscilloscope's circuits outside of the CRT and the deflection plates inside the CRT to which the deflection voltages are applied.

Figure 31–3 is an elementary block diagram of an analog oscilloscope. The block diagram is composed of a CRT and four system blocks. These blocks include the display system, vertical system, horizontal system, and trigger system. The CRT provides the screen on which waveforms of electrical signals are viewed. These signals are applied to the vertical input system. Depending on how the volts/div control is set, the vertical attenuator—a variable voltage divider—reduces the input signal

voltage to the desired signal level for the vertical amplifier. This is necessary because the oscilloscope must handle a wide range of signal-voltage amplitudes. The vertical amplifier then processes the input signal to produce the required voltage levels for the vertical deflection plates. The signal voltage applied to the vertical deflection plates causes the electron beam of the CRT to be deflected vertically. The resulting up-and-down movement of the beam on the screen, called the trace, is significant in that *the extent of vertical deflection is directly proportional to the amplitude of the signal voltage applied to the vertical, or V, input.* A portion of the input signal, from the vertical amplifier, travels to the trigger system to start or trigger a horizontal sweep. The trigger system determines *when* and *if* the sweep generator will

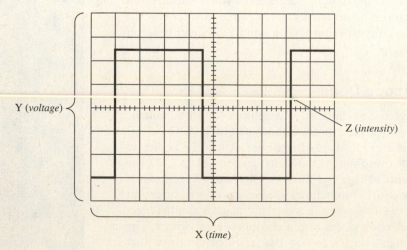

Figure 31–2. *X, Y,* and *Z* components of a displayed waveform.

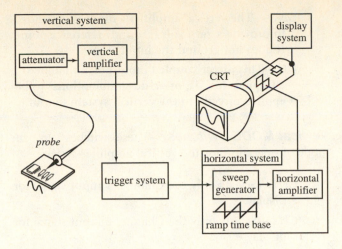

Figure 31–3. Analog oscilloscope block diagram.

be activated. With the proper LEVEL and SLOPE control adjustment, the sweep will begin at the same trigger point each time. This will produce a stable display as shown in Figure 31–4. The sweep generator produces a linear time-based deflection voltage. The resulting time-based signal is amplified by the horizontal amplifier and applied to the CRT's horizontal deflection plates. This makes it possible for the oscilloscope to graph a time-varying voltage. The sweep generator may be triggered from sources other than the vertical amplifier. External trigger input signals or internal 60-Hz (line) sources may be selected.

The display system includes the controls and circuits necessary to view the CRT signal with optimum clarity and position. Typical controls include intensity, focus, and trace rotation along with positioning controls.

Dual-Trace Oscilloscopes ·············

Most oscilloscopes have the ability to measure two input signals at the same time. These dual-trace oscilloscopes have two separate vertical amplifiers and an electronic switching circuit. It is then possible to observe two time-related waveforms simultaneously at different points in an electric circuit.

Operating Controls of a Triggered Oscilloscope ·····················

The type, location, and function of the front panel controls of an oscilloscope differ from manufacturer to manufacturer and from model to model. The descriptions that follow apply to the broadest range of general-use scope models.

Intensity. This control sets the level of brightness or intensity of the light trace on the CRT. Rotation in a clockwise (CW) direction increases the brightness. Too high an intensity can damage the phosphorescent coating on the inside of the CRT screen.

Focus. This control is adjusted in conjunction with the intensity control to give the sharpest trace on the screen. There is interaction between these two controls, so adjustment of one may require readjustment of the other.

Astigmatism. This is another beam-focusing control found on older oscilloscopes that operates in conjunction with the focus control for the sharpest trace. The astigmatism control is sometimes a screwdriver adjustment rather than a manual control.

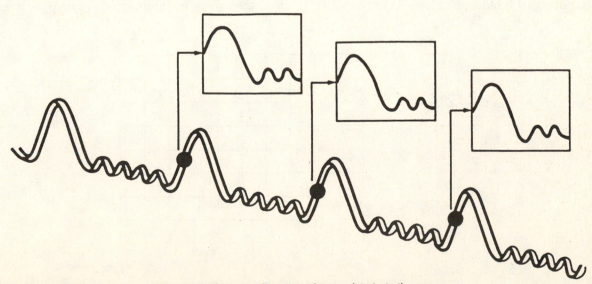

Figure 31–4. Triggering produces a stable display because the same trigger point starts the sweep each time. The SLOPE and LEVEL controls define the trigger points on the trigger signal. The waveform on the screen is all those sweeps overlaid in what appears to be a single picture.

Horizontal and Vertical Positioning or Centering.

These are trace-positioning controls. They are adjusted so that the trace is positioned or centered both vertically and horizontally on the screen. In front of the CRT screen is a faceplate called the *graticule,* on which is etched a grid of horizontal and vertical lines. Calibration markings are sometimes placed on the center vertical and horizontal lines on this faceplate. This is shown in Figure 31–5.

Volts/Div.

This control attenuates the vertical input signal waveform that is to be viewed on the screen. This is frequently a click-stop control that provides step adjustment of vertical sensitivity. A separate Volts/Div. control is available for each channel of a dual-trace scope. Some scopes mark this control Volts/cm.

Variable.

In some scopes this is a concentric control in the center of the Volts/Div. control. In other scopes this is a separately located control. In either case, the functions are similar. The variable control works with the Volts/Div. control to provide a more sensitive control of the vertical height of the waveform on the screen. The variable control also has a calibrated position (CAL) either at the extreme counterclockwise or clockwise position. In the CAL position the Volts/Div. control is calibrated at some set value—for example, 5 mV/div., 10 mV/div., or 2 V/div. This allows the scope to be used for peak-to-peak voltage measurements of the vertical input signal. Dual-trace scopes have a separate variable control for each channel.

Input Coupling AC-GND-DC Switches.

This three-position switch selects the method of coupling the input signal into the vertical system.

> *AC*—The input signal is capacitively coupled to the vertical amplifier. The dc component of the input signal is blocked.

> *GND*—The vertical amplifier's input is grounded to provide a zero volt (ground) reference point. It does not ground the input signal.

> *DC*—This direct-coupled input position allows all signals (ac, dc, or ac-dc combinations) to be applied directly to the vertical system's input.

Vertical MODE Switches.

These switches select the mode of operation for the vertical amplifier system.

> *CH1*—Selects only the Channel 1 input signal for display.

> *CH2*—Selects only the Channel 2 input signal for display.

> *Both*—Selects both Channel 1 and Channel 2 input signals for display. When in this position, ALT, CHOP, or ADD operations are enabled.

> *ALT*—Alternatively displays Channel 1 and Channel 2 input signals. Each input is completely traced before the next input is traced. Effectively used at sweep speeds of 0.2 ms per division or faster.

> *CHOP*—During the sweep the display switches between Channel 1 and Channel 2 input signals. The switching rate is approximately at 500 kHz. This is useful for viewing two waveforms at slow sweep speeds of 0.5 ms per division or slower.

> *ADD*—This mode algebraically sums the Channel 1 and Channel 2 input signals.

> *INVERT*—This switch inverts Channel 2 (or Channel 1 on some scopes) to enable a differential measurement when in the ADD mode.

Time/Div.

This is usually two concentric controls that affect the timing of the horizontal sweep or time-base

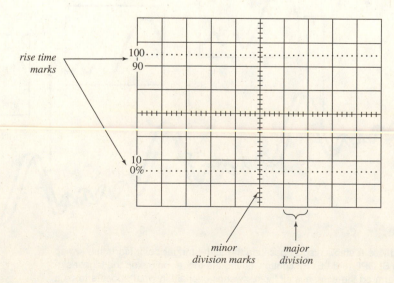

Figure 31–5. An oscilloscope graticule.

generator. The outer control is a click-stop switch that provides step selection of the sweep rate. The center control provides a more sensitive adjustment of the sweep rate on a continuous basis. In its extreme clockwise position, usually marked CAL, the sweep rate is calibrated. Each step of the outer control is therefore equal to an exact time unit per scale division. Thus, the time it takes the trace to move horizontally across one division of the screen graticule is known. Dual-trace scopes generally have one Time/Div. control. Some scopes mark this control Time/cm.

Vert. Pos.
This control is used to adjust the vertical position of the trace. Each channel of a dual-trace scope generally has its own Vert. Pos. control.

X-Y Switch.
When this switch is engaged, one channel of the dual-trace scope becomes the horizontal, or *X*, input, while the other channel becomes the vertical, or Y, input. In this condition the trigger source is disabled. On some scopes, this setting occurs when the Time/Div. control is fully counterclockwise.

Triggering Controls.
The typical dual-trace scope has a number of controls associated with the selection of the triggering source, the method by which it is coupled, the level at which the sweep is triggered, and the selection of the slope at which triggering takes place:

1. *Level Control.* This is a rotary control which determines the point on the triggering waveform where the sweep is triggered. When no triggering signal is present, no trace will appear on the screen. Associated with the level control is an Auto switch, which is often an integral part of the level rotary control or may be a separate push button. In the Auto position the rotary control is disengaged and automatic triggering takes place. In this case a sweep is always generated and therefore a trace will appear on the screen even in the absence of a triggering signal. When a triggering signal is present, the normal triggering process takes over.

2. *Coupling.* This control is used to select the manner in which the triggering is coupled to the signal. The types of coupling and the way they are labeled vary from one manufacturer and model to another. For example, ac coupling usually indicates the use of capacitive coupling that blocks dc; line coupling indicates the 50- or 60-Hz line voltage is the trigger. If the oscilloscope was designed for television testing, the coupling control might be marked for triggering by the horizontal or vertical sync pulses.

3. *Source.* The trigger signal may be external or internal. As already noted, the line voltage may also be used as the triggering signal.

4. *Slope.* This control determines whether triggering of the sweep occurs at the positive going or negative

going portion of the triggering signal. The switch itself is usually labeled positive or negative, or simply + or −.

Probes.
The input signal to the oscilloscope is connected to the vertical signal input jack using various types of connectors and a shielded coaxial cable. An insulated probe is used to connect the scope to the circuit being investigated. The probes themselves may contain circuits that affect the waveform viewed on the screen of the scope. The probe may produce a direct reading or may reduce the reading by some set ratio. Because the probe itself may introduce some distortions in the waveform being observed, an adjusting screw may be provided on the probe. This adjustment called probe compensation, is done on $10\times$ probes only.

SUMMARY

1. An oscilloscope can be used to measure dc, as well as low- and high-frequency ac voltages, waveforms, and time.

2. Oscilloscopes may be an analog or digital type.

3. An oscilloscope displays the amplitude of an ac waveform versus time.

4. A cathode-ray tube is the screen of an oscilloscope.

5. The purpose of the electron gun in a CRT is to emit and deflect an electron beam that strikes the phosphorescent coating on the inside of the screen and causes it to give off light.

6. The signal voltages applied to the vertical deflection plates of a CRT cause the beam to be deflected up and down.

7. The horizontal deflection plates receive the linear sweep voltage, which generates the time base.

8. The intensity control is used to set the brightness of the trace.

9. The focus control is used to narrow the beam into the sharpest trace. There may be an auxiliary astigmatism control for focusing.

10. The horizontal and vertical centering or positioning controls are used to position the trace on the CRT screen.

11. The etched faceplate in front of the CRT face that appears as vertical and horizontal graph lines is called the graticule. Linear calibration markers (height and width) are frequently etched on the graticule.

12. The Volts/Div. control is calibrated to measure the amplitude of signal waveforms along the vertical axis.

13. The Time/Div. control is calibrated to measure time along the horizontal axis.

14. The triggering controls determine the manner in which a trigger pulse is initiated to start the sweep generator.

15. The trigger can be run automatically, the mode that is frequently used.

16. The trigger circuit can be actuated by a signal from internal oscilloscope circuits.

17. The trigger circuit can also be actuated by an external signal voltage.

18. The trigger circuit can also be actuated by a power line—derived voltage from internal oscilloscope circuits.

S E L F   T E S T

Check your understanding by answering the following questions:

1. (True/False) Oscilloscopes are used for the observation of ac waveforms. _____

2. (True/False) Precise time measurements can be made with triggered scopes. _____

3. The waveforms seen on the screen of a CRT show the _____ versus _____.

4. The _____ deflection plates of a CRT are the signal plates; the _____ deflection plates are for the time-base voltage.

5. Of the controls on an oscilloscope, those that affect the height of the signal are called _____.

6. Those controls that affect the sharpness of the trace are called _____ and _____.

7. The etched faceplate in front of the face of the CRT is called the _____.

8. A frequently used triggering mode of a triggered oscilloscope is _____.

9. The controls that affect the up-and-down movement of the trace are called _____.

10. The height of the waveform displayed on the oscilloscope screen is directly proportional to the _____ of the waveform.

MATERIALS REQUIRED

Power Supply:
- Power source for oscilloscope
- 1 0–15 V dc variable supply

Instruments:
- Oscilloscope with triggered sweep, calibrated time base, calibrated vertical amplifier, internal voltage calibrator, and direct probe. The scope can be a single- or dual-trace oscilloscope.
- DMM

Miscellaneous:
- Operating or instruction manual for the oscilloscope
- Assorted leads for connecting to the oscilloscope terminals and jacks

NOTE: Before performing this experiment, read the operating manual completely. The manual should be conveniently available to you throughout the experiment.

PROCEDURE

1. Power to the oscilloscope should be **off.** Examine the front panel of the oscilloscope carefully, noting the types and functions of each control and jack. Examine the back of the scope for additional jacks, switches, and controls.

2. Tabulate the information on the scope in Table 31–1 (p. 241). Some of the information will be found in the operating manual. Other information can be found by inspection.

3. Turn the oscilloscope **on.** Set the time-base switch to the EXT or X-Y position. If you are using a dual-trace scope, set the control for channel 1 (or A) operation. After a brief warmup period, a bright spot should appear on the screen. If there is no spot, adjust the intensity control clockwise.

   CAUTION: Do not set the intensity too high and do not leave the spot on too long; the screen coating can be permanently damaged.

   If the spot still does not appear, it may be off screen. Use the vertical and horizontal positioning controls until the spot becomes visible on the screen.

4. If you have not already done so, adjust the applicable controls to produce a sharp, clear, centered spot on the screen. This will involve the vertical and horizontal positioning, focus, astigmatism (use only if an adjustable control is on the front panel), and intensity controls. Note that the focus and intensity controls (and the astigmatism control, if adjustable) interact with one another. Therefore, it is necessary to adjust both (or all three) to obtain the sharpest, thinnest point of light.

5. Set the time-base switch to automatic operation (this may involve simply disengaging the X-Y switch). Adjust the time-base (Time/Div.) control to 1 ms/div and the Trigger-level control to Auto and positive slope. A line (trace) should appear on the screen. Adjust the intensity control to produce a visible but not too intense trace. Use the horizontal positioning control to move the left end of the line to the leftmost graduation of the graticule. Adjust the trace rotation control, if needed, for a proper horizontal trace.

6. Set the vertical input coupling switch, for channel 1, to the ground (GND) position and adjust the vertical position control to establish a zero volt reference position in the center of the screen. Then, change the vertical coupling switch to the AC position.

7. Connect the vertical input of the oscilloscope to the calibrated square-wave output voltage of your oscilloscope. This value will not be the same for each type of scope, but typically will be about 500 mV peak to peak.

8. Adjust the Volts/Div. control to obtain three to five divisions peak to peak on your screen.

9. Adjust the variable Volts/Div. control out of the CAL position and note the effect on the viewed waveform. Return the variable Volts/Div. control to the CAL position.

10. Increase the Time/Div. control one setting faster. (For example, if the control was set to 1 ms/div, set it to 0.5 ms/div.). Observe the effect of this change on the waveform.

11. Adjust the variable Time/Div. control out of its CAL position and observe the effects on the viewed waveform. Return the control to the CAL position.

12. Change the vertical input coupling switch to the DC or direct position and note the effect on the waveform.

13. Turn off the oscilloscope. Have someone change all the control and switch settings. Repeat steps 3 through 6 with as little reference to the operating manual as possible.

14. Set your oscilloscope's control as follows:

Volts/Div.	1 V/div.
Time/Div.	1 ms/div.
Triggering	Auto
Slope	+
Vert. Coupling	DC

15. Establish a zero volt reference position on the center of the oscilloscope's graticule.

16. With a variable dc power supply in the off position, connect your scope probe's ground lead to the negative output of the dc supply and the scope's positive lead to the positive output of the variable supply. Also,

connect a DMM across the supply's output. This is shown in Figure 31–6.

17. Adjust the dc output voltage to 1V as indicated on the DMM. Observe the change in trace position on your scope. Adjust the output voltage to 2 and 3V respectively, noting the deflection change on the scope.

18. Turn **off** the power supply. Disconnect the supply and turn **off** the oscilloscope.

Optional Activity

Using a list of oscilloscope manufactures provided by the instructor, find the respective websites to research the various types of oscilloscopes they produce. Report on any unique features described by the manufacturers.

ANSWERS TO SELF TEST

1. true
2. true
3. amplitude; time
4. vertical; horizontal
5. Volts/Div. (or Volts/cm)
6. focus; astigmatism
7. graticule
8. automatic
9. vertical positioning, or centering
10. amplitude (voltage)

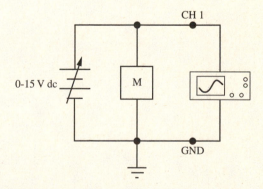

Figure 31–6. Circuit used for step 16.

Name _____ Date _____

TABLE 31–1. Oscilloscope Features, Functions, and Controls

Instrument No. _____ Manufacturer _____ Model _____

Features (Check all that apply)

Single trace _____	Calibrated time base _____
Dual trace _____	Calibrated vertical control _____
Triggered sweep _____	Two-channel +/− capability _____
Auto sweep _____	10× probe _____
XY operation _____	Direct probe _____
Sweep magnification _____	Other features (describe) _____
Internal voltage calibrator _____	_____

Lists of Controls and Their Functions

Control	*Function*

QUESTIONS

1. If the intensity control was adjusted too high when the Time/Div. was set to the X-Y position, what might happen to the scope?

_____

_____

_____

_____

2. What is the effect of changing either the Volts/Div. or Time/Div. controls out of the CAL position?

_____

_____

_____

_____

3. When observing the calibrated square-wave output signal, what effect did the ac- and dc vertical coupling switch have on the waveform?

_____

_____

_____

_____

4. Is there any relationship between the number of cycles of the waveform displayed and the setting of the Time/Div. control? Discuss fully.

_____

_____

_____

_____

_____

5. Refer to the data in Table 31–1. What controls affect the following, and how?
   (a) The height of the displayed waveform
   (b) The brightness of the waveform
   (c) The sharpness of the waveform
   (d) The position of the waveform on the screen
   (e) Turning the sweep generator on and off.

_____

_____

_____

_____

_____

_____

_____

_____

_____

_____

_____

_____

_____

SIGNAL-GENERATOR OPERATION

BASIC INFORMATION

Signal generators provide ac voltages over selected frequency ranges. They are also capable of producing various types of waveforms. Although no single generator can provide the entire range of frequencies and waveforms, they can produce frequencies lower than 1 Hz to many thousands of megahertz. Waveforms can be sine, sawtooth, or square. The voltage output of signal generators is usually low, because signal generators are normally used for testing and experimenting rather than as power supplies. A particular type of signal generator, called a *function*, or *waveform*, *generator*, produces sine, sawtooth, and square waves over a wide range of frequencies but at a relatively low voltage.

Signal generators are often identified by the frequency range they are designed to produce. Thus an AF signal generator is designed to provide frequencies in the audio (so-called sound) range, from several hertz to 20 kHz. The signal itself is a simple sine wave. The output voltage may be several volts to as much as 20 or more volts. Because signal generators are primarily test instruments, the accuracy and stability of their output is critical. The two signal generators covered in this experiment are the AF sine-wave generator and the function generator.

AF Signal Generator

Although AF signal generators are designed to operate in the audio-frequency range (up to about 20 kHz), most commercial generators have ranges in the hundreds of kilohertz. Because the signal generator is often used to provide a test signal for troubleshooting radio receivers, its frequency range extends to that of the standard FM and AM broadcast bands. Signal generators of this type are also equipped to generate modulating signals. To prevent extraneous signals in the air from affecting the signal-generator output, a shielded cable is used between the signal generator and the device to which it is connected.

In this experiment we focus on the waveform and frequency of the signal generator as well as its output voltage. We will observe all three on an oscilloscope while operating the controls of the generator.

The position and function of generator controls vary from manufacturer to manufacturer, but in general, the following are found on all generators.

1. *On-Off Switch.* Signal generators are bench-type instruments that require an ac source to operate (generally 120 V, 60 Hz). The on-off

switch applies power to the circuits of the generator. Some models of portable signal generators are battery operated.

2. *Output control.* This control is used to adjust the voltage output or amplitude of the generator. The output terminal of the generator is generally made to accept a shielded cable. Most outputs are unmetered; to determine the exact output level, an external voltmeter or oscilloscope is required.

3. *Range control.* In most signal generators, frequencies are grouped in a number of ranges spanning the entire frequency spectrum of the generator. For example, one manufacturer produces a generator with a frequency output of 10 Hz to 1 MHz in five decade bands: 10 Hz–100 Hz, 100 Hz–1 kHz, 1 kHz–10 kHz, 10 kHz–100 kHz, 100 kHz–1 MHz. Range controls can be click-stop rotary dials or pushbutton selector switches.

4. *Frequency control.* This is normally a continuously variable rotary dial that fine-tunes the exact frequency within the range set by the range control.

Some AF signal generators also contain modulation switches (AM, FM, or PM), and modulation level controls.

Function Generator

The purpose of the function generator is to produce signals for test and troubleshooting purposes. The signals serve as stimuli for tracing the behavior of parts of circuits or entire systems. In addition, the generated signal may in turn trigger other signals whose behavior is being studied or observed on an oscilloscope. The controls on a function generator are similar to those on an AF signal generator.

1. *On-Off Switch.* The power source for most function generators is a 120-V, 60-Hz supply, although some portable models are battery operated. The on-off switch controls power to the circuits of the generator.

2. *Waveform control.* Function generators are capable of producing two or more types of waveforms. The most common types of waveforms are sine, sawtooth, and square waves. These controls are sometimes labeled function controls.

3. *Output-level control.* This controls the voltage output, or amplitude, of the waveform.

4. *Range switch.* As in the case of the AF signal generator, the total frequency range of the generator is divided into frequency bands selected by a switch or series of switches.

5. *Frequency control.* This serves to select the exact frequency within the frequency band set by the range switch. It, in effect, fine-tunes the frequency of the output waveform.

6. *DC Offset control.* This control adds a positive or negative dc component to any of the waveforms produced by the generator.

7. *dB Attenuator.* This control allows reduction of the output signal, below the normal output, by a specific amount in terms of decibels. Output reduction is normally between -20 dB to -60 dB.

Some function generators have the ability to output specific signals for TTL and CMOS digital circuits. Also, function generators may have the ability to produce modulated output waveforms when an external input signal is applied to it.

An important characteristic of a signal generator is its internal resistance, r_i. This is shown in Figure 32–1. Typical values of internal resistance are 50 Ω and 600 Ω. When a load resistance, R_L, is placed across the generator's output terminals it will load the generator. If this load resistance is not at least 10 times greater than r_i, the output voltage will drop significantly due to the voltage divider action of r_i and R_L. Therefore, when using signal generators be sure that the load resistance will not load down the generator.

SUMMARY

1. A signal generator supplies very accurate and stable ac voltages of variable frequencies and output (voltage) levels.

2. An AF signal generator provides ac voltages in the audio (sound) range: 10 Hz to 20 kHz.

3. Commercial AF generators, however, often have ranges extending to 100 kHz and beyond.

4. The output of the typical AF generator is a sine wave, although some commercial generators also produce square waves.

5. The output (voltage) of the AF generator is generally unmetered. An external voltmeter or oscilloscope is used to set the output level.

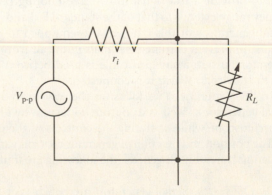

Figure 32–1. Internal resistance, r_i, of a single generator.

6. The following are the typical controls and functions of an AF signal generator:

- On-off power switch
- Frequency range control switch or switches or rotary click-stop dial to select frequency range
- Frequency dial to set the exact frequency within the range set by the range control
- Output level control used to set the output voltage of the generator

7. AF signal generators provide the test signals used to test and troubleshoot radio circuits. These generators sometimes contain modulation functions for checking AM and FM circuits.

8. Function (or waveform) generators produce signals of variable frequencies in different waveforms.

9. The typical function generator can produce sine, sawtooth, and square waves.

10. The following are the typical controls and functions of a function generator:

- On-off power switch
- Waveform selector switch
- Switch or switches or rotary dial to select the frequency range
- Frequency dial to set the exact frequency within the range set by the range switch
- DC offset control to add a positive or negative dc voltage to the ac output voltage
- Output level control to set the voltage level of the output

11. AF signal generators and function generators differ in features and design from one manufacturer to another. It is important to have convenient access to the operating manual of the equipment being used.

Check your understanding by answering the following questions.

1. Audio-frequency signal generators are often used to provide a signal when testing or troubleshooting _____ circuits.

2. The audio frequencies are in the range of _____ Hz to _____ kHz.

3. (True/False) The typical AF signal generator uses a built-in voltmeter to set the output voltage level. _____

4. A _____ control is used to set the band of frequencies that the AF generator can produce.

5. Fine-tuning of the exact output frequency of the generator is set by a _____ switch.

6. Another name for a function generator is a _____ generator.

7. A dc _____ control adds a negative or a positive voltage to the ac output of the function generator.

8. The most common waveforms provided by a function generator are the _____, _____, and _____ waveforms.

MATERIALS REQUIRED

Instruments:
- AF signal generator (sine-wave output)
- Function generator (with sine-, sawtooth-, and square-wave output)
- Triggered oscilloscope (single- or dual-trace) with direct probe
- Operating manuals for the above instruments
- DMM

Miscellaneous:
- Resistance decade box

PROCEDURE

Before turning on the power to any of the instruments used in this experiment, study the manufacturer's operating manual supplied with the instrument. Note especially the location and function of each switch, dial, terminal, and control on the face of the instrument. Review the procedure that follows and compare the names of the controls referred to in the procedure with the actual equipment you will be using. Your instructor will review this information with you before you turn on power.

A. AF Sine-Wave Signal-Generator Operation

A1. Connect the output of the signal generator to the vertical input of the oscilloscope. (If a dual-trace scope is used, make all connections to channel 1.) Set the output (voltage) level control to the middle of its full range, and the frequency range switch or control to the 10–100 kHz range. Turn **on** the signal generator.

A2. Turn **on** the scope. Set the scope for automatic triggering. If you are using a dual-trace scope, make sure it is set to display only the channel 1 signal. Adjust the signal generator to deliver a 100-kHz sine wave to the scope. Adjust the Volts/Div. control of the scope until the wave on the screen is 6 divisions peak-to-peak and centered vertically. This may also require an adjustment of the output level control of the signal generator. Adjust the Time/Div. control of the scope until two complete sine waves are displayed on the screen (further adjusting of the Volts/Div. and output level controls may be necessary).

A3. Slowly reduce the output level of the signal generator while observing the effect this has on the height of the wave displayed on the screen. Now slowly increase the output level beyond the setting of step A2 and observe the effect this has on the height of the wave displayed on the screen. Reset the output level control to restore the wave to its original 6-division peak-to-peak height.

A4. Change the frequency range switch and frequency control on the signal generator to produce a 1 kHz signal. Readjust the Time/Div. control to display two cycles across the entire width of the screen and centered vertically on the screen. Adjust the Volts/Div. control so that the sine waves are about 6 divisions peak-to-peak. In Table 32–1, record the setting of the Time/Div. control and the number of divisions spanned by the sine waves.

A5. Increase the frequency output of the generator to 2 kHz and observe the effect on the display. Do not adjust or change the setting of the Time/Div. control. In Table 32–1, record the number of cycles displayed and the number of divisions spanned by the cycles.

A6. Decrease the frequency output of the generator to 500 Hz. In Table 32–1, record the number of cycles displayed and the number of divisions spanned by the cycles. After recording the values, turn **off** the scope and the signal generator.

B. Function-Generator Operation

B1. Connect the output of the function generator to the vertical input of the scope. Set the function generator to the sine-wave output and 1 kHz; set the output level in the middle of its full range. Turn **on** the function generator.

B2. Turn **on** the scope and adjust the Time/Div. control to produce one cycle across the width of the screen. Adjust the Volts/Div. control to produce a display about 6 divisions peak-to-peak. Sketch the waveform displayed on the screen in Table 32–2.

B3. Switch the frequency to 2 kHz. Do not change or adjust the Time/Div. control. Center the display vertically and horizontally. Readjust the height to 6 divisions peak-to-peak if necessary. Sketch the waveform displayed on the screen in Table 32–2.

B4. Switch back to 1 kHz output (do not readjust or change the Time/Div. control). Set the function switch to produce a sawtooth-wave output. If necessary, adjust the Volts/Div. control to produce a display 6 divisions peak-to-peak. Sketch the sawtooth-wave displayed on the screen in Table 32–2.

B5. Increase the sawtooth-wave frequency to 2 kHz and repeat step B3.

B6. Switch back to 1 kHz and set the function switch to produce a square-wave output. Do not change or readjust the Time/Div. control. Center the waveform on the screen. Sketch the waveform displayed on the screen in Table 32–2.

B7. Repeat step B3 for the square-wave output. After you have completed this step, turn **off** the scope and function generator.

B8. Connect a DMM, along with an oscilloscope, across the output of the function generator. Place the scope's vertical input coupling switch in the dc position and Volts/Div. control to 5 V/div.

B9. With the function generator's square-wave function button still pushed in, push the sine-wave function button slightly so that all of the function buttons are in their out position. This will disable all output waveforms.

B10. Next, turn on the generator's dc offset control. While measuring with the DMM and monitoring the scope, adjust the offset control for its maximum positive and negative output dc values. Record the maximum positive and negative values in Table 32–3.

B11. Return the dc offset control to the off or zero volt position. Set the output waveform control to a sine-wave output. With the scope's Volts/Div. control set to 1v/div., adjust the output of the generator to 4 divisions peak-to-peak.

B12. Connect a resistance decade box, set to 10 kΩ, across the generator's output to form the circuit of Figure 32–1. Decrease the load resistance until the generator's output has dropped to 2 divisions peak to peak. Remove the resistance decade box and measure its resistance. This value of R_L will be equal to the generator's internal resistance, r_i. Record this value in Table 32–3.

B13. Turn **off** function generator and oscilloscope. Disconnect equipment.

ANSWERS TO SELF TEST

1. radio
2. 10; 20
3. false
4. range
5. rotary dial
6. waveform
7. offset
8. sine; sawtooth; square

TABLE 32–1. AF Signal-Generator Operation

Frequency Setting of Signal Generator, Hz	Number of Cycles Displayed on Screen	Time/Div. Setting of Scope, Time Units/Div.
1 k	2	
2 k		
500		

TABLE 32–2. Function-Generator Operation

	Sine Wave	Sawtooth Wave	Square Wave

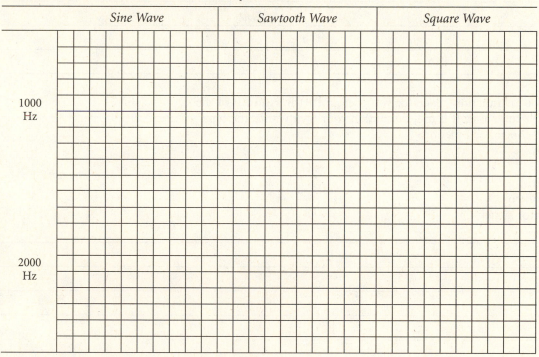

TABLE 32–3 Function Generator Characteristics

dc Offset		Internal Resistance
+, v	−, v	r_i, Ω

QUESTIONS

1. Discuss, in your own words, the relationship between the number of cycles of the waveform displayed on the screen and the frequency setting of the AF signal generator (with a constant Time/Div. control setting). Refer to your data for this experiment.

_____

_____

2. In procedure step A3, the output level of the signal generator was first decreased and then increased beyond the level of step A2. Discuss the effect of the decrease and increase on the number of cycles and amplitude of the waveform displayed.

_____

_____

_____

_____

_____

3. Explain how to find the internal resistance, r_i, of a generator.

_____

_____

_____

_____

_____

4. State whether or not the function generator you used in Part B had each of the following controls or functions. If it did, explain, in your own words, the type and purpose of the control or function.
   (a) On-off switch
   (b) Range control
   (c) Frequency control
   (d) Output-level control
   (e) Waveform-selector switch

_____

_____

_____

_____

_____

_____

_____

_____

_____

_____

EXPERIMENT 33

OSCILLOSCOPE VOLTAGE AND FREQUENCY MEASUREMENT

BASIC INFORMATION

OBJECTIVES

1 To check the oscilloscope and probes for proper measurement

2 To accurately measure dc and ac voltage values with the oscilloscope

3 To accurately measure time and frequency with the oscilloscope

Safety

An oscilloscope is essentially a voltage measuring instrument, much like the DMM, that can graphically display what it is measuring. Because voltage values greater than 30 volts are often measured with an oscilloscope, proper safety procedures must be followed.

Proper grounding is essential when preparing to take measurements or work on a circuit. A properly grounded oscilloscope can protect you from a hazardous shock and prevent serious damage to the oscilloscope and the circuits that you are working on. If a high voltage contacts the case of an ungrounded oscilloscope, you could receive a shock when you touch any part of the case.

To ground the oscilloscope, connect it to an electrically neutral reference point such as earth ground. This is normally accomplished by plugging in the oscilloscope's three-pronged power cord into an outlet grounded to earth ground. Grounding is also necessary for your measurements to be accurate. Therefore, your oscilloscope must share the same ground as the circuits you are testing. Portable oscilloscopes, such as shown in Figure 33–1 (p. 248), have insulated cases and controls. They do not require the separate connection to earth ground.

Oscilloscope Probes

The ground lead of the scope probe provides the proper grounding method for most measurements. This ground lead is normally connected to a known ground in the circuit being tested, such as the metal chassis. If this ground lead is connected to a "hot" point in a circuit, damage to the circuit or the oscilloscope is highly possible.

Oscilloscope probes are normally either a 1× or 10× type. A 1× probe is commonly called a direct probe because it doesn't contain any attenuation circuitry. The input signal on the probe's tip goes directly into the scope's input. Without this attenuation circuitry, more interference is introduced to the circuit being tested. The 1× probe does have the advantage of measuring very small signal levels. The 10× probe contains a small internal circuit that attenuates or reduces the signal being measured,

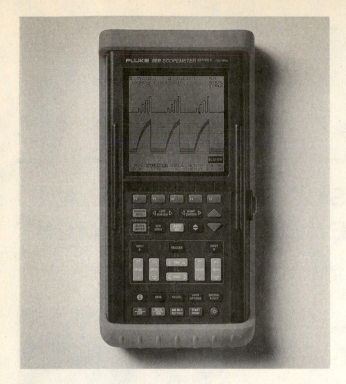

Figure 33–1. Portable scopemeter.

Probe Compensation

In Figure 33–2, the scope probe's parallel adjustable capacitor and the input capacitance of the oscilloscope's vertical input form another special voltage divider. Before using a 10× probe, first check to see if the probe's electrical properties are balanced with the oscilloscope's electrical properties. The adjustment made to produce this balance is called compensating the probe. Probe compensation is accomplished by connecting the 10× probe to the oscilloscope's calibrated square-wave output signal and then adjusting the probe's compensating adjustment for a correct flat square-wave as shown in Figure 33–3. An under compensated probe results in a rounded square-wave due to poor high frequency response while an over compensated probe causes overshoot due to excessive high frequency response. A 10× probe should be checked every time the oscilloscope is setup for taking measurements.

Voltage Measurements

The oscilloscope is normally used to make two basic measurements. These measurements are amplitude and time. After making these essential measurements, other values can be computed. To make a basic peak-to-peak voltage measurement, the following steps should be followed:

1. Preset the oscilloscope's controls to their proper positions and obtain a baseline trace.
2. Verify that the 10× probes are properly compensated and measuring correctly.
3. Set the VERTICAL MODE switch to display the channel being used and apply the ac signal.
4. With the Volts/Div. Variable control in the CAL detent position, set the Volts/Div. switch to display 5 to 6 divisions of the waveform.
5. Adjust the TRIGGER LEVEL and SLOPE controls to obtain a stable display.
6. Set the Time/Div. switch to display 2 to 3 cycles of the waveform.

minimizes circuit loading, and provides improved high frequency measurement as compared to a 1× probe. Figure 33–2 shows the circuitry of a typical 10× probe.

The voltage divider consisting of the probe's 9 MΩ resistor and the oscilloscope's 1 MΩ vertical input resistance attenuates or reduces the input signal by a factor of 10. Therefore 1/10th of the input signal is actually applied to the vertical system block. This voltage divider enables measuring higher voltage values than possible with a 1× probe. Also, as seen by the circuit being measured, the input resistance of the scope has been increased from 1 MΩ to 10 MΩ. This reduces circuit loading by the oscilloscope, similar to voltmeter circuit loading previously studied.

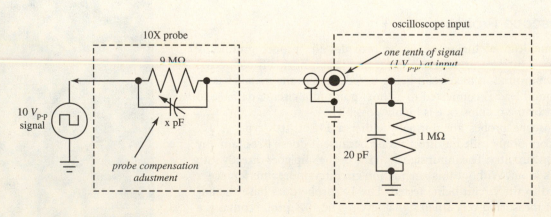

Figure 33–2. Typical 10× oscilloscope probe divider network.

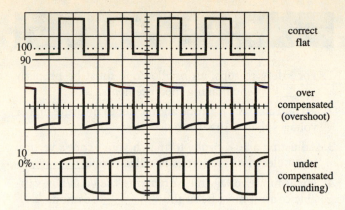

Figure 33–3. Probe compensation.

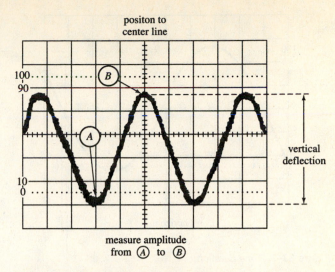

Figure 33–4. Peak-to-peak waveform voltage.

7. Use the VERTICAL POSITION control to align the negative peak of the waveform with one of the horizontal graticule lines.

8. Position the display horizontally so that one of the positive peaks lines up with the center vertical graticule line.

9. Measure the vertical deflection from its positive peak to its negative peak.

10. Calculate the peak-to-peak, $V_{p\text{-}p}$, voltage using the formula:

$V_{p\text{-}p}$ = Vertical Deflection Division × Volts/Div. Switch Setting (depending of 10× or 1× probe being used)

As an example:
> If the peak-to-peak deflection is 4.6 divisions with the Volts/Div. setting at 2V/div., $V_{p\text{-}p}$ = 4.6 div. × 2V/div. = 9.2 $V_{p\text{-}p}$.

This is shown in Figure 33–4.

The same method is also used to measure dc voltage. The AC-GND-DC switch is placed in the dc position and the probe is connected to a point of dc voltage in the circuit. The ground lead of the scope is connected to the ground of the circuit. The number of divisions the trace rises above or below the zero volt reference setting is a measure of the + or − dc voltage.

For measuring a combination of an ac signal riding on a dc component, place the AC-GND-DC switch in the dc position. This will couple the entire signal into the scope's vertical input. If only the changing or ac portion of a combination ac-dc waveform needs to be measured, place the coupling switch in the ac position to block the dc component.

Time and Frequency Measurement · · · · · ·

Time measurements include measurements such as a waveform's period (time for one complete cycle), pulse width, rise time, and timing of pulses. These measurements are taken using the horizontal scale of the oscilloscope. Since frequency is the reciprocal of period, once you measure the

waveform's period, its frequency is equal to one divided by the period. As an example, if the measured period is 16.67 ms, then, the frequency is 1/16.67 ms or 60 Hz.

To make a time or frequency measurement, the following steps should be followed:

1. Preset the oscilloscope's controls to their proper positions and obtain a baseline trace.

2. Verify that the scope probes are measuring correctly and properly compensated if 10×.

3. Set the VERTICAL MODE switch to display the channel being used and apply the ac signal.

4. Adjust the TRIGGER LEVEL and SLOPE controls to obtain a stable display.

5. With the Time/Div. variable control in the CAL position, adjust the Time/Div. switch to display one complete period of the waveform.

6. Position the display to place the measurement points on the center graticule line as shown in Figure 33–5 (p. 250).

7. Measure the horizontal distance between the time-measurement points.

8. Calculate the time duration by:

> Time Duration = no. of Horizontal Divisions × Time/Div. switch setting.

9. The corresponding frequency of this waveform can be found by:

> Frequency = 1/(Time Duration)

Other special pulse measurements can also be made with an oscilloscope. Consult the scope's instruction or operation manual for these procedures.

SUMMARY

1. An oscilloscope is essentially a graphical voltage measuring instrument.

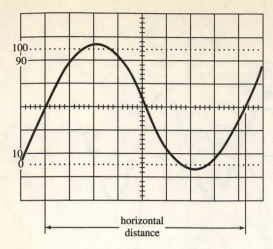

Figure 33–5. Time duration.

2. Proper scope grounding is necessary to take accurate measurements and to prevent hazardous conditions.

3. Your oscilloscope must share the same ground point as the circuits being tested.

4. A 10× probe contains built in attenuation that must be properly adjusted or compensated.

5. Before accurate amplitude or time measurements can be accomplished, the variable Volts/Div. and Time/Div. controls must be in the CAL position.

6. The vertical input coupling switch could be used to either pass or block the dc component of a waveform.

7. Once a waveform's period is measured, its frequency can be calculated.

$$F = 1/(\text{Time period})$$

PROCEDURE

A. DC Voltage Measurement

A1. Preset the oscilloscope's controls to their proper position. Turn **on** the scope and obtain a baseline trace.

A2. Verify proper operation of the oscilloscope and probes by measuring the scope's calibrated output voltage. If using a 10× probe, follow the proper procedures for probe compensation.

A3. Using a dc variable power supply and DMM, set the output voltage to the values specified below. Next, use the oscilloscope to measure these dc values. Sketch the scope's waveform trace on the graticule shown in Table 33–1 (p. 253). Record the corresponding Volts/Div. settings and measured voltage in the provided spaces. Also, on each sketch indicate the zero volt reference point.
a. 3V   b. 9V   c. −5V

MATERIALS REQUIRED

Power Supplies:
- Variable 0-15 V dc, regulated

Instruments:
- Oscilloscope
- Function Generator
- DMM

B. AC Voltage Measurement

B1. Connect the oscilloscope to the output of a function generator. Set the generator's output to the sine-wave function at a frequency of 100 Hz. For each of the following specified values, set the scope's controls to display two cycles of the waveform. Sketch the waveform as seen on the graticule and record the Volts/Div. and measured V_{p-p} value in Table 33–2. On each sketch note the zero volt reference point.
a. 2 V_{p-p}
b. 12 V_{p-p}
c. 2 V_{p-p} on 4V dc

C. Time and Frequency Measurement

C1. Set the function generator for a sine-wave output of 5 V_{p-p}. Adjust the generator for each of the following specified frequencies and measure the time (period) of each cycle. Sketch the waveforms as seen on the graticule in Table 33–3 (p. 254). Indicate, in the provided spaces, the Time/Div. setting, number of divisions, period and frequency for each sketch.

a. 1 kHz

b. 15 kHz

c. 100 Hz

D. Risetime Measurement

D1. Set your function generator's output for a 5 V_{p-p} 1 kHz square-wave. Next, measure and record the square-wave's risetime value. Record this value in the provided space and demonstrate this measurement technique to your instructor.

D2. Turn **off** the function generator and oscilloscope. Disconnect all equipment.

Optional Activity ·

This activity requires the use of electronic simulation software. Using electronic simulation software, measure various output amplitudes, frequencies, and waveforms from the waveform generator with an oscilloscope. Explore using the dual trace capability of the oscilloscope.

ANSWERS TO SELF TEST

1. grounded

2. false

3. compensation

4. CAL

5. time period

6. true

TABLE 33–1. DC Voltage Measurement

DC Input Voltage		
3V	9V	−5V
Volts/Div		
Voltage		

TABLE 33–2. AC Voltage Measurement

AC Input Voltage, $V_{\text{p-p}}$		
2V	12V	2V on 4V DC
Volts/Div		
Voltage		

TABLE 33–3. Time and Frequency Measurement

AC Input		
5V p-p 1 kHz	5V p-p 15 kHz	5V p-p 100 Hz

	5V p-p 1 kHz	5V p-p 15 kHz	5V p-p 100 Hz
Time/Div			
# of Div.			
Period			
Frequency			

QUESTIONS

1. When taking ac voltage measurements with a scope, what Volts/Div. setting will produce the most accurate reading?

_____

_____

_____

_____

_____

_____

_____

2. Under what circumstances would a 10× probe be preferred over a 1× probe?

_____

_____

_____

_____

_____

_____

_____

3. Why is it important to check a scope's control settings and probe compensation each time the scope is used?

_____

_____

_____

_____

_____

_____

_____

4. Why must the vertical input coding switch be set to the dc position when measuring a combination ac-dc waveform?

_____

_____

_____

_____

_____

_____

_____

PEAK, RMS, AND AVERAGE VALUES OF AC

BASIC INFORMATION

Generating an AC Voltage

When a conductor cuts lines of magnetic force, a voltage is induced in the conductor. This is the principle of the electric generator. We will apply this principle to the generation of an ac voltage.

Consider a single loop of wire so arranged on a shaft that it can be rotated through the magnetic field that exists between the north (N) and south (S) poles of a magnet, as in Figure 34–1. As the loop of wire turns, it cuts across the magnetic lines of force. This induces a voltage in each side of the loop. The direction (polarity) of the induced voltage is determined by the direction of the magnetic field and the direction of motion of the conductor.

In Figure 34–1 the loop is shown rotating in a counterclockwise direction. The left side of the loop is traveling down through the magnetic field, whereas the right side of the loop is traveling up through the field. The polarity of the voltage induced in one side is therefore opposite to that induced in the other side. But the two sides of the loop are actually in series, so that the two induced voltages are additive.

The voltage induced in the loop is proportional to the rate at which the lines of force are cut by the loop. Figure 34–2 (p. 258) is an end view

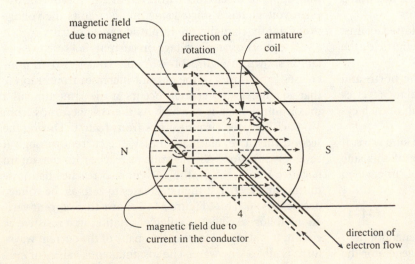

Figure 34–1. Simple electric generator.

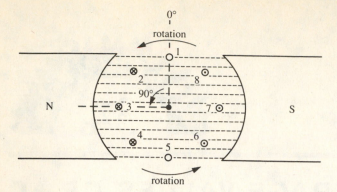

Figure 34–2. Rotation of the single loop in the magnetic field.

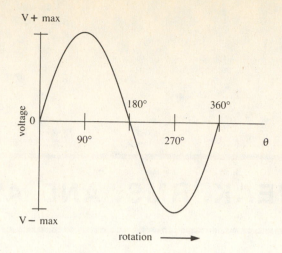

Figure 34–3. The variation of voltage with the rotation angle is a sine wave.

of the loop at various positions as it rotates between the north and south poles. Positions 1 and 5, 2 and 6, 3 and 7, 4 and 8 represent opposite sides of the same loop.

At position 1 the side of the loop is momentarily moving parallel to the lines of force. No lines are being cut, therefore, no voltage is being induced in this loop and its other side (position 5). As the loop moves to position 2 (and 6), it begins cutting lines of force at an oblique angle, and a voltage is induced in the sides of the loop. If the ends of the loop were connected to a load, current would flow in the loop. The symbol $\otimes$ shown in position 2 indicates that electron current would flow into the page; the symbol $\odot$ in position 6 indicates that current would flow out of the page (as would be expected, because 2 and 6 are in series). As the loop continues to rotate, the angle at which it cuts the lines of force becomes more nearly perpendicular until at position 3 and 7 the loop is crossing the lines of force at a right angle. Since the loop is assumed to be rotating at a constant speed, positions 3 and 7 are where the greatest number of lines of force are being cut per unit of time, and positions 1 and 5 are where the least lines of force are being cut per unit of time. It follows then that at positions 3 and 7 the highest voltage is being induced, and at positions 1 and 5 the lowest voltage is being induced (0 V in this case). Past position 3 the loop begins cutting the lines obliquely until at position 5 the loop is not cutting any lines.

If the voltage induced in the loop were plotted against the position of the loop within the magnetic field, the graph would resemble the one in Figure 34–3.

A complete rotation of the loop past the north and south poles is called a cycle. One cycle is equal to 360 electrical degrees. The positive or the negative half of each cycle is called an alternation.

The voltage induced in the loop will be a sine wave. If we call the peak of the instantaneous value of this sine wave V_M, then the voltage V at any instant of time, when the loop has rotated θ degrees, is

$$V = V_M \sin \theta \qquad (34\text{–}1)$$

The loop represents only one coil of an ac generator (also called an *alternator*) armature. In a practical generator the loop ends are brought out and connected to slip rings. The

typical ac generator also contains more than two poles. The armature can be rotated mechanically by a steam turbine, a gasoline engine, a water wheel, or even a hand crank. In the large ac generators used by power companies, the armature is stationary and built around the frame of the machine, and the magnetic poles are mounted on a shaft and rotated within the armature. The effect of wires cutting a magnetic field is the same whether the wire moves across a stationary field or a moving field cuts across a stationary wire. In both cases, a voltage is induced in the wire.

AC Voltage and Current

The generator in Figure 34–1 produces an ac voltage. When an ac voltage is connected to a resistive load, *alternating current* flows in the circuit. Because a cycle of ac voltage consists of one *positive* and one *negative* alternation, current in the circuit will flow in one direction during the positive alternation and in the opposite direction during the negative alternation. Ohm's and Kirchhoff's laws apply to a resistive ac circuit in the same way that they apply to dc circuits. However, in a dc circuit the applied voltage has a *single* value. In ac circuits the voltage is constantly changing in amplitude and polarity.

The effect of an ac voltage on current in a resistive circuit is shown in Figure 34–4. As the voltage $V_M \sin \theta$ increases from 0 to V_M, the current increases from 0 to I_M; that is, a maximum current occurs at the same time as a maximum voltage is reached. As the voltage drops from V_M back to 0, the current drops from I_M to 0. During the second (negative) alternation of V, the current through R reverses direction. As the voltage reaches its maximum negative value $(-V_M)$, the current also reaches its maximum value $(-I_M)$; it then decreases to zero as the voltage decreases to 0. Thus, Figure 34–4 shows that current variations follow exactly the voltage variation in a resistive ac circuit. It is also evident that the shape of the current waveform is also a sine wave. The instantaneous values of current in an ac waveform can be given by the formula

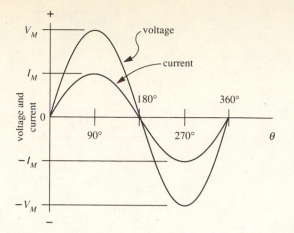

Figure 34-4. Voltage and current are in phase in a resistive circuit.

$$I = I_M \sin \theta \qquad (34\text{-}2)$$

We can also say that in a resistive ac circuit, current and voltage are *in phase*. There are ac circuits that contain capacitors and inductors in which current and voltage are not in phase. These are considered in later experiments.

Peak, RMS, and Average Values · · · · · · ·

The amplitude of a dc voltage can be identified by a single value. Can we specify an ac voltage by a single value? Besides peak-to-peak, there are actually three different values that can be used—namely, the *peak,* the *rms,* and the *average* value. Each identifies a different characteristic of the voltage, but they are all related. Given one value, the other three can be easily calculated.

Peak Value. If we say that the peak of a sinusoidal voltage is 100 V, we mean that the voltage reaches a maximum of +100 V on the positive alternation and −100 V on the negative alternation. From the peak value we can calculate the instantaneous level of voltage at any angle θ in the cycle using formula (34-1).

The peak value does *not* produce the same power as the same dc value, because the ac voltage varies constantly in amplitude, whereas the dc voltage maintains a constant level.

RMS Value. There is a value of ac voltage that will produce the same power as the equivalent dc level. This is the rms, or *root mean square,* value. Thus, if we say that the rms value of an ac voltage is 100 V, we mean that it will deliver the same power as 100 V dc. The rms value is the *square root* of the *mean* (average) of the sum of the *squares* of the instantaneous values of voltage in an ac alternation. That is,

$$V_{\text{rms}} = \sqrt{\frac{V_1^2 + V_2^2 + \cdots + V_n^2}{n}}$$

where $V_1, V_2, \ldots, V_n$ are successive instantaneous values of $V_M \sin \theta$. It can be shown that

$$V_{\text{rms}} = 0.707 V_M \qquad (34\text{-}3)$$

and that

$$V_M = 1.414 V_{\text{rms}} \qquad (34\text{-}4)$$

So we have two unique but related values of V that describe the amplitude of an ac voltage. The rms values (also called *effective* values) are more popularly used than peak values in giving the amplitude of an ac voltage. That is why ac voltmeters are calibrated in rms rather than peak values.

Electric power is a square function of voltage (V^2/R) or current ($I^2 \times R$). That is why the rms value, which is derived from a square value (V_1^2, etc.), is comparable to the equivalent value of dc voltage.

Average Value. There is yet another method of identifying an ac voltage, that is, by using the *average* value of the waveform. Voltage V_{ave} is also dependent on the peak value and the rms value and may be determined from these values by the formulas

$$V_{\text{ave}} = 0.636 V_M = 0.899 V_{\text{rms}} \qquad (34\text{-}5)$$

The peak voltage V_M can also be specified in terms of V_{rms} and V_{ave}. Thus,

$$V_M = 1.414 V_{\text{rms}} = 1.572 V_{\text{ave}} \qquad (34\text{-}6)$$

Alternating current is also identified in any one of four ways, peak-to-peak ($I_{p\text{-}p}$), peak (I_M), rms (I_{rms}), and average (I_{ave}). The formulas for current are identical to the formulas for voltage except that I replaces V.

Measurement of AC Voltages and Current

In the design of ac circuits, voltage and current measurements are usually made in rms values. An analog meter responds to average values, but the ac scale is calibrated in rms values. Normally a VOM does not have an ac current function. Current measurements can be made with a digital VOM that has both ac voltage and current scales. These are also calibrated in rms values. Some meters have both rms and peak value scales, but these may be used only in the measurement of *sinusoidal* voltages or currents. Some digital multimeters are referred to as "True RMS" meters. These meters have the ability to measure the rms value for nonsinusoidal waveforms. Depending on the meters internal circuitry, the value can be read directly on the meter's display or it might be necessary to perform a conversion formula to get the correct reading. DMMs and analog ac meters have an upper frequency limit when measuring ac voltages and currents. It is important to read a meter's instruction manual to find its characteristics, specifications, and measurement techniques.

An oscilloscope is normally used for measuring peak ac voltages or peak-to-peak values.

Differential Measurement · · · · · · · · · · · · · ·

As previously discussed, voltage is the difference in potential between two points. Many times one of these two points is measured with respect to the circuit's ground or earth ground. A single point measurement, V_B, measured with respect to ground is shown in Figure 34–5. Connecting an oscilloscope to take this measurement is very simple and straightforward. But what happens when it is necessary to take a voltage measurement across R_1, points A and B, where neither point is at ground? If the scope is connected to take the measurement as shown in Figure 34–6, the scope's ground and supply ground will effectively short out R_2, leaving only R_1, 10 Ω, to limit the circuit current. This could possibly cause damage to the circuit, supply or oscilloscope.

To properly and safely take a voltage measurement, where neither point is at ground, use the oscilloscope's differential measurement technique and connect the scope as shown in Figure 34–7. Note that both channels of the oscilloscope are used along with the scope's ground, which is connected to earth ground. Place the oscilloscope's vertical MODE switch to the ADD mode. Set both Vertical Volts/Div. switches to the same settings. Be sure that the Variable controls are in the CAL positions. Invert Channel #2 (some scopes have an invert button for Channel #1 instead) and establish the vertical zero volt reference position. Make the proper connections for each channel and ground. Both probes must be the same types (either 1X or 10X). The scope will now display the algebraic addition of Channel 1's voltage in respect to ground plus Channel 2's inverted voltage in respect to ground. The scope will therefore produce a differential measurement across two points, neither of which is at a ground reference point.

SUMMARY

1. An ac generator produces a sinusoidal voltage.

2. The formula for the instantaneous value V of a sinusoidal voltage is

$$V = V_M \sin \theta$$

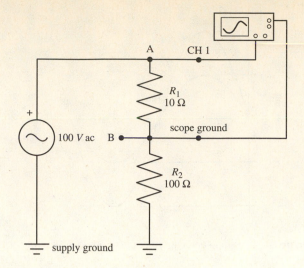

Figure 34–6. Improper voltage measurement.

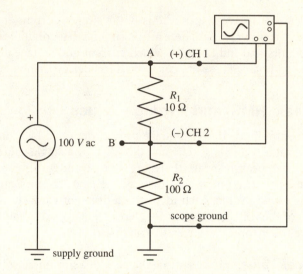

Figure 34–7. Differential voltage measurement.

where V_M is the peak, or maximum voltage, of the waveform and θ is the angle the waveform has reached at the instant it is measured.

3. The amplitude of a sinusoidal ac waveform may be specified in four ways: (a) its peak value V_M; (b) its root mean square value V_{rms}; (c) its average value V_{ave}; and its (d) peak-to-peak value.

4. The rms value is comparable to the equivalent value dc voltage because the *power* that each can produce is the same.

5. The four values of a sinusoidal voltage are interdependent, as their formulas indicate. Thus,

$$V_{rms} = 7.707 V_M$$

$$V_{ave} = 0.636 V_M = 0.899 V_{rms}$$

$$V_{p\text{-}p} = 2 V_M$$

6. These formulas may also be written as

$$V_M = 1.414 V_{rms} = 1.572 V_{ave} = 0.5 V_{p\text{-}p}$$

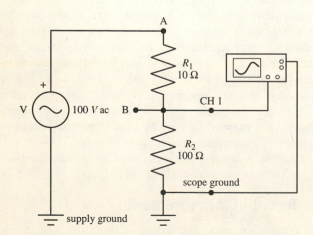

Figure 34–5. Single point voltage measurement.

7. The ac scale of an analog voltmeter and ammeter is calibrated in rms values, although the meter movement responds to average values.

8. When an ac voltage is given without specifying whether it is rms, peak, or average, the rms value is meant.

9. The relationships among rms, peak, and average values of alternating current are the same as those for ac voltage, with I substituted for V in the voltage formulas.

10. Ohm's and Kirchhoff's laws apply to resistive ac circuits just as they do to dc circuits. Thus, the rms value of current in a resistor, across which V_{rms} is applied, may be calculated from the formula

$$I_{rms} = \frac{V_{rms}}{R}$$

Similarly, I_M may be calculated for the resistor R by substituting in the formula

$$I_M = \frac{V_M}{R}$$

SELF TEST

Check your understanding by answering the following questions:

1. An ac voltage changes constantly in _____ and _____.

2. The shape of the voltage waveform delivered by an ac generator is a _____.

3. (True/False) The rms and average values of a sinusoidal voltage or current may be calculated from its peak value. _____

4. The V_M of a sinusoidal voltage is 25 V. The rms value of that voltage is _____ V.

5. The rms value of a sinusoidal current is 5 mA. The peak value of that current is _____ mA.

6. In question 5, the average value of current is _____ mA.

7. The rms voltage measured across a 47-Ω resistor is 15 V. The rms current in that resistor is _____ mA.

8. The peak current in the resistor in question 7 is _____ mA.

9. The ac voltmeter is usually calibrated in _____ (peak/rms/average) values.

10. The peak value of an ac voltage may be measured with a(n) _____.

MATERIALS REQUIRED

Power Supplies:
- 120-V, 60-Hz power source
- Isolation transformer
- Variable-voltage autotransformer (Variac or equivalent)

Instruments:
- Function generator
- DMM
- Oscilloscope
- AC ammeter (or additional DMM)

Resistors (½-W, 5%):
- 2 33
- 1 47
- 2 1 kΩ
- 1 1.5 kΩ

Miscellaneous:
- SPST switch
- Line cord set consisting of polarized plug, fuse, and on-off switch

PROCEDURE

CAUTION: You will be dealing with voltages that are potentially lethal. Pay strict attention to all electrical safety rules when performing this experiment.

A. 60-Hz Supply

A1. With the line cord unplugged and its switch **off,** connect the circuit of Figure 34–8 (p. 262). Set the variable voltage transformer to its lowest voltage; S_1 is **open.**

A2. Plug the line cord into a 120-V, 60-Hz outlet; turn the line switch **on; close** S_1. Adjust the output voltage of the variable transformer to 35 V. Maintain this voltage throughout Part A. Record this voltage under "Voltage, rms, Measured" in Table 34–1 (p. 265). (The voltmeter will read the voltage in rms units.)

A3. Measure the voltage across R_1, R_2, and R_3 with the voltmeter, and record the values under "Voltage, rms, Measured" in Table 34–1. **Open** S_1; line cord switch **off.**

A4. Connect an oscilloscope across the output of the variable-voltage transformer. Line cord switch **on;** S_1 **closed.** Connect your scope's ground lead to

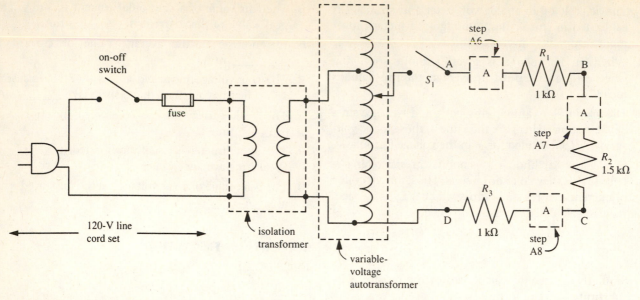

Figure 34–8. Circuit for procedure step A1. The ac ammeter is added in steps A6, A7, and A8.

point D and use the differential measurement technique for the scope's measurements. Measure the peak voltage across the transformer output terminals with the oscilloscope and record the value in Table 34–1 under the "Voltage, peak, Measured" column.

A5. Repeat step A4, using the oscilloscope to measure the peak voltage across each of the resistors R_1, R_2, and R_3. Turn **off** power whenever you change circuit connections. Record the values in Table 34–1. Switch line cord **off;** open S_1.

A6. With power to the resistors **off,** break the circuit between A and B and insert an ammeter in the circuit as shown in Figure 34–8 to measure the current through R_1. Set the range of the ac ammeter at 20 mA or above. Turn line cord switch **on; close** S_1. Record the ammeter reading in the "Current, rms, Measured" column in Table 34–1. S_1 **open;** line cord switch **off.**

A7. Break the circuit between B and C and insert an ammeter as in step A6 to measure the current through R_2. Line cord switch **on;** S_1 **closed.** Record the ammeter reading in Table 34–1 in the "Current, rms, Measured" column. S_1 **open.** Line cord switch **off.**

A8. Break the circuit between C and D and insert an ammeter as in step A6 to measure the current through R_3. Line cord switch **on;** S_1 **closed.** Record the ammeter reading in Table 34–1 in the "Current, rms, Measured" column. S_1 **open;** line cord switch **off;** line cord **unplugged** from outlet.

A9. Calculate the rms current that should be in each resistor. Record your answers in Table 34–1, in the "Current, rms, Calculated" column.

A10. Calculate the rms voltage across each resistor. Record your answers in Table 34–1 in the "Voltage, rms, Calculated" column.

A11. Calculate the peak voltage across each resistor. Record your answers in Table 34–1 in the "Voltage, peak, Calculated" column.

B. 1 kHz Supply

B1. With the Function generator **off** and S_1 **open,** connect the circuit of Figure 34–9.

B2. Turn **on** the generator. Set the frequency to 1 kHz. (The oscilloscope can be used to verify the value.) With an ac voltmeter connected across its output, adjust the sine-wave output to 5 V (or its highest output voltage, but pick an even value to keep calculations simple). Record the value in Table 34–2 (p. 265) in the "Voltage, rms, Measured" column.

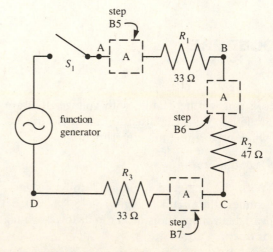

Figure 34–9. Circuit for procedure step B1. The ac ammeter is added in steps B5, B6, and B7.

B3. With the voltmeter measure the rms voltage across each resistor, R_1, R_2, and R_3. Record the values in Table 34–2 in the "Voltage, rms, Measured" column.

B4. Using the oscilloscope, measure the peak voltage across the function generator, R_1, R_2, and R_3. Record the values in Table 34–2 in the "Voltage, peak, Measured" column. **Open** S_1.

B5. Break the circuit between A and B and insert an ac ammeter as in Figure 34–9 to measure the current in R_1. **Close** S_1. Record the ammeter reading in Table 34–2 in the "Current, rms, Measured" column. **Open** S_1.

B6. Break the circuit between B and C and insert an ac ammeter as in Figure 34–9 to measure the current in R_2. **Close** S_1. Record the ammeter reading in Table 34–2 in the "Current, rms, Measured" column. **Open** S_1.

B7. Break the circuit between C and D and insert an ac ammeter as in Figure 34–9 to measure the current in R_3. **Close** S_1. Record the ammeter reading in Table 34–2 in the "Current, rms, Measured" column. **Open** S_1; turn **off** the function generator.

B8. Calculate the rms voltage across each resistor. Record your answers in Table 34–2 in the "Voltage, rms, Calculated" column.

B9. Calculate the peak voltage across each resistor. Record your answers in Table 34–2 under the "Voltage, peak, Calculated" column.

B10. Calculate the rms current in each resistor. Record your answers in Table 34–2 under the "Current, rms, Calculated" column.

ANSWERS TO SELF TEST

1. amplitude; polarity
2. sine wave
3. true
4. 17.7
5. 7.07
6. 4.50
7. 319
8. 451
9. rms
10. oscilloscope

Optional Activity ·

Obtain and study the instruction or operating manuals for the meters used in this experiment. Determine the DMM's frequency range and, if used, the frequency range of the ac ammeter. Determine each of the meter's ability to accurately measure nonsinusoidal waveforms. Write a report on the practical applications of these meters.

EXPERIMENT 34

TABLE 34–1.   60-Hz Voltage and Current

	Voltage, rms, V		Voltage, peak, V		Current, rms, mA	
	Measured	Calculated	Measured	Calculated	Measured	Calculated
Variable transformer output						
R_1						
R_2						
R_3						

TABLE 34–2.   1 kHz Voltage and Current

	Voltage, rms, V		Voltage, peak, V		Current, rms, mA	
	Measured	Calculated	Measured	Calculated	Measured	Calculated
Function generator output						
R_1						
R_2						
R_3						

QUESTIONS

1. Explain, in your own words, the relationship between peak, rms, and average values of an ac voltage.

_____

_____

_____

_____

2. Refer to your data in Table 34–1. Compare the measured peak voltage with the calculated value. Explain any differences.

_____

_____

_____

_____

3. When using the oscilloscope to measure the voltage drops of Figure 34–9, which measurements required a differential measurement technique and explain why.

_____

_____

4. Formula 34–3 can be used to find V_{rms} when V_M is known:

$$V_{rms} = 0.707V_M$$

The constant 0.707 is, therefore, equal to the ratio V_{rms}/V_M. Using your measured data in Table 34–1 for V_{rms} and V_M, calculate the average value of V_{rms}/V_M for the three resistor measurements. Compare your answer with the constant 0.707; explain any difference. (Show all calculations.)

_____

_____

_____

_____

_____

_____

_____

_____

_____

5. Repeat the process of Question 4 using your data in Table 34–2. Compare your answer with the constant 0.707; explain any difference. Also compare your calculated value with your value for Question 4. Explain any difference.

_____

_____

_____

_____

_____

_____

_____

_____

_____

_____

6. Do the data in the current column of Tables 34–1 and 34–2 verify the application of Ohm's law to resistive ac circuits? Refer to data to substantiate your answer.

_____

_____

_____

_____

_____

_____

_____

CHARACTERISTICS OF INDUCTANCE

BASIC INFORMATION

Inductance and Reactance of a Coil

Resistors limit the amount of current in dc as well as in ac circuits. In addition to resistors, reactive components, such as inductors and capacitors, impede currents in ac circuits.

In the circuit of Figure 35–1 a sine wave of voltage V produces a sine-wave current through the coil. This changing current creates a varying magnetic field about the coil. The periodic buildup and collapse of the magnetic field causes magnetic flux to cut across the windings of the coil, inducing a voltage V' in these windings. By Lenz' law this voltage is in opposition to the source voltage V and can therefore be termed a *counter emf*. When V is positive, V' is negative but less than V. The numerical value of the resultant emf that powers the circuit is $|V| - |V'|$, where $|V|$ and $|V'|$ are the absolute values of V and V', respectively. Because of the counter emf the alternating current in the coil is less than it would be if V were a dc source and a steady-state current were flowing. This property of the coil that opposes a changing current is termed the *inductance* of the coil.

The unit of inductance is the *henry* (abbreviated H), named in honor of the American physicist Joseph Henry. The number of henrys in an inductor can be measured by means of an instrument called an *inductance bridge*, LC meter, or Z-meter. The symbol used to denote inductance is L. The amount of opposition offered by an inductor is called *inductive reactance* and is denoted by the symbol X_L. Inductive reactance is measured in ohms.

The inductive reactance of a coil is not constant but is a variable quantity. It depends on the inductance L and the frequency f of the voltage source. Inductive reactance may be calculated from the formula

$$X_L = 2\pi fL \qquad (35-1)$$

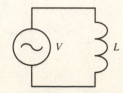

Figure 35–1. The sine wave of voltage will produce a current sine wave in the inductance, L.

OBJECTIVES

1 To observe experimentally the effect of an inductance on current in a dc and ac circuit

2 To verify experimentally the formula for inductive reactance:

$$X_L = 2\pi fL$$

3 To measure phase shift with an oscilloscope

where

- π is the constant 3.14,
- f is the frequency in hertz,
- L is the inductance in henrys.

Some important facts may be deduced from formula (35–1). The first is that X_L varies directly with frequency—that is, it is directly proportional to f. This variation may be shown graphically by a specific example.

Problem. Draw a graph of X_L versus f for an inductor whose inductance is $L = 1.59$ H.

Solution. The inductive reactance X_L is found for each of a series of frequencies. Thus, when $f = 0$ Hz,

$$X_L = 2\pi f L = 2(3.14)(0)(1.59) = 0 \ \Omega$$

When $f = 100$ Hz,

$$X_L = 6.28(100)(1.59) = 1 \text{ k} \ \Omega$$

Similarly, calculations show that when

$$f = 200 \text{ Hz, } X_L = 2 \text{ k}\Omega$$

$$f = 300 \text{ Hz, } X_L = 3 \text{ k}\Omega$$

$$f = 400 \text{ Hz, } X_L = 4 \text{ k}\Omega$$

$$f = 500 \text{ Hz, } X_L = 5 \text{ k}\Omega$$

Figure 35–2 shows the variation of X_L versus f for $L = 1.59$ H. It is a straight-line graph, so we call the relationship between X_L and f a *linear* one.

A second fact that may be seen from formula (35–1) is that for direct current, that is, for $f = 0$, $X_L = 0$. This fact is consistent with the definition of inductance, as that characteristic of a coil that opposes a *change* in current. A steady direct current produces no change. Inductance has no effect on a direct current with a constant voltage; $X_L = 0$ for such a direct current.

Formula (35–1) also defines the relationship between X_L and L. If f is kept constant, X_L increases or decreases as L is increased or decreased, respectively. Again, if X_L is plotted against L, a straight line results.

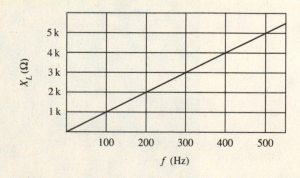

Figure 35–2. As frequency increases, inductive reactance increases.

Resistance of a Coil

An inductor consists of a number of turns of insulated wire wound around a core. There may be only several turns, or there may be thousands of turns. The more turns there are on a particular core, the greater is the inductance of the coil. The diameter of the wire used in winding a coil depends on the maximum current the coil will be required to carry. The larger the diameter of the wire, the greater the current capabilities of the coil. A smaller-diameter wire will carry less current. If more current is permitted to flow in a coil than that for which it is rated, the coil will overheat and may destroy the insulation around the windings, thus shorting windings together. Or an overheated coil may burn out, that is, the wire may open. In either event the coil will become defective.

Many circuits require relatively low currents. Very fine diameter wire is normally used in the construction of the coils used in these circuits. Since the resistance of a wire is directly proportional to its length, the resistance of inductors consisting of many turns of fine wire will be relatively high. Thus, the characteristics of inductance *and* resistance are associated with an inductor. Although the resistance of a coil cannot be separated from the inductance, a schematic symbol of a coil with inductance L and resistance R (Figure 35–3) shows these two characteristics as though they were separate and acting in series. The resistance can be measured with an ohmmeter. For practical reasons in solving circuit problems a coil is usually represented by its series-equivalent form (Figure 35–3).

The resistance associated with the coil suggests that one method of troubleshooting an inductor is to place an ohmmeter across the two terminals and compare the measured with the rated resistance. If they are the same, it may be assumed that the winding is continuous—that is, it is neither open nor shorted. If the resistance is infinite, the inductor is open. If the resistance is very much lower than the rated value, say, 20 Ω instead of 500 Ω, it may be assumed that the coil winding is shorted. In practice, resistance checks of an inductor are made to determine if it is continuous or open. Special devices are used to check coils for shorted turns.

Measurement of X_L

The inductive reactance of a coil (X_L) may be determined by measurement. In the circuit of Figure 35–4(a) a sinu-

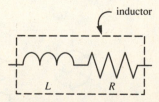

Figure 35–3. Schematic series-equivalent form of an inductor shows separate symbols for the resistive and inductive characteristics.

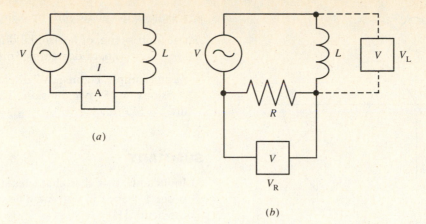

(a)

(b)

Figure 35–4. Circuits used to determine X_L.

soidal voltage V causes a current I to flow. Ohm's law for ac circuits states that the voltage V_L across the coil is $V_L = I \times Z$. The symbol Z stands for *impedance*. This is the term used to express the total opposition to current in ac circuits. If the R of the coil is low compared with X_L, $Z = X_L$ and

$$V_L = I \times X_L \qquad (35\text{–}2)$$

This formula may be written in the form

$$X_L = \frac{V_L}{I} \qquad (35\text{–}3)$$

Formula (35–3) can be used to solve for X_L of a coil if frequency f is given. This is accomplished using an ac voltmeter to measure the voltage V_L across L and measuring the current I with an ac ammeter. The measured values of V_L and I are then substituted in formula (35–3), and X_L is calculated.

Note that a slight error is introduced by this procedure, for L has resistance in addition to inductance. However, if the value of X_L is appreciably higher than R of the coil, $(X_L) > (10R)$ may be ignored. This is usually considered when X_L is greater than $10R$.

An alternative method for measuring the current I does not require the use of an ammeter. Instead of the meter A, a resistor R of known value is placed in series with L, as in Figure 35–4(b). In a series circuit, current is the same everywhere. Hence, the current in R is the same as the current in L. The voltage V_R is then measured across R. Current is next determined by substituting the measured value of V_R and the known value of R in the formula

$$I = \frac{V_R}{R} \qquad (35\text{–}4)$$

Next, V_L is measured and X_L calculated by substituting the values of V_L and I in formula (35–3).

Phase Relationships

Ohm's law states that as the voltage across a resistor varies, the current through the resistor varies directly. That is, if the voltage is doubled, the current doubles; if the voltage decreases by a third, the current decreases by a third. This is true whether the source is dc or ac.

In the case of a sinusoidal alternating current, at any instant the voltage and current obey Ohm's law, and at the next instant the change in voltage is accompanied by a like change in current. When the instantaneous voltage is zero, the instantaneous current is zero. When v is maximum, i is maximum (where v and i are the instantaneous values of voltage and current, respectively). Figure 35–5 shows this relationship graphically. In the case of ac resistive circuits, we say the voltage and current are *in phase*.

When an inductor is connected in an ac circuit, the voltage across the inductor and the current through the inductor are not in phase. For a pure inductor (that is, a device having only inductance) the voltage and current are out of phase by 90°. By convention we say the current *lags* the voltage by 90°. For example, when the voltage reaches its maximum instantaneous value, the current is zero. Not until one-fourth time period later will the current reach its maximum value, and by that time the voltage is equal to zero. In this experiment the oscilloscope will be used to show graphically, as in Figure 35–6 (p. 270), how the voltage and current sine waves are shifted in time.

In a circuit composed of both resistance and inductance, the degree of phase shift depends on the amount

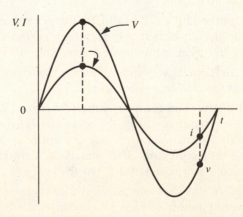

Figure 35–5. In a resistive circuit, voltage and current are in phase.

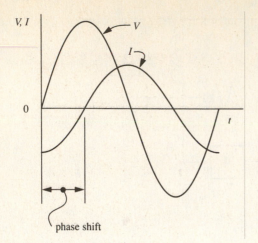

Figure 35–6. In a nonresistive circuit, voltage and current are not in phase.

of resistance as compared to the amount of inductive reactance. If the resistance is 10 times the inductive reactance, the X_L, the phase angle will be 6°. If X_L is 10 times the resistance value, the phase angle will be 84°. If the reactive component is equal to the resistive component, then the phase angle is 45°.

Phase Measurement

Phase measurements can be made with a dual trace oscilloscope when the signals are of the same frequency. To make this measurement, the following procedure can be used:

1. Preset the scope's controls and obtain a baseline trace. Set the Trigger Source to whichever input is chosen to be the reference input. Channel 1 is often times used as the reference, but Channel 2 as well as External Trigger or Line could be used.

2. Set both Vertical Input Coupling switches to the same position, depending on the type of input.

3. Select both Vertical MODE; then select either ALT or CHOP, depending on the input frequency.

4. Although not necessary, set both Volts/Div. and both Variable controls so that both traces are approximately the same height.

5. Adjust the TRIGGER LEVEL to obtain a stable display. Typically set so that the beginning of the reference trace begins at approximately zero volts.

6. Set the Time/Div. switch to display about one full cycle of the reference waveform.

7. Use the Position controls, Time/Div. switch, and Variable time control so that the reference signal occupies exactly 8 horizontal divisions. The entire cycle of this waveform represents 360° and each division of the graticule now represents 45° of the cycle.

8. Measure the horizontal difference between corresponding points of each waveform on the horizontal graticule line as shown in Figure 35–7.

9. Calculate the phase shift by using the formula:

Phase Shift = (no. of horizontal difference divisions) × (no. of degrees per division)

As an example, Figure 35–7 displays a difference of 0.6 divisions at 45° per division. The phase shift = (0.6 div.) × (45°/div.) = 27°.

SUMMARY

1. Inductance L is that characteristic of a coil that opposes a change in current. The unit of inductance is the henry (H).

2. The amount of opposition that an inductance offers to a changing current is called inductive reactance (X_L). Inductive reactance is measured in ohms.

3. The inductive reactance of a coil is not a constant but is directly proportional to the frequency f of the changing current. It is also directly proportional to the value of the inductance L.

4. Inductive reactance may be calculated from the formula

$$X_L = 2\pi f L$$

where

X_L is in ohms,

f is in hertz,

L is in henrys.

5. Every coil has some resistance R associated with it. Only the resistance of an inductance will limit direct current in a coil. The inductance of a coil will not affect direct current.

6. The continuity of a coil winding may be determined by measuring the resistance of the coil. If the resistance measured is infinite, the coil winding is open.

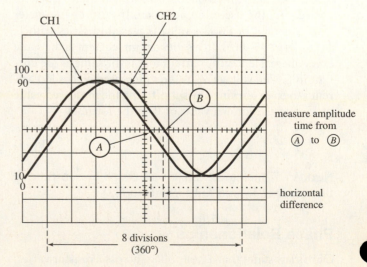

Figure 35–7. Oscilloscope phase shift measurement.

7. The ac voltage across a coil V_L is equal to the product of the alternating current in the coil and the inductive reactance of the coil; that is

$$V_L = I \times X_L$$

8. If the ac voltage V_L across a coil and the alternating current I in the coil are known, the inductive reactance of the coil may be calculated from the formula

$$X_L = \frac{V_L}{I}$$

This formula assumes that the resistance R of the coil is low compared with the reactance of the coil.

9. A dual trace oscilloscope can be used to measure the phase difference or displacement of the voltage and current.

SELF TEST

Check your understanding by answering the following questions:

1. (True/False) Inductance of a coil is measured in ohms. _____

2. The _____ of a coil is directly _____ to the _____ of the voltage source, if the L of the coil remains constant.

3. If $L = 2.5$ H and $f = 1$ kHz, $X_L = $_____ Ω.

4. If the frequency f is doubled while L is held constant, X_L is _____.

5. The graph of X_L versus f is a _____ if L is held constant.

6. The resistance R of a coil is associated with the resistance of the _____ of the coil.

PROCEDURE

A. Effect of Inductance on Direct and Alternating Current

A1. Measure the resistance of the large choke (7–10 H). Record the value in Table 35–1 (p. 275). Also record the value of the resistor equal to the dc resistance of the choke.

A2. With the power supply **off**, S_1 **open**, and S_2 in the Ⓐ position, connect the circuit of Figure 35–8 (p. 272). Adjust the variable power supply V_{PS} to its lowest value. Set the meters to read dc voltage and current.

A3. Power **on; close** S_1. Increase V_{PS} until the ammeter A reads 20 mA. Do not change this voltage set-

7. If the ac voltage V_L across a coil is 6 V and the current I in the coil is 0.02 A, then $X_L = $_____ Ω.

8. If $f = 60$ Hz and $X_L = 1$ kΩ, then $L = $_____ H.

MATERIALS REQUIRED

Power Supplies:
- 120-V, 60-Hz source
- Variable-voltage autotransformer (Variac or equivalent)
- Isolation transformer
- Variable 0–15 V dc, regulated

Instruments:
- Oscilloscope
- Variable-frequency function generator
- DMM
- AC ammeter (25-mA range)
- DC ammeter (25-mA range)

Resistors:
- 1 3.3 kΩ (½-W, 5%)
- 1 resistor equal to the dc resistance of the choke

Inductors:
- 1 large choke (7–10 H) (Magnetek-Triad #C8X 7-H, 240-Ω, 75-mA; Magnetek-Triad #C3X 10-H, 500-Ω, 50-mA; or equivalent)
- 1 100-mH choke

Miscellaneous:
- SPDT switch
- SPST switch
- Polarized line cord set with on-off switch and fuse

ting, because it will be used in the next step. Record the voltage in Table 35–1 under "Direct-Current Source."

A4. Throw S_2 to position Ⓑ. If necessary, adjust the voltage to equal that in step A3. Measure the current through the inductance and record this value as well as the voltage in Table 35–1 under "Direct-Current Source." **Open** S_1; turn V_{PS} **off**.

A5. With the line cord unplugged, the on-off switch **off**, S_1 **open** and S_2 in position Ⓐ, connect the circuit of Figure 35–9 (p. 272). The choke and resistor are the same as used in the circuit of Figure 35–8. Set the

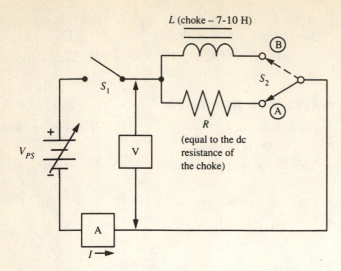

Figure 35–8. Circuit for procedure step A2.

meters to measure ac voltage and current. Adjust the autotransformer to its lowest value.

A6. Plug the line cord into a 120-V source and turn the line cord switch **on. Close** S_1. Increase the voltage of the autotransformer until the ammeter reads 20 mA. Maintain the setting of the autotransformer for the next step. Record the reading of the voltmeter under "Alternating-Current Source" in Table 35–1.

A7. Throw S_2 to position Ⓑ. If necessary, adjust the voltage to equal that of step A6. Measure the current through the inductance and record this value and the voltage in Table 35–1 under "Alternating-Current Source" column.

A8. Increase the voltage of the autotransformer until the ammeter reads 25 mA. Record the voltage in Table 35–2 (p. 275). Decrease the voltage in steps and

record it at 20 mA, 15 mA, and 10 mA in Table 35–2. **Open** S_1; turn line cord switch **off**; unplug cord.

A9. Calculate V/I for each voltage and current in Table 35–2. Record your answers in the table. Calculate the average V/I and record your answer in the table. Calculate X_L for the inductance of the choke you are using and 60-Hz frequency. Record your answer in the table.

B. Effect of Frequency on Inductance

B1. With the Function Generator in the off position, connect the circuit of Figure 35–10.

B2. Examine Table 35–3. Using the oscilloscope, you will take a series of measurements of peak-to-peak voltage around the circuit at the frequencies noted in the table.

B3. Turn **on** the Function Generator and adjust the sine-wave output to 10 V p-p. (If the generator cannot produce a voltage that high, choose as high a voltage as possible but one that will make calculations simple.)

B4. Complete Table 35–3 by measuring and recording the output voltage of the generator, the voltage across the inductance, and the voltage across the resistor.

B5. Calculate the current in the circuit for each frequency using the formula

$$I = \frac{V_R}{R}$$

where

    I is the current in the circuit,

    V_R is the voltage across the resistor,

    R is the measured resistance of the resistor.

Record your answers in Table 35–3.

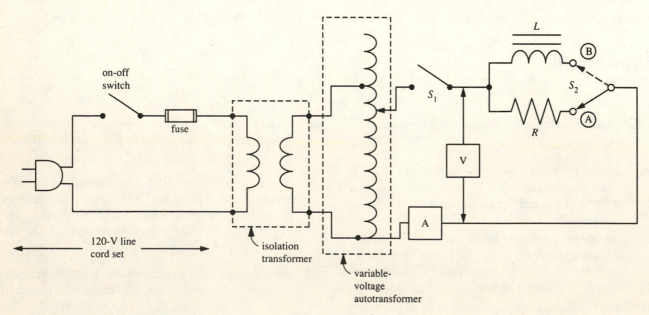

Figure 35–9. Circuit for procedure step A5.

272   EXPERIMENT 35

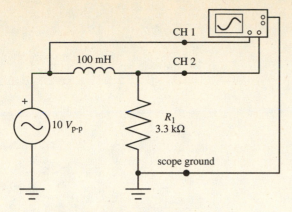

Figure 35–10. Circuit for procedure step B1.

B6. Calculate the inductive reactance for each of the frequencies, using your measured values for the voltage across the inductor V_L, the voltage across the resistor V_R, and the measured resistance of the resistor R. Since

$$X_L = \frac{V_L}{I_L} \text{ and } I_L = I_R = \frac{V_R}{R}$$

$$X_L = \frac{V_L}{\frac{V_R}{R}} = \frac{V_L}{V_R} \times R$$

where

X_L is inductive reactance,

V_L is the voltage across the inductance,

I_L is the current in the inductance.

Record your answers in Table 35–3.

B7. Calculate the inductive reactance of the inductor using the formula

$$X_L = 2\pi f L$$

where

f is the frequency of the generator,

L is the rated value of the inductor (100 mH in this experiment).

Record your answers in Table 35–3.

C. Effect of Inductance on Phase Shift

C1. Use the circuit of Figure 35–10 to measure and observe phase shift. Establish Channel 1, connected at the input, as your reference point.

C2. Set the Function Generator for an output of 10 $V_{\text{P-P}}$ at 2 kHz. Measure the phase shift of the output voltage across R_1, with reference to the input, with Channel #2. The output voltage is a result of the current flowing through R_1. Record your measurement in the provided space marked "Phase Shift Measurement."

C3. Repeat step C2, using an input frequency of 10 kHz. Record your measurement as in step C2.

ANSWERS TO SELF TEST

1. false
2. inductive reactance; proportional; frequency
3. 15.7k
4. doubled
5. straight line
6. winding
7. 300
8. 2.65

Name _____ Date _____

TABLE 35–1. Effect of Inductance on Direct and Alternating Current

Resistance (Measured), Ω	Position of S_2	Component in Circuit	Direct-Current Source		Alternating-Current Source	
			Voltage, V	Current, mA	Voltage, V	Current, mA
Resistor	Ⓐ	R		20 mA		20
Choke	Ⓑ					

TABLE 35–2. Inductive Reactance X_L at 60 Hz

Current, mA	25 mA	2.0 mA	15 mA	10 mA	Average $\frac{V}{I}$ at Line Frequency, Ω
Voltage, V					
$\frac{V}{I}$ Calculated					
Choke X_L Calculated (rated)					

TABLE 35–3. Effect of Frequency on Inductance

Frequency f, Hz	Sine-Wave Generator Voltage, V, $V_{p\text{-}p}$	Voltage across Inductance V_L, $V_{p\text{-}p}$	Voltage acsoss Resistor V_R, $V_{p\text{-}p}$	Current I Calculated, mA	Inductive Reactance X_L, Ω	
					$\frac{V_L}{V_R} \times R$	$2\pi f L$
2						
3						
5						
7						
8						
9						
10						

Phase Shift Measurement: Phase Shift @ 2 kHz _____   Measured resistance of Inductor _____
Phase Shift @ 10 kHz _____

QUESTIONS

1. Explain, in your own words, the effect of inductance on current in a dc circuit. Is the effect the same in an ac circuit? If not, explain the difference.

_____

_____

_____

_____

2. Refer to your data in Table 35–1. Is there any significant difference between the effect of L and R on current in a dc circuit as compared to the effect of L and R on current in an ac circuit? Cite specific data in your table.

_____

_____

_____

_____

3. Refer to your data in Table 35–2. Is X_L the same for each calculated value of V/I? If not, explain any differences.

_____

_____

_____

_____

4. Refer to your data in Table 35–3. Are the calculated values of I peak values or rms values? Explain.

_____

_____

_____

_____

5. On a separate 8½ × 11 sheet of graph paper, plot a graph of current I_L through the inductance versus the voltage V_L across the inductance using your data in Table 35–2. The horizontal (x) axis should be V_L and the vertical (y) axis should be I_L. Label the graph and the axes.

6. On a separate 8½ × 11 sheet of graph paper (do not use the sheet from Question 5), plot a graph of frequency f versus inductive reactance X_L using your data in Table 35–3. Choose an appropriate scale to show the relationship between X_L and f. The horizontal axis should be frequency and the vertical axis should be inductive reactance.

TRANSFORMER CHARACTERISTICS

INTRODUCTORY INFORMATION

Ideal Transformer

A transformer is a device for coupling ac power from a source to a load. The source of power is connected to the primary winding, the load to the secondary winding. In the process, certain transformations take place which are related to the construction, and to the primary to secondary turns ratio, of the transformer.

Transformers find many uses in electronics. There are power transformers, impedance matching transformers, audio transformers, radiofrequency transformers, and the like.

A conventional transformer consists of two or more windings on a core, magnetically coupled (Figure 36–1). An ac voltage applied across the input or primary winding causes current in the primary. This sets up an expanding and collapsing magnetic field which cuts the turns of the secondary winding, inducing an ac voltage in the secondary. When a load is connected across the secondary, current flows in the load.

The core around which the primary and secondary are wound may be iron, as in the case of low-frequency power and audio transformers. An air core may be employed for coupling higher-frequency circuits. The core material and the geometry of the windings determine the characteristics of coupling.

An alternating current in a winding of an iron-core transformer magnetizes the core first in one direction and then in the other. It is this moving magnetic field in the core which cuts the windings of the other coil, inducing a voltage in it.

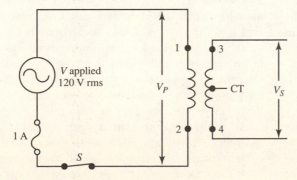

Figure 36–1. Stepdown transformer without load.

In an ideal iron-core transformer, that is, one in which there are no power losses, 100 percent of the source (primary) power would be delivered to the load. Formula (36–1) shows this power relationship.

$$V_p \times I_p = V_s \times I_s \tag{36-1}$$

Here V_p and I_p are the primary voltage and current, respectively, while V_s and I_s are the secondary voltage and current. In a lossless transformer, the ratio between the primary voltage and the voltage induced in the secondary is the same as the ratio (a) between the number of turns N_p of the primary and number of turns N_s of the secondary.

$$\frac{V_p}{V_s} = \frac{N_p}{N_s} = a \tag{36-2}$$

If $a = 1$, there are as many turns in the primary as there are in the secondary, and the voltages appearing across the primary and secondary are equal. This type of 1:1 transformer is called an *isolation transformer.*

If a is greater than 1, a lower voltage appears across the secondary than across the primary. This is called a voltage stepdown transformer.

If a is less than 1, a higher voltage appears across the secondary than across the primary. This is a voltage stepup transformer.

Formula (36–1) for an ideal transformer may also be written as

$$\frac{V_p}{V_s} = \frac{I_s}{I_p} \tag{36-3}$$

From Eqs. (36–2) and (36–3) we have

$$\frac{V_p}{V_s} = \frac{I_s}{I_p} = a \tag{36-4}$$

The last formula states that current and voltage in the windings of a transformer are inversely related. Therefore, a voltage stepup transformer is also a current stepdown transformer, whereas a voltage stepdown is a current stepup transformer.

As shown in Figure 36–1, this transformer has a center-tap, (CT), connected to its secondary. The center-tap provides the ability for the transformer to output ½ of the normal secondary output voltage. When the transformer's secondary load is connected from 3 to CT or CT to 4, the transformer's effective turns ratio has been changed. The split secondary also allows the transformer's secondary to produce two voltages equal in amplitude, but 180° out of phase in respect to the CT.

Power Losses in a Transformer · · · · · · · · ·

The ideal transformer does not exist because there are power losses which do not permit 100 percent transfer of power from the source to the load. One such loss is related to the resistance of the windings and is the I^2R or heating loss. Thus, there are I^2R losses associated with the

resistance of the primary winding (primary loss) and also with the resistance of the secondary (secondary loss). When there is no load on the secondary winding, there is no current and hence no power loss in the secondary. However, there is current and hence some I^2R loss in the primary.

Eddy currents are present in the core of an iron-core transformer. A circulating eddy current is induced in the iron core by the changing magnetic field. Eddy currents heat the transformer and act like an I^2R loss. Eddy currents thus rob the source of power and represent another power loss.

Another loss associated with a transformer core is *hysteresis loss*. The hysteresis effect results from the fact that magnetic lines of force lag behind the magnetizing force that causes them. Hysteresis can be understood by considering the fact that when a magnetizing force is removed from an iron-core magnetic circuit, a portion of the flux remains within the iron. This residual magnetism can be removed by applying to the iron a magnetizing force opposite in direction to that of the initial force. The energy required to demagnetize the iron acts as a core loss, which is associated with the reversal of magnetizing current in the winding.

One other loss must be mentioned. This is associated with the magnetic leakage which exists in a transformer. Not all the lines of magnetic force will link the turns of the secondary winding. The lines of force thus lost to the magnetizing circuit constitute magnetic leakage.

Effect of Load Current on Primary Current ·

Primary current in the transformer of Figure 36–2 depends on the load current in the secondary, as is evident from the formula

$$\frac{I_s}{I_p} = a = \frac{N_p}{N_s}$$

or

$$I_p = \frac{I_s}{a} \tag{36-5}$$

Therefore, as load current I_s increases due to a decrease in load resistance R_2, primary current must increase. The increase in primary current is explained by the assumption that the change in load impedance (as is evidenced by an increase in load current) appears also as a reflected impedance in parallel with the primary. Since power in the primary and secondary was assumed to be equal, and since power can be dissipated only in a resistance, the reflected impedance must be the resistance R_1 (Figure 36–2).

In addition to the primary current I_p resulting from the reflected load impedance, the primary supplies the current for the iron-core losses and the magnetizing current. The phasor sum of these two currents (the magnetizing current is out of phase with the voltage) is called

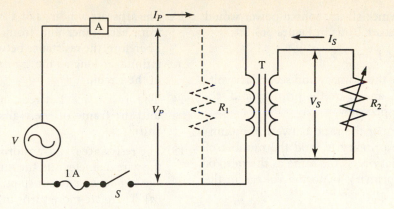

Figure 36–2. Primary current I_p increases when secondary current I_s increases, due to reflected impedance (R_1) from the secondary to the primary.

the "exciting current." The exciting current is about 3 to 5 percent of the rated output of the transformer. This explains why primary current will normally measure more than that predicted by Formula (36–5). Note that when a vector is used to represent time-varying voltages and currents the term phasor is used. Phasors are discussed more fully in subsequent experiments.

Resistance-Testing Transformer Windings

Resistance (ohmmeter) tests of the individual windings of the small transformers used in electronics are used to determine the continuity of the windings. An ohmmeter test also establishes the resistance of each winding. The technician then compares the measured resistance with the rated value to determine if a suspected transformer is in fact defective.

The following considerations may be helpful in analyzing the results of resistance measurements.

Winding Measures Infinite Resistance. This winding is "open." The break may be at the beginning or end of the winding, where the connection is made to the terminal leads. This type of break can be easily repaired by resoldering the leads to the winding. If the discontinuity is elsewhere, the transformer must be replaced.

Winding Resistance "Very" High. A winding whose resistance measures "very" high compared with its rated value may be open, or there may be a cold-solder joint at the terminal connections. If the condition cannot be corrected, the transformer must be replaced.

Winding Resistance "Very" Low Compared with Rated Value. Turns of the winding must be "shorted" somewhere on the transformer, or the winding may be shorted to the frame. However, a small difference in resistance between the rated and measured values may be insignificant. For example, if a primary rated at 120 Ω measures 100 Ω, the difference may be attributable to the

inaccuracy of the meter. In cases of doubt, other tests (which will not be discussed here) are indicated.

Resistance between Windings. The windings of transformers, other than autotransformers, are insulated from each other. There should be infinite resistance between insulated windings, as long as the transformer is not connected in a circuit. If the insulation between two windings breaks down there will be a measurable resistance between these windings, signifying a defective transformer.

Factors which Determine Resistance of a Winding. The resistance of a winding depends on the diameter of the wire and the number of turns. Thus, resistance varies inversely as the square of the diameter and directly as the number of turns. Large diameters are required for high-current windings, smaller diameters for windings which carry less current. The primary of a voltage stepdown transformer has more windings than the secondary. Moreover, a voltage stepdown is also a current stepup transformer. Hence the secondary must carry more current than the primary. Therefore, the resistance of the primary will be higher than that of the secondary. How much higher depends on the transformer ratio and on the diameter of the two wires.

SUMMARY

1. A transformer consists of two or more windings on a core. The core is iron for low-frequency applications or air for higher frequencies.

2. When an ac source is applied across the primary winding, a voltage is induced in the secondary. If there is a load on the secondary, current also flows in that winding.

3. A transformer is therefore a device which couples ac power from a source connected to the primary, to a load connected to the secondary winding.

4. In an ideal transformer all the source power would be delivered to the load. Thus in Figure 36–2

$$V_p \times I_p = V_s \times I_s$$

where V_p and V_s are the primary and secondary voltages, respectively; I_p and I_s are the primary and secondary currents, respectively.

5. In an ideal transformer the ratio between the number of turns in the primary N_p and the number of turns in the secondary N_s is the same as the ratio of the voltage in the primary V_p to the voltage in the secondary V_s. Thus

$$\frac{N_p}{N_s} = a = \frac{V_p}{V_s}$$

6. In an ideal transformer the ratio of current in the secondary I_s to current in the primary I_p is also a. Thus

$$\frac{N_p}{N_s} = a = \frac{I_s}{I_p}$$

7. A voltage stepdown transformer is one in which the secondary voltage is lower than the primary. A voltage stepdown transformer is also a current stepup transformer. In this type of transformer $a > 1$.

    NOTE: The symbol $>$ means greater than; $<$ means less than.

8. If $a < 1$, the transformer is a voltage stepup, current stepdown device.

9. The power losses which occur in a transformer include:
    (a) I^2R or heating losses due to the resistance of the windings.
    (b) Eddy current losses present in an iron-core transformer.
    (c) Hysteresis losses, which result from the lag between the magnetic lines of force and the magnetizing force which induces them.
    (d) Magnetic leakage, which occurs because not all the magnetic lines of force link the turns of the secondary winding.

10. An increase in I_s will result in an increase in I_p and a resulting decrease in primary impedance. The decrease in primary impedance is caused by the impedance reflected from the secondary into the primary. The reflected impedance is the resistance R_1 (Figure 36–2) which acts in parallel with the primary winding.

11. The primary current supplies the power losses associated with a transformer. Therefore, because there are losses associated with every transformer, $I_p > I_s/a$.

12. Ohmmeter tests of the resistance of individual windings of a transformer are used to determine the continuity of the windings.

13. The separate windings of a transformer are insulated from each other and from the transformer frame. Therefore, the resistance between windings should be infinite, as long as the transformer is *not* connected in the circuit.

14. The resistance between an individual winding and the frame of the transformer should also be infinite.

15. The resistance of a transformer winding depends on the wire diameter and the number of turns. In a voltage stepdown, current stepup transformer:
    (a) There are more turns in the primary than in the secondary.
    (b) The secondary wire diameter is larger than that of the primary wire.
    (c) The resistance of the secondary is lower than the primary.

SELF TEST

Check your understanding by answering these questions:

1. In a particular ideal transformer, the primary has 10 times as many turns as the secondary. The voltage across the primary is 120 V. The voltage across the secondary must be _____ V.

2. In the transformer of question 1, the current in the secondary is 3 A. The current in the primary must therefore be _____ A.

3. In the transformer of question 1, the turns ratio N_p/N_s is _____.

4. There is no *ideal* transformer for all practical purposes. _____ (true/false)

5. The power losses in a transformer include those due to:
    (a) _____ of the windings
    (b) _____
    (c) _____
    (d) _____

6. An increase in secondary current causes a(n) _____ (increase/decrease) in primary current.

7. The impedance reflected from the secondary into the primary is a _____ (resistance/inductance).

8. The power losses in a transformer are supplied by the _____.

9. In an iron-core power transformer which is not connected in a circuit, the resistance measured between the primary and secondary windings is 10 Ω. This resistance _____ (is/is not) normal.

10. The primary and secondary windings must be _____ in the transformer in question 9.

MATERIALS REQUIRED

Power Supplies:
- Isolation transformer
- Variable-voltage autotransformer (Variac or equivalent)

Instruments:
- DMM
- Multirange ac Milliammeter

Resistors:
- 1 each 100 Ω, 75 Ω, 50 Ω, and 25 Ω Power Resistors (25 W or above) or Decade Power Rheostat

Transformer:
- 120V primary to 25.2 V secondary with CT

Miscellaneous:
- SPST Switch; Fused line cord

PROCEDURE

....

Transformer Turns Ratio, a:

1. Connect the circuit of Figure 36–1.

   CAUTION: Be certain that the line voltage (120 V ac) is isolated and applied to the *primary* of the voltage stepdown transformer. The secondary is not loaded.

2. Measure the voltage applied to the primary, V_p, and each secondary voltage (3 to 4, 3 to CT, and 4 to CT). Record your data in Table 36–1 (p. 283).

3. Calculate and record in the table, each of the turns ratios a.

Effect of Load on Primary Current

4. Connect an ac milliammeter A in the primary circuit as in Figure 36–2. The ac milliammeter should be set on the highest range, then adjusted to the proper range. V is the 120 V of the 60-Hz line. R_2 is a decade rheostat acting as the load resistor.

5. Measure and record the primary current I_p, the voltage across the primary V_p, and the voltage across the full secondary V_s, in turn for each value of load resistance R_2 shown in Table 36–2. A DMM is used to measure V_p and V_s.

6. Calculate and record the secondary current ($I_s = V_s/R_2$). Substitute the value of I_s in the equation $I_{p1} = I_s/a$. Calculate and record I_{p1}. (The calculated value I_{p1} is the current which would appear in the primary of an ideal transformer.)

7. Subtract I_{p1} from I_p (the current measured in the primary) and record in Table 36–2 under column labeled "ΔI."

8. Calculate and record the power P_s dissipated by the load resistor R_2 for each value of R_2. Also calculate the transformer's power efficiency when R_2 is 100 Ω and 25 Ω.

$$\% \text{ efficiency} = \frac{P_{out}}{P_{in}} \times 100.$$

   CAUTION: In all cases, be certain that the wattage rating of R_2 is high enough for the application.

Resistance Measurement of the Windings

9. Remove power and disconnect the transformer primary and secondary from the circuit.

10. Measure the resistance of the primary and secondary with an ohmmeter, and record in Table 36–3 (p. 283). Also measure and record the resistance between windings and between each winding and the frame.

ANSWERS TO SELF TEST

1. 12
2. 0.3
3. 10:1`
4. true
5. (a) resistance or heating; (b) eddy current; (c) hysteresis; (d) magnetic leakage
6. increase
7. resistance
8. input power source
9. is not
10. shorted

EXPERIMENT

36

TABLE 36–1. Transformer Turns Ratio

V_p, V	$V_s(3\text{–}4)$, V	$V_s(3\text{–CT})$, V	$V_s(4\text{–CT})$, V	$a = \dfrac{V_p}{V_s(3\text{–}4)}$	$a = \dfrac{V_p}{V_s(3\text{–CT})}$

TABLE 36–2. Effect of Load Current on Primary Characteristics

R_2, Ω	I_p, mA	V_p, V	V_s, V	$I_s = \dfrac{V_s}{R_2}$ mA	$I_{p1} = I_s/a$, mA	$\Delta I = I_p - I_{p1}$, mA	Power P_s
100 Ω							
75 Ω							
50 Ω							
25 Ω							

% efficiency: R_L 100Ω = R_L 25Ω =

TABLE 36–3. Resistances in a Transformer

Resistance, Ω							
Primary	Secondary			Primary to Secondary		Primary to Frame	Secondary to Frame
$R_{1\text{–}2}$	$R_{3\text{–}4}$	$R_{3\text{–CT}}$	$R_{CT\text{-}4}$	$R_{1\text{–}3}$	$R_{2\text{–}4}$	$R_{1\text{–Frame}}$	$R_{3\text{–Frame}}$

QUESTIONS

1. Is the transformer you used a voltage stepup or a current stepup? Use your experimental data to support your answer.

_____

_____

_____

_____

2. How is primary current affected by current in the secondary?

_____

_____

_____

3. As the transformer became loaded closer to its rated current value, what happened to its power efficiency?

_____

_____

_____

_____

_____

_____

_____

_____

4. What is the relationship between the resistance of a winding and the
   (a) Current rating of that winding?
   (b) Number of turns of the winding?

_____

_____

_____

_____

_____

_____

_____

_____

_____

_____

5. In step 2, why were the measured secondary voltages higher than the rated values for your transformer?

_____

_____

_____

_____

_____

_____

_____

_____

EXPERIMENT 37

INDUCTANCES IN SERIES AND PARALLEL

BASIC INFORMATION

Inductance in Series

Just as in the case of resistors, inductors may be connected in series. The effect of series-connected inductors is to offer greater opposition to alternating currents than an individual inductor can, acting alone. Measurement shows that series-connected inductors, where no mutual coupling exists, act like an equivalent single inductor whose inductance L_T is the sum of the individual inductances. That is,

$$L_T = L_1 + L_2 + L_3 + \cdots + L_n \qquad (37\text{--}1)$$

Verification of the total inductance of series-connected inductors, as given by formula (37–1), may be obtained by measuring L_T directly on a LCR meter or Capacitor/Inductor Analyzer as shown in Figure 37–1. Inductance may also be measured indirectly, as in this experiment.

Consider the circuit of Figure 37–2 (p. 286). The ac source voltage V_{ac} is supplying current to R and L. The value of X_L may be determined, by measuring I and the voltage V_L across L and substituting these values in the formula

$$X_L = \frac{V_L}{I} \qquad (37\text{--}2)$$

After X_L has been calculated, L may be found from the formula

$$L = \frac{X_L}{2\pi f} \qquad (37\text{--}3)$$

Figure 37–1. LCR meter. *(Courtesy of BK Precision.)*

OBJECTIVES

1 To test inductors using a LCR meter

2 To verify experimentally that the total inductance L_T for two inductors L_1 and L_2 connected in series, where no mutual coupling exists, is:

$$L_T = L_1 + L_2$$

3 To verify experimentally that L_T for inductors connected in parallel, where no mutual coupling exists, is:

$$\frac{1}{L_T} = \frac{1}{L_1} + \frac{1}{L_2}$$

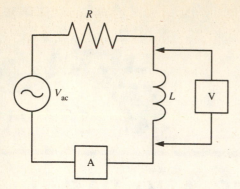

Figure 37–2. A series *RL* circuit may be used to determine the inductive reactance, X_L, of an inductor.

where

L is the required inductance in henrys,

X_L is the calculated value of inductive reactance in ohms,

f is the frequency of the ac source in hertz.

In the circuit of Figure 37–2 the current I may be measured directly with an ac ammeter. Another method is to measure the resistance of R and the voltage V_R across R. The current may then be calculated from the formula

$$I = \frac{V_R}{R} \qquad (37-4)$$

This method for finding inductance can be applied to a single inductor or to a number of inductors connected in series. Naturally, the same current will be in each inductor and in the series resistor.

Inductance in Parallel

Inductors may be connected in parallel in the same way resistors are connected in parallel. Two or more inductors connected in parallel, where no mutual coupling exists, act as an equivalent single inductor whose inductance L_T is given by the general formula

$$\frac{1}{L_T} = \frac{1}{L_1} + \frac{1}{L_2} + \frac{1}{L_3} + \cdots + \frac{1}{L_n} \qquad (37-5)$$

Thus, the total inductance of parallel-connected inductances is less than the smallest inductance in the circuit.

Two Parallel Inductors

Two inductors connected in parallel are equivalent to a single inductor L_T whose value can be found using the formula

$$L_T = \frac{L_1 \times L_2}{L_1 + L_2} \qquad (37-6)$$

Similar Inductors Connected in Parallel

Two or more parallel inductors with the same inductance rating and with no mutual coupling can be replaced by an equivalent inductor whose value can be found using the formula

$$L_T = \frac{L}{n}$$

where

L_T is the total inductance,

L is the value of each of the individual inductances that make up the parallel combination,

n is the number of parallel inductances.

Measuring Parallel Inductances

Verification of the total inductance L_T of parallel-connected inductors as given by formulas (37–5), (37–6), and (37–7) may be made by measuring across the parallel combination with a LCR Meter or Capacitor/Inductor Analyzer.

Inductance can also be measured indirectly by measuring the total current delivered to the parallel circuit and the voltage across the parallel-connected inductors. Inductive reactance can be found using the formula

$$X_L = \frac{V_L}{I}$$

After X_L has been calculated, L_T can be found from the formula

$$L = \frac{X_L}{2\pi f} \qquad (37-8)$$

where

L_T is the total inductance in henrys,

X_L is the inductive reactance in ohms,

f is the frequency of the source in hertz.

In the circuit of Figure 37–3 the current delivered to the parallel combination is being measured by the ammeter A. The voltage across the combination is V_L. The voltage across the resistor is V_R. The formula

$$I = \frac{V_R}{R}$$

gives the value of the total current drawn by the parallel inductances. Application of formulas (37–2) and (37–8) will result in a value for L_T.

SUMMARY

1. The total inductance L_T of series-connected inductors, where no mutual coupling exists, is given by the formula

$$L_T = L_1 + L_2 + L_3 + \cdots + L_n$$

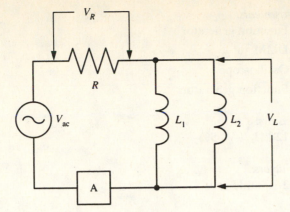

Figure 37–3. Parallel-connected inductances.

where $L_1, L_2, L_3, \ldots, L_n$ are inductors connected in series. This is the same as the formulas for R_T of resistors connected in series.

2. The inductance of coils and chokes may be measured directly with a LCR meter or capacitor/inductor analyzer.

3. The inductance of coils and chokes may also be determined indirectly by measuring the ac voltage V_L across the coil and the alternating current I in the coil. From these measurements X_L can be calculated, for $X_L = V_L/I$. Knowing X_L, it is possible to calculate L by the formula

$$L = \frac{X_L}{2\pi f}$$

4. The total inductance L_T of two or more parallel-connected inductances with no mutual coupling is given by the formula

$$\frac{1}{L_T} = \frac{1}{L_1} + \frac{1}{L_2} + \frac{1}{L_3} + \cdots + \frac{1}{L_n}$$

which is the same as the formula for parallel-connected resistors.

5. The total inductance of parallel-connected inductances can be measured directly using a LCR meter or capacitor/inductor analyzer.

6. The total inductance can be measured indirectly by measuring the current delivered to the parallel combination and the voltage across the combination. The inductive reactance can be calculated from these measurements using the formula

$$X_L = \frac{V_L}{I}$$

Once X_L is determined, the value of L_T can be found using the formula

$$L = \frac{X_L}{2\pi f}$$

Check your understanding by answering the following questions:

1. Three chokes, 4.2, 2.5, and 8 H, are connected in series. No mutual coupling exists. Their total inductance is _____ H.

2. The ac voltage V_L across a coil L is 22 V. The frequency is 60 Hz. The alternating current in the coil is 0.025 A.
   (a) $X_L =$ _____ Ω
   (b) $L =$ _____ H

3. A choke L and a 1 kΩ resistor R are connected in series as in Figure 37–4. The voltage across the resistor is 50 V, and the voltage across the choke is 40 V. The frequency of the applied voltage is 60 Hz.
   (a) The current I in the circuit = _____ A
   (b) X_L of the choke = _____ Ω
   (c) L of the choke = _____ H

4. Three chokes, 4.2, 2.5, and 8 H, are connected in parallel with no mutual coupling. The total inductance is $L_T =$ _____ H.

5. Four inductors having no mutual coupling are connected in parallel. Each inductor is 4.2 H. The total inductance is $L_T =$ _____ H.

6. In the circuit of Figure 37–5 (p. 288), $V_R = 20$ V, $V_L = 30$ V, $f = 1$ k Hz, and $R = 10$ kΩ.
   (a) The current delivered by the voltage source is _____ A
   (b) The inductive reactance of the circuit is _____ Ω
   (c) The total inductance is $L_T =$ _____ H

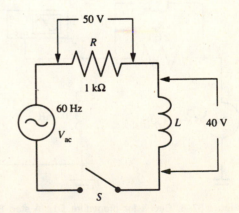

Figure 37–4. Circuit for Self-Test question 3.

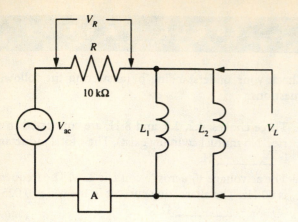

Figure 37–5. Circuit for Self-Test question 6.

Instruments:
- Function generator
- DMM
- Oscilloscope
- Function generator

Resistors:
- 12 kΩ, ½-W, 5%

Inductors:
- 2 100 mH coils

MATERIALS REQUIRED

Power Supplies:
- Function generator

PROCEDURE

A. Inductor Testing

A1. For each of the supplied inductors, determine what its coded value is based on the inductor coding system being used. Some inductors will have their respective value printed or stamped right on them. Others will use a color coding system or a 3-digit numbering system. Record these coded values in Table 37–1 (p. 291).

A2. Use a LCR Meter or Capacitor/Inductor Analyzer to measure the inductance of each inductor. Record the values in the Measured Value column of Table 37–1.

A3. Using a DMM, measure and record each of the inductor's dc resistance values in Table 37–1.

B. Behavior of Inductors Connected in Series

B1. Connect the circuit shown in Figure 37–6. The circuit will be used to determine the inductance value with indirect measurements. The inductors are designated L_1 and L_2. The inductance is one of the two being used in this experiment.

B2. Adjust the ac voltage to 5 V. Using the DMM and oscilloscope, measure the rms voltages across inductor L_1 and the resistor. Record the values in Table 37–2 (p. 291). Turn off the function generator. Disconnect inductor L_1.

B3. Calculate the current in the circuit and the inductance of inductor L_1. Record your answers in Table 37–2.

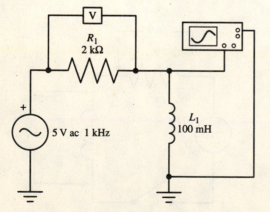

Figure 37–6. Circuit for procedure B1. In step B4, L_2 is substituted for L_1.

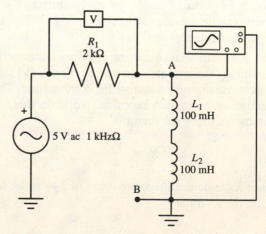

Figure 37–7. Circuit for procedure B6.

B4. Connect inductor L_2 in the circuit of Figure 37–6. Check to verify that $V_{ac} = 5$ V. If necessary, adjust the voltage. Repeat step B2. Record the measured values for V_{L2} and V_R in Table 37–2. Turn off the generator, but do not disconnect inductor 2.

B5. Repeat step B3 for inductor L_2.

B6. Break the circuit of Figure 37–6 and connect inductor L_1 in series with inductor L_2 as in Figure 37–7. (Separate L_1 and L_2 as much as possible to avoid coupling action between the inductors.)

B7. Measure the voltage across the combination of the two inductors L_1 and L_2—that is, across A and B, and across the resistor. Record the values in Table 37–2.

B8. Calculate the current in the circuit and the inductance of the series combination of L_1 and L_2. Record your answers in Table 37–2.

C. Behavior of Inductors Connected in Parallel

C1. With the function generator **off** and the circuit from step B7, disconnect L_2 from the circuit and reconnect it in parallel with L_1 as in Figure 37–8. (Separate L_1 and L_2 far enough to prevent coupling.)

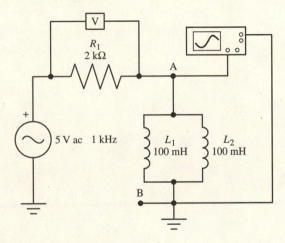

Figure 37–8. Circuit for procedure C1.

C2. Measure the voltage across the parallel combination of L_1 and L_2 (across A and B) and the voltage across the resistor V_R. Record the values in Table 37–2 for the parallel combination. Turn **off** the function generator and oscilloscope. Disconnect the meter and inductors.

C3. Calculate the total current in the circuit and the total inductance L_T of the parallel combination of L_1 and L_2. Record your answers in Table 37–2.

ANSWERS TO SELF TEST

1. 14.7

2. (a) 880; (b) 2.33

3. (a) 0.05; (b) 800; (c) 2.12

4. 1.31

5. 1.05

6. (a) 0.002 m; (b) 15 k; (c) 2.39

Name _____ Date _____

TABLE 37–1. Testing Inductors

Coil	Coded Value, mH	Measured Value, mH	Resistance, Ω
L_1			
L_2			

TABLE 37–2. Determining the Total Inductance of Inductors in Series and in Parallel

Inductor	Voltage across Inductor(s) V_L, V_{ac}	Voltage across Resistor V_R, V_{ac}	Total Current in Circuit I, mA	Inductance L, mH	Total Inductance L_T, mH
1					✕
2					✕
1 and 2 in series				✕	
1 and 2 in parallel				✕	

QUESTIONS

1. If an inductor developed a shorted turn, what effect would this have when being tested with a DMM and LCR Meter?

_____

_____

_____

_____

_____

_____

_____

_____

2. Explain the process for finding the inductance of an inductor experimentally.

_____

_____

_____

_____

_____

_____

_____

3. Are the values of inductance determined from procedure steps B3 and B5 equal to the rated values of the two inductors? If not, explain the discrepancy.

_____

_____

_____

_____

_____

_____

_____

_____

_____

4. Using formulas (37–1) and (37–5) and the values of L_1 and L_2 from Table 37–2, calculate L_T. Compare this with the value of L_T determined from V_L and I in Table 37–2. Explain any difference.

_____

_____

_____

_____

_____

_____

_____

_____

_____

5. Compare the formulas for calculating the total inductance of series and parallel inductances with the formulas for calculating the total resistance of series and parallel resistances.

_____

_____

_____

_____

_____

_____

_____

_____

_____

RC TIME CONSTANTS

BASIC INFORMATION

Capacitance and Capacitors

Capacitance is created when two conductors are separated by a noncon-ductor, or *dielectric.* This description would appear to be applicable to many electrical conditions, and, in fact, it is. For example, two wires sep-arated only by air (a dielectric) exhibit capacitance that under certain sit-uations can create problems in a circuit. The symbol for capacitance is *C.*

The unit of capacitance is the *farad,* named after the English scientist Michael Faraday. The abbreviation for farad is F. Since the farad is such a large unit of measurement a device with 1 F of capacitance would be very big, so farads are usually expressed in smaller units using metric pre-fixes. Typical capacitance values used in electronic circuits are given in millionths of a farad (a microfarad, or μF). Other typical units are pico-farads (pF) and nanofarads (nF).

Charging and Discharging a Capacitor

In electric circuits capacitors are used for many purposes. For example, they are used to store energy, to pass alternating current while blocking direct current, and to shift the phase relationship between current and voltage. They are also used as components in filter and resonant circuits. In this experiment they will be used in timing circuits.

A capacitor can store a charge of electrons over a period of time. The process of building up the charge of electrons in the capacitor is known as *charging.* To charge a capacitor it is necessary to have a voltage across the capacitor. Figure 38–1(*a*) (p. 294) shows a circuit containing a dc volt-age source and a capacitor. A three-position switch connects the capaci-tor to either a series circuit containing a voltage source and a resistor or a series circuit containing the capacitor and resistor alone. A neutral po-sition disconnects the capacitor from both circuits. In Figure 38–1(*a*) the switch is in neutral position 0.

If the switch is moved to position c, as in Figure 38–1(*b*), a charge of electrons will flow from the voltage source to the bottom plate of the ca-pacitor. At the same time, an equal charge of electrons will flow to the voltage source from the top plate of the capacitor. This process contin-ues until the capacitor is fully charged. This point is reached when the voltage applied to the capacitor is equal and opposite to the voltage across the plates of the capacitor. The polarity of the charged capacitor is shown

OBJECTIVES

1 ▶ To determine experimen-tally the time it takes a ca-pacitor to charge through a resistance

2 ▶ To determine experimen-tally the time it takes a capacitor to discharge through a resistance

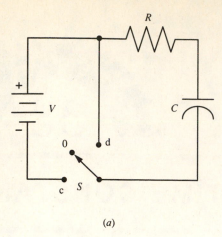

(a)

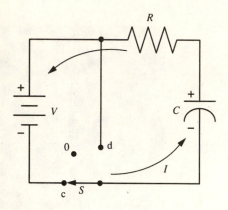

(b)

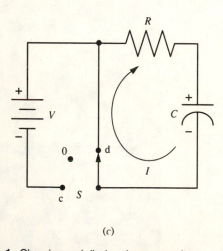

(c)

Figure 38–1. Charging and discharging a capacitor.

in Figure 38–1(b). Since the voltage across the capacitor is equal to the voltage applied to the capacitor by the source, there is no further movement of charges. If the switch in the circuit is now moved to the neutral position 0, the capacitor will remain fully charged.

With the capacitor fully charged, the switch is moved to position d as shown in Figure 38–1(c). This places the capacitor in a complete circuit containing the capacitor and a resistor. The charge of electrons can now travel from the bottom plate through the resistor to the top plate of

the capacitor until the charges on both plates are equal. At this point the capacitor is said to be fully discharged. A voltmeter placed across the capacitor would read 0 V. Because current is defined as the movement of charges, the charging and discharging of a capacitor involves the flow of current. As noted in Figure 38–1(b) and (c), the direction of current flow during the charging process is opposite to the direction of flow during the discharging process.

The Charge on a Capacitor

The size of a capacitor, given in farads (more commonly in microfarads or picofarads), is known as its capacitance.

The relationship between the charge Q on a capacitor and the capacitance of the capacitor C is given by the formula

$$Q = C \times V \tag{38-1}$$

where

Q is the charge in coulombs,

C is the capacitance in farads,

V is the voltage across the capacitor in volts.

Time Required to Charge a Capacitor

The circuit of Figure 38–1(a) will be used to demonstrate another characteristic of the charging and discharging process. Figure 38–2(a) indicates two voltages of particular interest in these processes, V_R, the voltage across the resistor R, and V_C, the voltages across the capacitor C.

The capacitor is in a completely discharged state, and the switch is in position 0. Now the switch is moved to position c and the voltmeter across C is observed. At the instant the switch is moved to position c, the meter reads 0 V, but as C was charging, the meter indicated an increasing voltage. This demonstrates that a capacitor does not charge instantaneously; it takes *time* to charge.

Figures 38–2(b) through (e) show how current and voltage in the circuit of Figure 38–2(a) behave. At the instant the switch is moved to position c, the charge of electrons rushes to the bottom plate of the capacitor. The only object impeding this flow of current is the resistor. At that instant, the capacitor appears merely as a short circuit. Therefore, at that instant, the resistor develops the full voltage drop equal to the applied voltage V. The value of current at that instant is equal to $(V/R) = I_c$. Since C appears as a short circuit, the voltage across C, V_C, is zero.

As C charges, the voltage across C increases and opposes the voltage of the source V. This reduces I_c, and as a result, V_R decreases. Finally, when C is fully charged, its voltage is equal and opposite to V; therefore, $I_C = 0$, and $V_R = 0$.

Figure 38–2(b) shows that the voltage across AB is a constant V once the switch is moved to position c. In Figure 38–2(c), an almost instantaneous inrush of current I_C flows when

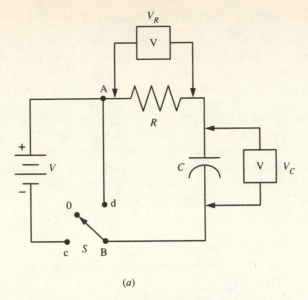

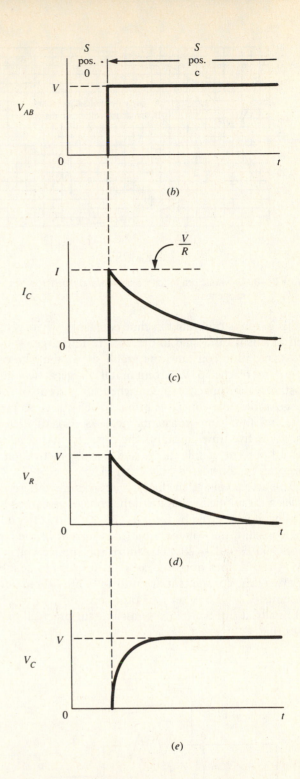

Figure 38–2. Capacitor C charging through an RC circuit.

the switch is moved to position c. However, as the capacitor charges, I_C decreases until it is zero, when C is fully charged.

Figure 38–2(c) shows the effects of the inrush of I_C and its decrease to zero. The voltage drop across R, V_R, goes from V to zero, as in Figure 38–2(d).

The voltage across C as seen in Figure 38–2(e) builds from zero until at its fully charged state it is equal to V.

The active voltage V_A causing current to flow is the difference between the applied voltage V and the voltage to which the capacitor has been charged to that point. This relationship is expressed by the formula

$$V_A = V - V_C \qquad \textbf{(38–2)}$$

When $V = V_C$, the capacitor is fully charged, and $V_A = 0$. Figures 38–2(b) through (e) show these conditions.

The circuit shown in Figures 38–1(a) and 38–2(a) is known as an RC (resistor-capacitor) circuit. The time it takes the capacitor to charge to any particular level is given in a unit expressed in time *constants*. Time constants can be calculated from the formula

$$t = R \times C \qquad \textbf{(38–3)}$$

where

t = time constant in seconds

R = resistance in ohms

C = capacitance in farads

It can be shown (though the derivation is beyond the scope of this book) that in one time constant, a capacitor charges to approximately 63.2 percent of the voltage applied across it. Thus, in an RC circuit containing $R = 1\ M\Omega$ and $C = 1\ \mu F$, the time constant will be

$$t = R \times C$$
$$= 1 \times 10^6 \times 1 \times 10^{-6}$$
$$= 1\ s$$

This means that after 1 s the capacitor will have developed 63.2 percent of the applied voltage. If the applied voltage was 100 V, a voltmeter across C would read approximately 63.2 V 1 s after the voltage was applied.

Charge Rate of a Capacitor

Figure 38–3 (p. 296), curve A, is a graph showing the rise of voltage across a capacitor during charging. The capacitor is charging through an RC circuit. The horizontal axis

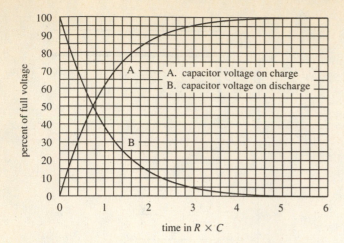

Figure 38–3. Universal graph for the charge and discharge of a capacitor.

Discharge Rate of a Capacitor · · · · · · · ·

Figure 38–3, curve B, shows the discharge of a capacitor in an *RC* circuit. The discharging process takes place after the capacitor is charged, as in Figure 38–2(*a*). In Figure 38–4(*a*) the switch is moved to position d, which results in a series circuit consisting of the capacitor *C* and the resistor *R*. Figures 38–4(*b*) through (*e*) show how current and voltage behave in the circuit of Figure 38–4(*a*).

At the instant the switch in Figure 38–4(*a*) is moved to position d, the voltage across AB is *V*. Closing the switch at position d short-circuits AB, and the voltage drops to zero, as shown in Figure 38–4(*b*).

in this graph is calibrated in time constants, whereas the vertical axis is calibrated in percent of full charge.

At one time constant, the graph shows that the capacitor has reached 63.2 percent of full charge; at two time constants, the capacitor has reached 86 percent of full charge; three time constants brings the charge to 95 percent, and four time constants yield 98 percent of full charge. At five time constants the capacitor has reached over 99 percent of full charge and for all practical purposes can be considered fully charged.

The active voltage in the *RC* circuit is the difference between the applied voltage and the voltage developed by the capacitor, as given by formula (38–2). Thus, after one time constant, the voltage producing current in the circuit is $100 - 63.2 = 26.8$ percent of the applied voltage. If $V = 100$ V, after one time constant it drops to 26.8 V; after two time constants it drops to 14 V; after three time constants, $V_A = 5$ V; after four time constants, $V_A = 2$ V; and finally at five constants we assume for practical purposes that $V_A = 0$.

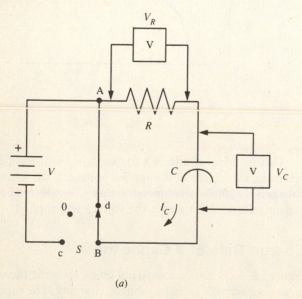

(*a*)

Figure 38–4. Capacitor *C* discharging through an *RC* circuit.

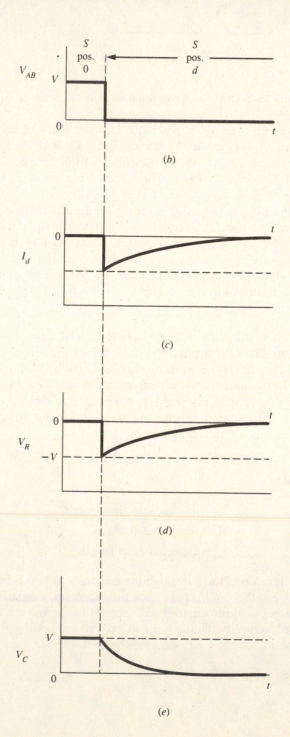

When the switch is in position 0, no current flows. At the instant the switch is moved to position d, an in-rush current I_d (the discharge current), impeded only by R, flows in the circuit produced by voltage V_C across the capacitor. As the capacitor discharges, V_C decreases and I_d decreases, as is shown in Figure 38–4(c). Note that this curve shows current flows in the direction opposite to the direction of I_C during the charging process.

With the switch in position 0, no current flows through R, and therefore the voltage drop across R is zero. With the switch in position d, current begins flowing at its maximum value, thus producing the maximum voltage drop across R equal to V_C. As V_C decreases, I_d decreases; therefore, the voltage drop across R decreases, as Figure 38–4(d) shows.

Finally, in Figure 38–4(e) we see the maximum voltage $V_C = V$ across the capacitor at its fully charged state. As soon as the switch is moved to position d, C begins discharging and the voltage across C decreases as shown in Figure 38–4(e) until $V_C = 0$.

SUMMARY

1. The charge in coulombs Q on a capacitor is equal to the capacitance C in farads times the voltage V in volts to which the capacitor has charged. Thus, $Q = C \times V$.

2. The time required to charge a capacitor to 63.2 percent (approximately) of the applied voltage is called a time constant.

3. When a capacitor of C farads is charging through a resistance of R ohms, the time constant t in seconds of the charging circuit is $t = R \times C$.

4. At any instant of time while a capacitor is charging, the active voltage V_A equals the difference between the charging source voltage V and the voltage across the capacitor; that is,

$$V_A = V - V_C$$

5. It takes five time constants for a capacitor to charge to 99+ percent (approximately) of the applied voltage. For practical purposes we say that a capacitor is fully charged after five time constants.

6. When a capacitor of C farads is discharging through a resistor of R ohms, the discharge time constant t in seconds is $t = R \times C$.

7. When discharging, a capacitor will lose 63.2 percent of its charge in one time constant.

8. In each succeeding time constant the capacitor will lose an additional 63.2 percent of the voltage still across it.

9. At five time constants, we say the capacitor is fully discharged.

10. The graph of the charge and discharge of a capacitor is shown in Figure 38–3, which represents the universal chart for the charge and discharge of a capacitor.

11. A time constant is a *relative*, not an absolute, measure of time.

S E L F   T E S T

Check your understanding by answering the following questions:

1. A 0.25-μF capacitor is charging through a 2.2-MΩ resistor in series with an applied voltage of 20 V. The circuit has a time constant of _____ s.

2. In one time constant the capacitor in question 1 will have charged to _____ V.

3. In _____ s the capacitor in question 1 will have charged to 17.2 V.

4. A 0.05-μF capacitor charged to 100 V is permitted to discharge through a 220-kΩ resistor. At the end of _____ ms, the voltage across the capacitor will have dropped to 37 V (approximately).

5. The discharge time constant in question 4 is _____ ms.

6. It takes _____ time constants to discharge a capacitor almost completely.

MATERIALS REQUIRED

Power Supply:
- Variable 0–15 V dc, regulated

Instrument:
- Electronic analog voltmeter or DMM (the meter must not be heavily damped)
- Function generator
- Oscilloscope

Resistors (½-W, 5%):
- 1 1-MΩ
- 1 10-kΩ
- 1 2-kΩ resistor

Capacitors:
- 1 1-μF 25 V Electrolytic
- 10.1-μF

Miscellaneous:
- 2 SPST switches
- Time (can be watch with sweep second hand or digital watch with seconds readout)

PROCEDURE

A. Discharging of a Capacitor

A1. With power **off** and S_1 and S_2 **open,** connect the circuit of Figure 38–5. Set the power supply to its lowest voltage. Observe the polarity of C.

A2. **Close** S_1 and increase the power supply voltage until the meter reads 12 V.

A3. Without changing the setting of the power supply, measure the output voltage of the supply. Record this value in Table 38–1 (p. 301).

A4. The circuit of Figure 38–5 is a simple series circuit consisting of a 1-MΩ resistor in series with the internal resistance of the meter R_{in}. Using Ohm's law, the value of R_{in} can be calculated.

$$R_{in} = \frac{12}{V_{PS} - 12} \times 1 \text{ M}\Omega$$

where 12 is the reading of the meter in step A2 and V_{PS} is the voltage across the power supply from step A3. Calculate R_{in} for your meter and record your answer in Table 38–1. Check the meter specifications or nameplate to determine the rated value for R_{in}. Record the value in Table 38–1.

A5. Using the rated values for R_{in} and C, calculate the time constant of a circuit containing only R_{in} and C. Record your answer in Table 38–1, under "Discharge Time Constant $R_{in}C$, Calculated, s."

    Table 38–1 has a column of time-constant multiples from 1 through 5 and 10. Calculate the elapsed time in seconds for each time constant. For example, if $RC = 10$ s, then after one time constant 10 s will have passed; after two time constants, another 10 s will have passed, for a total of 20 s elapsed time; after three time constants, 30 s, and so on. Record these times in the "Discharge Time, Seconds" column.

A6. Charge the capacitor by **closing** S_2 (S_1 is still **closed** from step A2). At the instant S_2 is closed, the capacitor acts as a short, and the meter reading will

drop. As the capacitor charges, the meter reading will increase until it stabilizes at 12 V, indicating the capacitor is fully charged.

NOTE: In the following steps the behavior of the capacitor during discharge will be recorded. Three discharge trials will be run. The voltage across the capacitor will be measured at intervals of the time constant as the capacitor discharges.

A7. **Open** S_1. This causes the capacitor to discharge through the meter. Using a timer that measures seconds accurately, measure the voltage across the capacitor at 1, 2, 3, 4, and 5 time constants, and at 10 time constants. Record the values in Table 38–1, under "1st Trial."

A8. **Close** S_1. The capacitor will recharge until the meter reads 12 V, at which point the capacitor is fully charged.

A9. Repeat step A7; record the voltages for the six time intervals under "2d Trial."

A10. Repeat step A8 to recharge the capacitor.

A11. With the capacitor fully charged, repeat step A7; record the voltages for the six time intervals under "3d Trial." **Open** S_1 and S_2; turn **off** power.

A12. Calculate the average measured voltage for each of the six time constants and record your answers in the "Average" column of Table 38–1. Also, calculate the voltage for each multiple of the calculated time constant $R_{in}C$. Record your answers in Table 38–1 under the "Calculated" column.

B. Charging of a Capacitor

B1. With power **off** and S_1 and S_2 **open,** connect the circuit of Figure 38–6. Set the power supply to its lowest voltage.

B2. **Close** S_1 and S_2 and increase the power supply voltage until the meter measures 12 V. The capacitor is fully discharged at this point.

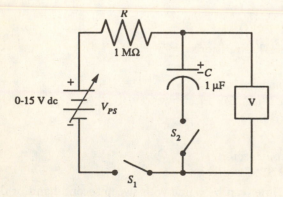

Figure 38–5. Circuit for procedure step A1.

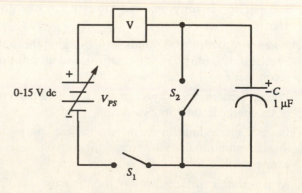

Figure 38–6. Circuit for procedure step B1.

B3. Complete the "Charge Time" column in Table 38–2 (p. 301) by entering the values in the "Seconds" column from the same column in Table 38–1. The discharge time constant is the same as the charge time constant, and both are equal to $R_{in}C$.

NOTE: In the following steps the behavior of the capacitor during charge will be recorded. Three charge trials will be run. For each trial the voltage measured by the meter (that is, the voltage across R_{in}) will be recorded at each multiple of the time constant.

B4. **Open** S_2. This causes the capacitor to charge through the meter. At the instant S_2 is opened, C acts as a short circuit, and the total voltage is across the meter. As the capacitor charges, the meter readings will decrease. Record the voltmeter measurements for each of the six time intervals in Table 38–2, under "1st Trial." At 10 times the time constant, the capacitor can be considered fully charged.

B5. **Close** S_2; this discharges the capacitor. When the meter measures 12 V, the capacitor is fully discharged.

B6. Repeat step B4. Record the values in Table 38–2, under "2d Trial."

B7. Repeat step B5 to discharge the capacitor fully.

B8. Repeat step B4. Record the values in Table 38–2, under "3d Trial." **Open** S_1 and S_2; turn **off** power.

B9. Calculate the voltage across the capacitor for each trial and each time interval. The voltage across the capacitor will be

$$V_C = V_{PS} - V_{R_{in}}$$

where

V_C is the voltage across the capacitor,

V_{PS} is the voltage of the power supply (12 V in this experiment),

$V_{R_{in}}$ is the voltage read by the meter.

Record your answers in Table 38–2.

B10. Calculate the average voltage across the capacitor for each time interval. Record your answer in the "Average" column.

B11. For each multiple of the time constant, calculate the voltage across the capacitor. Record your answers in Table 38–2 under "Calculated."

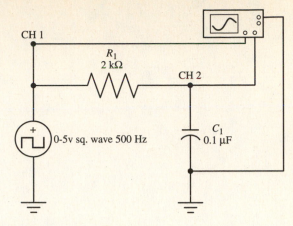

Figure 38–7. Square-wave input.

C. Square-Wave Input

C1. Connect the circuit shown in Figure 38–7. This circuit will be used to demonstrate RC time constants when a square-wave input is applied, instead of opening and closing switches.

C2. Adjust the output of the function generator for a 5 V_{P-P} square-wave (0 V to 5 V) at 500 Hz. Use the generator's dc offset control to shift the waveform so that its ($+$) peak is at five volts and its ($-$) peak is at zero volts.

C3. Calculate and record the time constant values and the percent charge for the time constant periods shown in Table 38–3 (p. 301).

C4. Calculate the expected voltage across C_1 for each respective time constant.

C5. Using the oscilloscope, measure and record the voltages across C_1 at each time constant in Table 38–3. Set the scope's Time/Div. to 0.2 ms/Div.

ANSWERS TO SELF TEST

1. 0.55

2. 12.6

3. 1.1

4. 11.0

5. 11.0

6. 5

TABLE 38–1. Discharging of a Capacitor

Power Supply Voltage V_{PS}, V	Internal Resistance of Meter R_{in}, Ω		Discharge Time Constant $R_{in}C$, Calculated, s	Voltage Across Capacitor, V_C, V
	Calculated	Rated		
				12

Discharge Time		Voltage Across Capacitor V_C, V				
Time Constants	Seconds	1st Trial	2d Trial	3d Trial	Average	Calculated
1						
2						
3						
4						
5						
10						

TABLE 38–2. Charging of a Capacitor

Charge Time		Voltmeter Reading $V_{R_{in}}$, V			Voltage Across Capacitor V_C, V				
Time Constants	Seconds	1st Trial	2d Trial	3d Trial	1st Trial	2d Trial	3d Trial	Average	Calculated
1									
2									
3									
4									
5									
10									

TABLE 38–3. Square-Wave Input

Time Constants	1	2	3	4	5
Time					
% Change					
V_C Calculated					
V_C Measured					

QUESTIONS

1. Explain, in your own words, the charging and discharging processes of a capacitor.

_____

_____

_____

_____

_____

_____

2. Explain, in your own words, what the time constant of an *RC* circuit is.

_____

_____

_____

_____

_____

_____

3. Discuss the key factors that probably limited the accuracy of your measurements in this experiment.

_____

_____

_____

_____

_____

4. On a separate 8½ × 11 sheet of graph paper, plot the following time constants versus voltage graphs using the same set of axes. The time constants should be on the horizontal (*x*) axis and the voltages across *C* should be on the vertical (*y*) axis. Label all axes and graphs.
   (a) Time constant versus the average measured value of the voltage across the capacitor from Table 38–1.
   (b) Time constant versus the calculated voltage across the capacitor from Table 38–1.
   (c) Time constant versus the average measured value of the voltage across the capacitor from Table 38–2.
   (d) Time constant versus the calculated voltage across the capacitor from Table 38–2.

5. Do the graphs plotted in Question 4 agree with the universal time-constant graph as shown in Figure 38–3? Discuss any differences.

_____

_____

_____

_____

REACTANCE OF A CAPACITOR (X_C)

BASIC INFORMATION

Reactance of a Capacitor

The capacitive reactance X_C of a capacitor is the amount of opposition it offers to current in an ac circuit. The unit of capacitance reactance is the ohm. However, like the X_L of a coil, the X_C of a capacitor cannot be measured with an ohmmeter. Rather, capacitive reactance must be measured indirectly from its effect on current in an ac circuit.

Capacitive reactance is dependent on frequency and is given by the formula

$$X_C = \frac{1}{2\pi fC} \tag{39-1}$$

where

X_C is in ohms,

C is the capacitance in farads,

f is the frequency in hertz.

The value of C in microfarads (μF) may be substituted directly in the formula

$$X_C = \frac{10^6}{2\pi fC} \tag{39-2}$$

Remember that the C in formula (39–2) is expressed directly in microfarads. For example, to find the reactance of a 0.1-μF capacitor at a frequency of 1 kHz, substitute in formula (39–2):

$$X_C = \frac{10^6}{(6.28)(1000)(0.1)} = 1592 \ \Omega = 1.59 \ \text{k}\Omega$$

Formulas (39–1) and (39–2) show that the higher the frequency, the lower the value of the reactance of a capacitor, and the lower the frequency, the higher the reactance. For direct current, $f = 0$; thus, X_C is infinite. This means that a capacitor appears as an open circuit to dc, after 5 time constants, and no direct current will flow through a capacitor.

The reactance of a capacitor may be determined by measurement. In the circuit of Figure 39–1 (p. 304) a sinusoidal voltage V causes a current I

OBJECTIVE

1 To verify experimentally the formula for capacitive reactance

$$X_C = \frac{1}{2\pi fC}$$

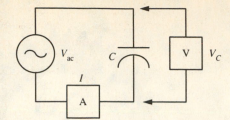

Figure 39–1. A sinusoidal voltage V causes current I to flow in the circuit. Since R in the circuit is very low compared to X_C, $Z = X_C$ and $V_C = V$.

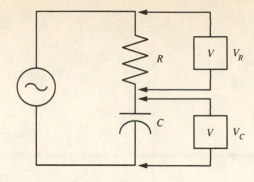

Figure 39–2. Series RC circuit.

to flow in the circuit. Ohm's law for ac circuits states that

$$I = \frac{V}{Z} \tag{39–3}$$

where

I is in amperes,

V is in volts,

Z (called impedance) is in ohms. The symbol is used to denote total ac opposition to current.

In Figure 39–1, if R in the circuit is very small compared with X_C, the reactance of C can be considered the impedance of the circuit. That is,

$$X_C = Z \tag{39–4}$$

Hence, the capacitive current I is given by the formula

$$I = \frac{V}{X_C} = \frac{V_C}{X_C} \tag{39–5}$$

Formula (39–5) may be written in the form

$$X_C = \frac{V_C}{I} \tag{39–6}$$

We use formula (39–6) in determining the reactance of a capacitor at a given frequency f by measuring the voltage V_C across C with an ac voltmeter and measuring the current in the circuit with an ac ammeter. The values of V_C and I are then substituted in formula (39–6), and X_C is calculated. Note in the experimental circuit of Figure 39–1 that the voltmeter V is connected directly across C and not across the voltage source V. Thus, we measure V_C and not V. The reason is that the ammeter A has resistance, across which there will be a voltage drop. If the resistance of the ammeter is high enough, its effect on impeding current would have to be considered if the applied voltage V were measured. By measuring V_C, formula (39–6) can be used to find X_C.

An alternative method for measuring alternating current does not require the use of an ac ammeter.

Consider the circuit of Figure 39–2. This is a series circuit. Thus, the current flowing through R and C is the same. With this circuit arrangement it is possible to determine the value of X_C experimentally by measuring the voltages V_C across C and V_R across R. To determine the current I in amperes in the circuit, the known value of R

in ohms and the measured value V_R in volts are substituted in Ohm's law.

$$I = \frac{V_R}{R} \tag{39–7}$$

Knowing I, it is possible to find X_C by substituting for I and V_C in formula (39–6). It is also possible to find X_C without calculating I by combining formulas (39–6) and (39–7) and solving for X_C. We get

$$X_C = \frac{V_C}{V_R} \times R \tag{39–8}$$

By substituting the measured values V_C, V_R, and R in formula (39–8), we can determine the value of X_C. Having found X_C experimentally, we can verify approximately the validity of the formula for X_C by comparing the experimental value of X_C with the formula value. This method requires that the capacitance C be known.

SUMMARY

1. The amount of opposition that a capacitor offers to current in an ac circuit is called capacitive reactance, X_C. Capacitive reactance is given in ohms.

2. The capacitive reactance of a capacitor is not constant but varies with frequency and capacitance.

3. Capacitive reactance is inversely proportional to frequency. As f increases, X_C decreases; as f decreases, X_C increases.

4. Capacitive reactance is inversely proportional to capacitance. As C increases, X_C decreases; as C decreases, X_C increases.

5. The capacitive reactance of a capacitor C in a circuit of frequency f may be calculated from the formula

$$X_C = \frac{1}{2\pi f C}$$

where X_C is in ohms, f is in hertz, and C is in farads. When C is given in microfarads (μF), the formula can be written as

$$X_C = \frac{10^6}{2\pi f C \text{ (in } \mu\text{F)}}$$

6. Capacitive reactance cannot be measured directly—only by its effect in an ac circuit. Thus, in Figure 39–1, we measure the voltage V_C across C and the current I in the circuit. Then, using Ohm's law for ac circuits, we can find X_C by substituting V_C and I in the formula

$$X_C = \frac{V_C}{I}$$

7. Another method of determining X_C is illustrated by the circuit of Figure 39–2. Here the voltages V_R and V_C are measured across R and C, respectively. Then X_C is calculated by substituting V_R and V_C in the formula

$$X_C = \frac{V_C}{V_R} \times R$$

SELF TEST

Check your understanding by answering the following questions:

1. The unit of capacitance is the _____. The unit of capacitive reactance is the _____.

2. The reactance of a capacitor depends on the _____ of the voltage source and the _____ of the capacitor.

3. The X_C of a capacitor whose value is 0.1 μF in a circuit where a 60-Hz signal is applied is _____ Ω.

4. If the frequency of the signal in question 3 is changed to 600 Hz, the X_C of the 0.1-μF capacitor is _____ Ω.

5. The voltage across a capacitor is 9.5 V, and the current in the capacitor is 10 mA. The capacitive reactance of the capacitor is _____ Ω.

6. A 1 k-Ω resistor and a 0.05-μF capacitor are connected in series. The voltage measured across the resistor is 5 V, and the voltage measured across the capacitor is 15 V. The X_C of the capacitor is _____ kΩ.

7. The frequency of the applied voltage in question 6 is _____ kHz.

MATERIALS REQUIRED

Power Supplies:
- Isolation transformer
- Variable-voltage autotransformer (Variac or equivalent)

Instruments:
- DMM
- Oscilloscope
- 0–5-mA ac ammeter
- LCR meter or capacitor/inductor analyzer

Resistor:
- 1 5.6 kΩ, ½-W, 5%

Capacitors (nonelectrolytic):
- 1 0.5-μF or 0.47-μF, 25-WV dc
- 1 0.1-μF, 100-WV dc

Miscellaneous:
- SPST switch
- Polarized line cord with on-off switch and fuse

PROCEDURE

1. Measure the capacitance of the two test capacitors using a LCR meter or capacitor/inductor analyzer. Record the values in Table 39–1 (p. 307) in the "Measured Value" column.

2. With the line cord unplugged, the line switch **off** and S_1 **open,** connect the circuit of Figure 39–3 (p. 306). Adjust the autotransformer to its lowest voltage setting.

3. Plug in the line cord; turn line switch **on** and **close** S_1. Increase the output voltage of the autotransformer until the ammeter measures 2 mA. Measure the voltage across the capacitor, V_C, and record the value in Table 39–1 for the 0.5-μF capacitor.

4. Repeat step 3 for a current of 3 mA.

5. Repeat step 3 for a current of 4 mA. After measuring V_C, **open** S_1 and disconnect the 0.5-μF capacitor from the circuit. Adjust the autotransformer to its lowest setting. The remaining circuit will be reused in step 7.

6. Calculate the capacitive reactance of the capacitor for each of the three currents (2 mA, 3 mA, and 4 mA). First, calculate X_C using the voltmeter and ammeter measurements and applying Ohm's law for ac circuits. Next, calculate X_C using the formula

$$X_C = \frac{1}{2\pi fC}$$

Record your answers in Table 39–1 for the 0.5-μF capacitor in the "Voltmeter-Ammeter Value" and "Reactance Formula Value" columns.

7. Connect a 0.1-μF capacitor in the circuit of Figure 39–3. Repeat steps 3 through 6, but use 1, 2, and 3 mA. Record all values in Table 39–1 for the 0.1-μF capacitor. After taking all measurements, **open** S_1. This circuit will be reused in step 9.

8. Use an ohmmeter to measure the resistance of the 5.6 kΩ resistor. Record your answer in Table 39–2 (p. 307).

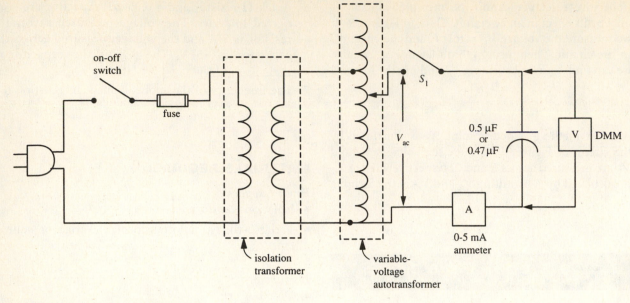

Figure 39–3. Circuit for procedure step 2.

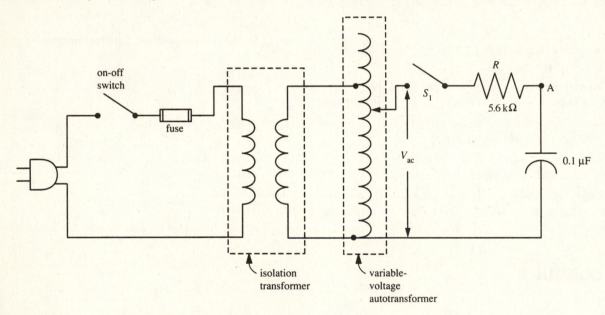

Figure 39–4. Circuit for procedure step 9.

9. With the circuit of step 7 and S_1 **open,** connect the 5.6 kΩ resistor in series with the 0.1-μF capacitor as shown in Figure 39–4 (p. 306).

10. **Close** S_1. Increase the output voltage of the auto-transformer to 10 V rms. With the oscilloscope calibrated for peak-to-peak voltage measurements, connect the hot lead of the scope to point A. Measure the peak-to-peak voltage across the capacitor and across the resistor. Use the differential measurement technique for V_R. Record the values in Table 39–2 for the 0.1-μF capacitor. **Open** S_1. Disconnect the capacitor. This circuit will be reused in step 12.

11. Calculate the capacitive reactance using the voltage-ratio formula (39–8). Record your answer in Table 39–2 in the "Capacitive Reactance (Calculated)" column.

12. With S_1 **open,** connect a 0.5-μF capacitor in the circuit of step 10.

13. Repeat step 10. Record the measurements of V_C and V_R in Table 39–2 for the 0.5-μF capacitor. After completing the measurements, **open** S_1, turn line switch **off;** unplug line cord.

14. Repeat step 11.

ANSWERS TO SELF TEST

1. farad; ohm
2. frequency; capacitance
3. 26.5 k
4. 2.65 k
5. 950
6. 3
7. 1.06

TABLE 39–1. **Voltmeter-Ammeter Method for Determining Capacitive Reactance**

Capacitance C, μF			Voltage across Capacitor V_C, V	Capacitive Reactance X_C (Calculated), Ω	
Rated Value	Measured Value	Current I, mA		Voltmeter-Ammeter Value	Reactance Formula Value
0.5 or 0.47		2			
		3			
		4			
0.1		1			
		2			
		3			

TABLE 39–2. **Voltage-Resistance Method for Determining Capacitive Reactance**

Capacitor Rated Value, μF	Resistor R, kΩ		Voltage across Capacitor V_C, V_{p-p}	Voltage across Resistor V_R, V_{p-p}	Capacitive Reactance X_C (Calculated) Voltage-Ratio Formula, kΩ
	Rated Value	Measured Value			
0.1	5.6				
0.5	5.6				

QUESTIONS

1. Explain, in your own words, two methods that can be used to determine capacitive reactance experimentally. Discuss any special conditions or restrictions in using these methods.

——
——
——
——
——
——
——
——
——
——
——
——

2. Refer to your data in Table 39–1. Should all values of X_C obtained using the voltmeter-ammeter measurements be the same? If so, explain why. If not, explain the difference.

_____

_____

_____

_____

_____

_____

_____

_____

_____

_____

3. Refer to your data in Table 39–2. Should the two values of X_C calculated using the voltage-ratio formula be the same? If not, explain the difference.

_____

_____

_____

_____

_____

_____

_____

_____

_____

4. If the capacitor you are testing, using Figure 39–3, is shorted what will be the result?

_____

_____

_____

_____

_____

_____

_____

CAPACITORS IN SERIES AND PARALLEL

BASIC INFORMATION

Total Capacitance of Series-Connected Capacitors · ·

In the series-connected circuit of Figure 40–1 the same line current I flows through C_1 and C_2. The effect of connecting these capacitors in series is to increase the total capacitive reactance and thus reduce the line current that would flow if either capacitor were in the circuit alone.

The total capacitive reactance of the series circuit X_{C_T} is the sum of the capacitive reactances of C_1, and C_2:

$$X_{C_T} = X_{C_1} + X_{C_2}$$

Substituting the formula for capacitive reactance gives

$$\frac{1}{2\pi f C_T} = \frac{1}{2\pi f C_1} + \frac{1}{2\pi f C_2}$$

The term $2\pi f$ cancels, leaving the formula for capacitances in series:

$$\frac{1}{C_T} = \frac{1}{C_1} + \frac{1}{C_2}$$

which in its general form is

$$\frac{1}{C_T} = \frac{1}{C_1} + \frac{1}{C_2} + \frac{1}{C_3} + \cdots + \frac{1}{C_n} \qquad (40\text{–}1)$$

Because the reactance of the series combination is greater than the reactance of either capacitor considered alone, the total capacitance C_T must be less than the capacitance of any of the individual series capacitances.

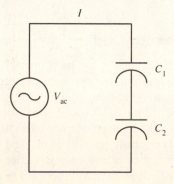

Figure 40–1. Series-connected capacitors.

The total capacitance C_T of capacitor combinations may be determined experimentally in a number of ways. The simplest is to connect the required combination and measure C_T with a capacitance meter. If a capacitor analyzer, such as shown in Figure 40–2 is used, C_T can be measured to a high degree of accuracy.

Another method, of finding total capacitance, is to determine the capacitive reactance X_{C_T} of the combination. Once X_{C_T} is determined, C_T may be found from the formula

$$C_T = \frac{1}{2\pi f X_{C_T}} \qquad (40\text{--}2)$$

This method is used in this experiment. The reactance of the capacitor combination may be found by measuring the current I_C and the voltage V_C across the combination. The total capacitive reactance X_{C_T} is then calculated from the formula

$$X_{C_T} = \frac{V_C}{I_C} \qquad (40\text{--}3)$$

The capacitive current may be measured with an ac ammeter. The ac voltage V_R across a series-connected resistor can be measured, and the series current, which is also the capacitive current I_C can be calculated from the formula

$$I_C = \frac{V_R}{R} \qquad (40\text{--}4)$$

Total Capacitance of Parallel-Connected Capacitors

Parallel-connected capacitors are frequently used in power and electronics circuits. The circuit in Figure 40–3 has two capacitors, C_1 and C_2, connected in parallel across the voltage source V_{ac}. Thus, the voltage across each capacitor is the same. The current supplied to this circuit divides, so that I_1 flows to C_1 and I_2 flows to C_2. The to-

Figure 40–2. Capacitor/Inductor Analyzer (*Courtesy of Sencore.*)

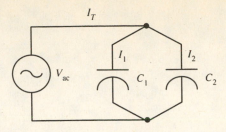

Figure 40–3. Parallel-connected capacitors.

tal current is therefore $I_T = I_1 + I_2$. The two capacitors could be replaced by a single equivalent capacitor drawing the total current I_T.

The flow of current is impeded by the capacitive reactance of each capacitor, which is given by the formula

$$I_1 = \frac{V_{ac}}{X_{C_1}}; \quad I_2 = \frac{V_{ac}}{X_{C_2}} \qquad (40\text{--}5)$$

The total current, in terms of V_{ac} and X_C, is

$$I_T = \frac{V_{ac}}{X_{C_1}} + \frac{V_{ac}}{X_{C_2}} \qquad (40\text{--}6)$$

$$= \frac{V_{ac}}{\frac{1}{2\pi f C_1}} + \frac{V_{ac}}{\frac{1}{2\pi f C_1}} = V_{ac}\,(2\pi f C_1) + V_{ac}\,(2\pi f C_2)$$

Dividing by V_{ac}, we obtain

$$\frac{I_T}{V_{ac}} = 2\pi f C_1 + 2\pi f C_2 \qquad (40\text{--}7)$$

But

$$\frac{I_T}{V_{ac}} = \frac{1}{X_{C_T}} = 2\pi f C_T$$

Substituting this expression in formula (40–7) gives

$$2\pi f C_T = 2\pi f C_1 + 2\pi f C_2$$

which, upon canceling $2\pi f$, becomes

$$C_T = C_1 + C_2$$

The general form of this formula for two or more capacitors connected in parallel is

$$C_T = C_1 + C_2 + C_3 + \cdots + C_n \qquad (40\text{--}8)$$

Capacitor Coding

Before a capacitor can be properly tested, the various capacitor-coding systems must be understood. Capacitor values are specified in either microfarads (μF) or picofarad (pF) units, although some capacitor testers will indicate values in nanofarads (nF). Large value capacitors, such as electrolytics, will be marked in whole numbers indicating their value in μF. Generally, non-electrolytics marked with whole numbers are in pF units, such as 220, 470, and 680.

Film-type capacitors often use a special coding system. If the code reads "104K," the first two numbers indicate the first two digits in its value. The third number, 4 in this case, indicates the multiplier or number of zeros added to the first two numbers. The K represents a tolerance of 10%. The value is in pF; therefore, this capacitor would be 100,000 pF or more commonly stated, 0.1 μF.

Disk-Ceramic capacitors may be marked with either whole numbers or decimal fractions. The whole numbers are in pF and fractional numbers in μF. The actual code system depends on the manufacturer, but codes similar to film-type capacitors are common. Also, often indicated on the capacitor will be the working volts dc (WVDC) and temperature range for the capacitor. Coding systems are also used for surface mount or chip capacitors. Capacitor codes can be found in the appendix of this book.

SUMMARY

1. In series-connected capacitors the current is the same in each of the capacitors. The total reactance X_{C_T} of series-connected capacitors is the sum of the reactances of the series capacitors in the circuit.

2. The total current I_T in a circuit consisting of series-connected capacitors is less than the current would be if any one of the capacitors were in the circuit by itself.

3. In a circuit consisting of two or more series-connected capacitors, the total capacitance C_T of the combination is

$$\frac{1}{C_T} = \frac{1}{C_1} + \frac{1}{C_2} + \frac{1}{C_3} + \cdots + \frac{1}{C_n}$$

4. The total capacitance of *series-connected capacitors* is calculated in the same way as the total resistance R_T of *parallel resistors*.

5. One method for experimentally determining the total capacitance C_T of series-connected capacitors is to measure the C_T of the combination with a LCR meter or capacitor/inductor analyzer.

6. Another method to determine C_T is to measure the total ac current I_T of the combination and the voltage V_C across the combination, and to calculate X_{C_T} using the formula

$$X_{C_T} = \frac{V_C}{I_T}$$

Knowing X_{C_T} and frequency f of the source, calculate C_T by substituting X_{C_T} and f in the formula

$$C_T = \frac{1}{2\pi f X_{C_T}}$$

7. In a circuit consisting of two or more capacitors connected in parallel, the total current I_T is equal to the sum of the branch currents.

8. The total current in a parallel circuit is greater than the branch current in any one of the capacitors in the circuit.

9. In a circuit consisting of two or more parallel-connected capacitors, the total capacitance C_T is equal to the sum of the individual branch capacitors,

$$C_T = C_1 + C_2 + C_3 + \cdots + C_n$$

This is similar to the formula for finding the total resistance of series-connected resistors.

SELF TEST

Check your understanding by answering the following questions:

1. In the circuit of Figure 40–1 $C_1 = 0.40$ μF and $C_2 = 0.05$ μF. The total capacitance of C_T of this combination is _____ μF.

2. If the frequency f of the applied voltage V is 100 Hz, then in the circuit of question 1,
   (a) $X_{C_1} =$ _____ kΩ
   (b) $X_{C_2} =$ _____ kΩ
   (c) $X_{C_T} =$ _____ kΩ

3. In the circuit of Figure 40–4 the frequency of the source is 1 kHz, and the capacitances C_1, C_2, and C_3, are equal. The voltage across the resistor $V_R = 4$ V, and the voltage across the three series-connected capacitors $V_{C_T} = 6$ V. Find the value of each of the capacitors: $C_1 = C_2 = C_3 =$ _____ μF.

4. In Figure 40–3 the current I_1 in C_1 is 40 mA, and the current I_2 in C_2 is 20 mA. The total current I_T in the circuit is _____ mA.

5. In Figure 40–3 $C_1 = 0.5$ μF, and $C_2 = 1.0$ μF. The total capacitance C_T is _____ μF.

6. The applied voltage V_{ac} in Figure 40–3 is 12 V. The total current in the circuit is 10 mA. Find the total reac-

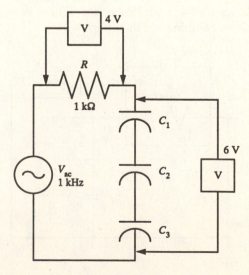

Figure 40–4. Circuit for Self-Test question 3.

tance and the total capacitance of the circuit at a frequency of 60 Hz.

(a) $X_{C_T} =$ _____ kΩ

(b) $C_T =$ _____ μF

MATERIALS REQUIRED

Instruments:
- Oscilloscope
- LCR meter or capacitor analyzer
- AC milliammeter or DMM

- A second DMM
- Function generator

Resistors (½-W, 5%):
- 1 10-kΩ

Capacitors:
- 1 0.022-μF 25-WV dc
- 2 0.1-μF, 25-WV dc
- 1 0.47-μF, 25-WV dc
- 1 470 pF Disk Ceramic 25-WV dc

PROCEDURE

If they are not already marked, label the capacitors 1 through 5 for identification during this experiment.

A. Total Capacitance of Capacitors in Series

A1. Using a LCR meter or capacitor analyzer, measure the capacitance of each of the capacitors in this experiment and record the values in Table 40–1 (p. 315). Also record their coded values.

A2. Connect the capacitors in series in the five combinations listed in Table 40–2 (p. 315). Measure the total capacitance of each of the combinations using your capacitor tester or meter. Record the values in Table 40–2 in the "Measured" column.

A3. Using the measured values of capacitance from Table 40–1, calculate the total capacitance of each of the five combinations in Table 40–2. Record your answers in the table in the "Calculated" column.

A4. With the function generator **off,** connect the circuit of Figure 40–5. Set the generator to its lowest output voltage. In this circuit a 0.47-μF capacitor is in series with a 0.1-μF capacitor. This is the $C_4 + C_5$ series combination in Table 40–2.

A5. Turn on the function generator. Increase the output of the generator until $V_{ac} = 5$ V rms at 200 Hz.

A6. Using the DMM, measure the voltage across the resistor V_R and across the two capacitors (that is, across AB), V_{C_T}. Use the oscilloscope, values converted to measure rms voltage, to verify your measurements. (Check with your instructor if there are significant differences in the measurements.) Record the values in Table 40–2 in the V_R and V_{C_T} columns. Turn **off** the function generator. This circuit will be reused in step A8.

A7. Calculate the current I_T in the circuit using Ohm's law. Use the measured values of voltage and resistance. Record your answer in Table 40–2 in the "I_T" column. Calculate capacitive reactance using your calculated I_T and the measured V_{C_T}. Record your answer in Table 40–2 in the "X_{C_T}" column. Calculate C_T using your calculated value of X_{C_T}. The line frequency f is = 200 Hz. Record your answer in Table 40–2 in the "C_T" column.

A8. Add a 0.1-μF capacitor in series with the two capacitors of step A4. The new circuit has a 0.47-μF and two 0.1-μF capacitors in series. This is the $C_3 + C_4 + C_5$ combination of Table 40–2.

A9. Turn on the function generator. Increase the voltage until $V_{ac} = 5$ V rms at 200 Hz. Measure V_R and V_{C_T} (across AB) as in A6. Record your values in Table 40–2. **Open** S_1. Turn **off** the generator. This circuit will be adapted for Part B.

A10. Calculate I_T, X_{C_T}, and C_T as in step A7. Record all answers in Table 40–2.

B. Total Capacitance of Capacitors in Parallel

B1. With the function generator set at its lowest output voltage, connect the circuit of Figure 40–6. In this case a 0.47-μF capacitor is in parallel with a 0.1-μF capacitor. This is the $C_4 + C_5$ parallel combination of Table 40–3 (p. 315).

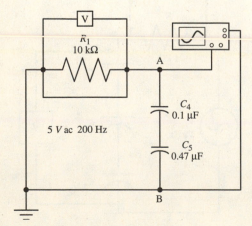

Figure 40–5. Circuit for procedure step A4.

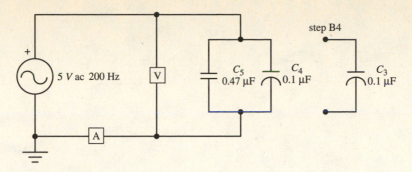

Figure 40–6. Circuit for procedure step B1.

B2. Turn **on** the function generator. Adjust the output of the function generator to 5 V_{ac} at 200 Hz. Measure the voltage across the parallel combination and the total current in the circuit. Record the values in Table 40–3. Turn **off** the generator.

B3. Applying Ohm's law for reactance, calculate X_{C_T} using the measured values of I_T and V_{C_T}. Record your answer in Table 40–3. Calculate total capacitance using your calculated value of X_{C_T}. The line frequency f is 200 Hz. Record your answer in Table 40–3, in the "Voltmeter-Ammeter Method" column. Applying the formula $C_T = C_1 + C_2 + \cdots + C_n$, calculate C_T using the measured values of C_4 and C_5 from Table 40–1. Record your answer in Table 40–3 in the "Formula Value" column.

B4. Add a third capacitor, 0.1-μF, in parallel as shown in Figure 40–6. In this case there are three capacitors connected in parallel: 0.47-μF and two 0.1-μF capacitors. This is the $C_3 + C_4 + C_5$ parallel combination of Table 40–3. Repeat step B2. After record-

ing the measurements in Table 40–3, turn **off** the generator.

B5. Repeat step B3 for the three-capacitor combination.

B6. With the $C_4 + C_5$ parallel combination disconnected from the circuit, measure the total capacitance with a LCR meter or capacitor analyzer. Record the value in Table 40–3 in the "Measured Value" column.

B7. Repeat step B6 with the $C_3 + C_4 + C_5$ parallel combination.

ANSWERS TO SELF TEST

1. 0.044
2. (a) 3.98; (b) 31.8; (c) 35.8
3. 0.32
4. 60
5. 1.5
6. (a) 1.2; (b) 2.2

TABLE 40–1. Measured Values of Capacitance

Capacitor Number	Rated Value	Measured Value	Coded Value
1	470 pF		
2	0.002 μF		
3	0.1 μF		
4	0.1 μF		
5	0.47 μF		

TABLE 40–2. Determining Total Capacitance of Capacitors in Series

Series Combination	Method 1		Method 2				
	Total Capacitance C_T, μF		Voltage across Resistor V_R, V_{ac}	Voltage across Series Combination V_{C_T}, V_{ac}	Current I_T, mA	Total Capacitive Reactance X_{C_T}, Ω	Total Capacitance C_T, μF
	Measured	Calculated					
$C_1 + C_2$							
$C_2 + C_3$							
$C_2 + C_3 + C_4$							
$C_4 + C_5$							
$C_3 + C_4 + C_5$							

TABLE 40–3. Determining Total Capacitance of Capacitors in Parallel

Parallel Combination	Total Current I_T, mA	Voltage across Parallel Combination V_{C_T}, V_{ac}	Total Capacitive Reactance (Calculated) X_{C_T}, Ω	Total Capacitance C_T, μF		
				Voltmeter-Ammeter Method	Formula Value	Measured Value
$C_4 + C_5$						
$C_3 + C_4 + C_5$						

QUESTIONS

1. Explain, in your own words, the effect of adding more capacitors in parallel on the total current of the parallel circuit, on the current through each branch capacitor, and on the voltage across each branch capacitor. Assume a constant-voltage source.

——

——

——

2. Explain, in your own words, the effect of adding more series capacitors in a series circuit on the total current of the series circuit, the current through each capacitor, and the voltage across each series capacitor. Assume a constant-voltage source.

_____

_____

_____

_____

3. Refer to your data for Method 1 in Table 40–2. Compare the measured values with the calculated values for the various combinations of series capacitors. Discuss any differences between measured and calculated values.

_____

_____

_____

_____

_____

4. Refer to your data for Method 2 in Table 40–2. Compare the measured values with the calculated values for the two series combinations used. Discuss any differences between measured and calculated values.

_____

_____

_____

_____

_____

5. Refer to Table 40–3. Which method appears to be most accurate for determining the value of capacitance? Discuss your reasons.

_____

_____

_____

_____

6. Compare the formulas for finding total capacitance for capacitors in series and in parallel with the formula for finding total resistance for resistors in series and in parallel.

_____

_____

_____

_____

_____

THE CAPACITIVE VOLTAGE DIVIDER

BASIC INFORMATION

AC Voltage Across a Capacitor

Ohm's law applied to ac circuits states that the current I in a circuit is equal to the applied voltage V divided by the total opposition to alternating current. The total opposition to current in ac is called *impedance*. The symbol for impedance is Z. Thus,

$$I = \frac{V}{Z} \tag{41-1}$$

In a circuit containing only capacitance C, such as in Figure 41–1, the total opposition to alternating current is the reactance X_C of capacitor C. Therefore, in the circuit of Figure 41–1, Z and X_C are the same, and

$$I = \frac{V}{X_C} \tag{41-2}$$

From formula (41–2) we see that the voltage drop V_C across capacitor C is equal to the product of the current I in the capacitor and the reactance X_C of the capacitor.

$$V_C = I \times X_C \tag{41-3}$$

Capacitive Voltage Divider

If an ac voltage is applied across capacitors connected in series, as in Figure 41–2 (p. 318), there is a voltage drop across each capacitor. This

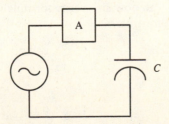

Figure 41–1. The total opposition to ac in this circuit is the capacitive reactance X_c caused by C.

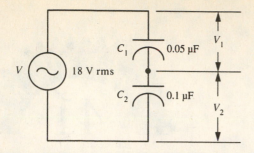

Figure 41-2. Voltage drops in a capacitive voltage divider.

voltage drop is equal to the current times the capacitive reactance. In a series circuit the current I is the same throughout the circuit, so the voltage across each capacitor is

$$V_1 = I \times X_{C_1} \tag{41-4}$$

$$V_2 = I \times X_{C_2}$$

The smaller the capacitor, the greater the reactance and, therefore, the greater the voltage drop across it in a capacitive voltage divider. In Figure 41-2 the reactance of a 0.05-μF capacitor is twice the reactance of a 0.1-μF capacitor; thus, $V_1 = 2V_2$. If the applied voltage V were 18 V in that circuit, then V_1 would be $= 12$ V and V_2 would be $= 6$ V; the sum of V_1 and V_2 would equal the applied voltage V.

Formula (41-4) can be written in terms of the respective capacitances:

$$V_1 = I \times X_{C_1} = I \times \frac{1}{2\pi f C_1} \tag{41-5}$$

$$V_2 = I \times X_{C_2} = I \times \frac{1}{2\pi f C_2}$$

The ratio of voltage drops and capacitances can be found by dividing V_1 by V_2.

$$\frac{V_1}{V_2} = \frac{\dfrac{1}{2\pi f C_1}}{\dfrac{1}{2\pi f C_2}} = \frac{2\pi f C_2}{2\pi f C_1}$$

The $2\pi f$ terms cancel, leaving

$$\frac{V_1}{V_2} = \frac{C_2}{C_1} \tag{41-6}$$

Using similar reasoning for series circuits with more than two capacitors, it can be shown that, in general,

$$\frac{V_a}{V_b} = \frac{C_b}{C_a}$$

where V_a is the voltage across capacitor C_a in series with capacitor C_b, and V_b is the voltage across capacitor C_b.

Now consider the total reactance X_{C_T} of series-connected capacitors,

$$V = I \times X_{C_T} \tag{41-7}$$

That is, the applied voltage V equals the product of I and the total reactance of the series-connected capacitors.

Since $V_1 = I \times X_{C_1}$, dividing V_1 by V leads to

$$\frac{V_1}{V} = \frac{I \times X_{C_1}}{I \times X_{C_T}} = \frac{X_{C_1}}{X_{C_T}} = \frac{\dfrac{1}{2\pi f C_1}}{\dfrac{1}{2\pi f C_T}}$$

and

$$\frac{V_1}{V} = \frac{C_T}{C_1}$$

Therefore,

$$V_1 = V \times \frac{C_T}{C_1} \tag{41-8}$$

Formula (41-8) states that the voltage drop across any capacitor C_1 in a series-connected capacitive voltage divider equals the product of the applied voltage and the ratio of the total capacitance C_T to the capacitance of C_1. Note that this is similar to the formula for voltage distribution in a series-connected resistive voltage divider, with the exception that in a resistive divider,

$$V_1 \text{ (across } R_1\text{)} = V \times \frac{R_1}{R_T}$$

That is, in a series capacitive divider the positions of C_1 and C_T are reversed, as compared with the positions of R_1 and R_T in a series-resistive divider.

It is now possible to apply formula (41-8) to the circuit of Figure 41-2. In that circuit

$$\frac{1}{C_T} = \frac{1}{C_1} + \frac{1}{C_2} = \frac{1}{0.05} + \frac{1}{0.1}$$

$$C_T = \frac{1}{30} = 0.0333 \ \mu F$$

Substituting $C_T = 0.0333$, $C_1 = 0.05$, and $V = 18$ V in formula (41-8) gives

$$V_1 = 18 \times \frac{0.0333}{0.05} = 12 \text{ V}$$

Similarly,

$$V_2 = 6 \text{ V}$$

These are the same results as were obtained in our earlier solution of this circuit.

We can now also verify formula (41-6), for according to that formula,

$$\frac{V_1}{V_2} = \frac{C_2}{C_1}$$

and

$$\frac{12}{6} = \frac{0.1}{0.05} = \frac{2}{1}$$

as expected.

SUMMARY

1. The voltage drop V_1 across a capacitor C_1 in a capacitive voltage divider is given by the formula $V_1 = I \times X_{C_1}$, where I is the current in the capacitor and X_{C_T} is the reactance of the capacitor.

2. In a series-connected capacitive voltage divider

$$\frac{V_a}{V_b} = \frac{C_b}{C_a}$$

where C_a and C_b are the capacitances of series-connected capacitors with voltage drops V_a and V_b, respectively.

3. The total current I in a capacitive voltage divider where the applied voltage is V and the total capacitance is C_T is

$$I = \frac{V}{X_{C_T}}$$

4. The voltage V_1 across a capacitor C_1 in a series-connected voltage divider with applied voltage V is

$$V_1 = V \times \frac{C_T}{C_1}$$

where C_T is the total capacitance of the series-connected capacitors.

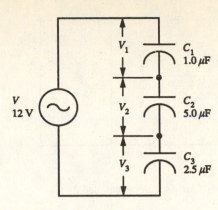

Figure 41–3. Circuit for Self-Test question 3.

2. In the circuit of Figure 41–2 $X_{C_1} = 1.5$ kΩ. Therefore,
   (a) $X_{C_2} = $ _____ Ω
   (b) $X_{C_T} = $ _____ Ω
   (c) $I = $ _____ mA

3. In the capacitive voltage divider of Figure 41–3, $C_1 = 1.0$ μF, $C_2 = 5.0$ μF, $C_3 = 2.5$ μF, and $V = 12$ V.
   (a) $V_1 = $ _____ V
   (b) $V_2 = $ _____ V
   (c) $V_3 = $ _____ V

MATERIALS REQUIRED

Instruments:
- DMM
- LCR meter or capacitor analyzer
- Function generator

Capacitors:
- 1   0.47-μF, 25-WV dc
- 2   0.1-μF, 25-WV dc
- 1   0.047-μF, 25-WV dc

SELF TEST

Check your understanding by answering the following questions:

1. In the circuit of Figure 41–1, $V = 10$ V, and $X_C = 500$ Ω. The current I in the circuit is _____ A.

PROCEDURE

If they are not already labeled, mark the capacitors with numbers 1 through 4 for identification.

1. Using a LCR meter or capacitor analyzer, measure the capacitance of each of the four capacitors used in this experiment. Record the values in Table 41–1 (p. 321).

2. With the function generator off, connect the circuit of Figure 41–4 (p. 320). Set the generator to its lowest output voltage.

3. Turn **on** the function generator. Increase the output voltage of the generator until $V_{ac} = 5$ V at 200 Hz. Maintain this voltage throughout this experiment.

4. Measure voltages across C_1 and C_2. Record the values in Table 41–2 (p. 321) for the $C_1 + C_2$ combination. Turn off the generator.

5. Connect another 0.1-μF capacitor in series with the 0.047-μF and 0.1-μF capacitors. This is combination $C_1 + C_2 + C_3$ in Table 41–2.

6. Turn on the generator. The input voltage should be 5 V ac at 200 Hz. Measure the voltage across each of the capacitors connected in series ($C_1, C_2,$ and C_3). Record the values in Table 41–2 for the $C_1 + C_2 + C_3$ combination.

7. Replace C_1 with C_4 so that the series combination consists of 0.1-μF, 0.1-μF, and 0.5-μF capacitors (combination $C_2 + C_3 + C_4$ in Table 41–2).

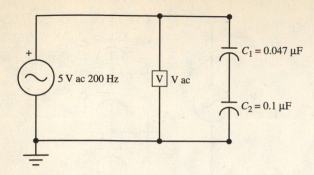

Figure 41–4. Circuit for procedure step 2.

8. Check V_{ac} and adjust to 5 V if necessary. Measure the voltage across each capacitor in the series combination. Record the values in Table 41–2 for the $C_2 + C_3 + C_4$ combination. Turn **off** the generator.

9. Using the measured values for C_1, C_2, C_3, and C_4, calculate the total capacitance for each of the three series combinations in Table 41–2. Record your answers in the table.

10. Using your calculated values for C_T, calculate the voltage across each capacitor using the voltage-ratio formula. Record your answers in Table 41–2 in the "Calculated Voltage" columns.

ANSWERS TO SELF TEST

1. 0.02
2. (a) 750; (b) 2.25 k; (c) 8;
3. (a) 7.5; (b) 1.5; (c) 3.0

TABLE 41–1.   Measured Values of Capacitance

Capacitor Number	1	2	3	4
Related value, μF	0.047	0.1	0.1	0.5
Measured value, μF				

TABLE 41–2.   Verifying the Voltage-Ratio Formula for Series-Connected Capacitors

Series Combination	V_{ac}	Measured Voltages, V V_1	V_2	V_3	V_4	Total Series Cap. (Cal.) C_T, μF	Calculated Voltage, V V_1	V_2	V_3	V_4
$C_1 + C_2$	5									
$C_1 + C_2 + C_3$	5									
$C_2 + C_3 + C_4$	5									

QUESTIONS

1. Explain, in your own words, the relationship between the voltage across a capacitor in a capacitive voltage divider and the voltage applied to the voltage divider.

_____

_____

_____

_____

_____

_____

_____

_____

2. In reference to Figure 41–4, if C_2 decreased in value what would happen to the capacitor voltage drops.

_____

_____

_____

_____

_____

_____

_____

_____

3. Refer to your data in Table 41–2. Do the measured values verify the relationship stated in Question 1? Cite specific data to support your answers.

_____
_____
_____
_____
_____
_____
_____
_____
_____
_____
_____
_____
_____

4. What conclusion can you make about the sum of the voltages V_1 through V_4 in a series-connected capacitive voltage divider? Refer to specific data in Table 41–2 to support your answer.

_____
_____
_____
_____
_____
_____
_____
_____
_____
_____
_____
_____
_____
_____

IMPEDANCE OF A SERIES *RL* CIRCUIT

BASIC INFORMATION

Impedance of a Series *RL* Circuit

The total opposition to alternating current in an ac circuit is called *impedance Z*. Ohm's law applied to ac circuits states that

$$I = \frac{V}{Z}$$

$$V = I \times Z$$

$$Z = \frac{V}{I}$$

Consider the circuit of Figure 42–1. If we assume that the inductance L through which alternating current flows has zero resistance, the current is impeded only by X_L. That is, $Z = X_L$. In this case, if $L = 8$ H and $f = 60$ Hz,

$$X_L = 2\pi fL = 6.28(60)(8) = 3014\ \Omega = 3.014\ \text{k}\Omega$$

How much current I will there be in the circuit if $V = 10$ V? Applying Ohm's law, we obtain

$$I = \frac{V}{X_L} \tag{42–1}$$

where I is in amperes, V is in volts, and X_L is in ohms. Therefore,

$$I = \frac{10\ \text{V}}{3.014\ \text{k}} = 3.32\ \text{mA}$$

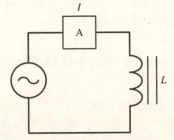

Figure 42–1. The current *I* in a circuit containing only inductance is limited by the inductive reactance X_L caused by *L*.

OBJECTIVES

1 To verify experimentally that the impedance Z of a series *RL* circuit is given by the formula

$$Z = \sqrt{R^2 + X_L^2}$$

2 To study the relationship between impedance, resistance, inductive reactance, and phase angle

If there is resistance R associated with the inductance L or if L is in series with a resistor of, say, 3 kΩ (Figure 42–2), the current will be less than 3.32 mA. How much current will flow, assuming the same X_L as previously computed? If an ammeter were placed in the circuit of Figure 42–2, it would measure 2.351 mA. That means the impedance of the circuit is

$$Z = \frac{V}{I} = \frac{10 \text{ V}}{2.35 \text{ mA}} = 4.253 \text{ k}\Omega$$

Obviously this is less than the arithmetic sum of R and X_L (which is 6.014 kΩ). Even though they have the same unit—ohms—resistance and reactance cannot be added arithmetically.

The current in R is out of phase with the voltage in X_L by 90°. Time-varying (ac) currents and voltages are often represented by *phasors*. Because I and V are related to Z we use the term phasor also when discussing Z.

The phasor diagram in Figure 42–3 shows R and X_L as phasors 90° apart. By convention R is shown at 0° (the horizontal line to the right). Inductive reactance X_L is represented by the vertical line in the upward direction, and capacitive reactance is represented by a vertical line in the downward direction. The impedance Z of a circuit is the *phasor sum* of R and X. Thus, in Figure 42–3 the phasor sum of R and X_L is the line Z. The angle θ is called the *phase angle* between R and Z.

The phasors can be added graphically by drawing R and X_L to some convenient scale and then completing the rectangle so that two sides are equal to R and two sides are equal to X_L. The diagonal of the rectangle is equal to Z. If this line is measured using the same scale as R and X_L, the value of Z can be found directly.

Note that the phasors R, X_L, and Z form a right triangle. In a right triangle, the sides are related by the following rule (referred to in mathematics texts as the pythagorean theorem):

The sum of the squares of the two sides of a right triangle is equal to the square of the longest side (called the *hypotenuse* of the triangle).

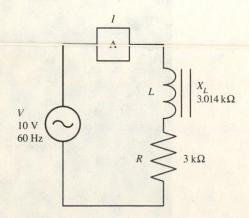

Figure 42–2. The impedance Z of a circuit containing R in series with L is greater than that of a circuit with L alone.

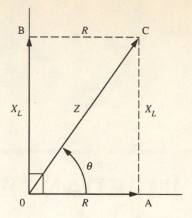

Figure 42–3. Phasor diagram of an *RL* series circuit. Impedance Z is the phasor sum of R and X_L.

In terms of the R, X_L, Z triangle in Figure 42–3, this relationship can be written as the formula

$$Z^2 = R^2 + X_L^2$$

To find Z, we take the square root of both sides of the formula.

$$Z = \sqrt{R^2 + X_L^2} \qquad \textbf{(42–2)}$$

If we go back to the circuit of Figure 42–2, formula (42–2) can be used to find the circuit Z and verify the measured I.

$$Z = \sqrt{(3 \text{ k})^2 + (3.014 \text{ k})^2}$$

$$Z = 4.253 \text{ k}\Omega$$

With 10 V applied, the current in the circuit of Figure 42–2 is

$$I = \frac{V}{Z} = \frac{10 \text{ V}}{4.253 \text{ k}\Omega}$$

$$I = 2.351 \text{ mA}$$

which verifies the measured values previously given.

Formula (42–2) and the illustrative problem point up a very important difference between the mathematics of ac circuits and that of dc circuits. In a dc circuit consisting of series-connected resistors, the total opposition to current R_T is simply the arithmetic sum of R_1, R_2, etc. In an ac circuit consisting of a series-connected resistor and inductance, the total opposition to current, the impedance Z, is *not* the arithmetic sum of R and X_L, but is the *phasor sum* of R and X_L.

Alternative Method for Solving Series *RL* Circuits

Certain relationships between the sides of a right triangle and its angles can be used to solve problems in ac circuits.

If the two sides of the triangle representing R and X_L are known, the third side of the triangle, Z, can be found without using formula (42–2). The angle between R and Z

is known as the phase angle. The quotient of X_L divided by R is known as the *tangent* of the phase angle, written as

$$\tan \theta = \frac{X_L}{R}$$

In the previous problem, $X_L = 3.014 \text{ k}\Omega$ and $R = 3 \text{ k}\Omega$. Thus,

$$\tan \theta = \frac{3.014 \text{ k}}{3 \text{ k}} = 1.005$$

A table of trigonometric functions or the trigonometric functions on a calculator can then be used to find the value of θ. The calculator method makes use of the $\tan^{-1}$ key. Once 1.005 is keyed into the calculator, press the $\tan^{-1}$ key or sequence of keys ($\tan^{-1}$ requires a two-key operation on most scientific calculators). The answer displayed will be the value of the angle θ. In the problem given

$$\theta = \tan^{-1} (1.005)$$

the displayed answer will be 45.143°. This is the value of the phase angle between R and Z.

The quotient of R divided by Z is known as the *cosine* of the phase angle:

$$\cos \theta = \frac{R}{Z}$$

or solving for Z.

$$Z = \frac{R}{\cos \theta} = \frac{3 \text{ k}\Omega}{\cos 45.143} = 4.253 \text{ k}\Omega$$

Remember that for practical purposes the answers given by the calculator can be rounded to three decimal places.

The following problem is solved using a calculator throughout without rounding until the final answer.

Problem. A series RL circuit has $R = 40 \ \Omega$ and $X_L = 25 \ \Omega$. Find the impedance of the circuit.

Solution. Draw the phasor diagram (Figure 42–4).

$$\theta = \tan^{-1} \left(\frac{X_L}{R} \right) = \tan^{-1} \left(\frac{25}{40} \right) = 32.0°$$

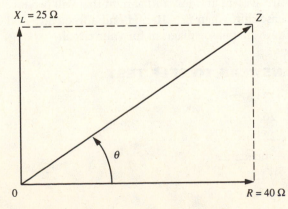

$X_L = 25 \ \Omega$

Z

θ

0

$R = 40 \ \Omega$

Figure 42–4. Phasor diagram for series RL problems.

$$Z = \frac{40}{\cos 32.0} = 47.2 \ \Omega$$

Using formula (42–2)

$$Z = \sqrt{40^2 + 25^2} = 47.170 \ \Omega$$

which agrees with the previous answer.

It should be noted that many scientific calculators can also solve formula (42–2) with just a few keystrokes. The student should refer to the instruction manual supplied with the calculator being used.

SUMMARY

1. In an ac circuit the total opposition to current is called the impedance of the circuit. The symbol for impedance is Z. The unit of impedance is the ohm.

2. Ohm's law extended to ac circuits states that the current I equals the ratio of the applied voltage V and impedance Z. Thus,

$$I = \frac{V}{Z}$$

3. In a series RL circuit, the impedance Z is the phasor sum of R and X_L, where X_L is out of phase with R by an angle of 90° (Figure 42–3).

4. The numerical value of Z can be found from the impedance right triangle in Figure 42–3 using the formula

$$Z = \sqrt{R^2 + X_L^2}$$

5. In Figure 42–3 if θ is the phase angle between Z and R, then

$$\theta = \tan^{-1} \left(\frac{X_L}{R} \right)$$

6. Impedance Z may be calculated if θ and R are known. Thus,

$$Z = \frac{R}{\cos \theta}$$

where the angle θ is calculated from the formula in item 5 above.

S E L F   T E S T

Check your understanding by answering the following questions:

1. In the series RL circuit (Figure 42–2), $X_L = 100 \ \Omega$ and $R = 200 \ \Omega$. $Z = $ _____ Ω.

2. In the circuit of question 1, $\theta = $ _____ °.

3. In the circuit of question 1, $\dfrac{R}{\cos \theta} = $ _____ Ω.

4. The ratio R/cos θ, where θ is the phase angle between Z and R, gives the _____ of the circuit of Figure 42–2.

5. In a series RL circuit, $R = 45\ \Omega$, $X_L = 45\ \Omega$, and the applied voltage $V = 10$ V. The current I in the circuit is _____ A.

6. In the circuit of Figure 42–2, $V_R = 15$ V. The current in the inductor L is _____ mA.

MATERIALS REQUIRED

Instruments:
- DMM
- Function generator
- Capacitor/inductor analyzer or LCR meter

Resistors:
- 1 3.3 kΩ, ½,-W 5%

Inductor:
- 1 47 mH
- 1 100 mH

PROCEDURE

....

1. Using a capacitor/inductor analyzer or LCR meter, measure the 47 mH and 100 mH inductors to verify their values. Record the measured values in Table 42–1.

2. With the function generator's power switch in the **off** position, connect the circuit shown in Figure 42–5.

3. Turn the function generator on and adjust its output for a value of 5 V_{p-p} at a frequency of 5 kHz using the oscilloscope. Record this input value in Table 42–1 in the V_{in} column.

4. Measure the V_{p-p} values across the resistor and inductor. Remember to use the ADD Mode and INVERT button of your oscilloscope to measure across L_1. Record these values in the Table 42–1.

5. Using the measured voltage across R_1 and its resistance value, calculate and record the series circuit current. Since the resistor and inductor are in series, this calculated current for R_1 is the same for L_1.

6. Using the measured inductor voltage drop and its series current value, calculate and record L_1's inductive reactance.

7. Next, use Ohm's law and the series reactance equation (42–2) to solve for the circuit's impedance. Record both values in Table 42–1.

8. Replace the 47 mH inductor with the 100 mH inductor measured in step #1.

9. Repeat steps 2 through 7, recording all values in the respective 100 mH row of Table 42–1.

10. Examine Table 42–2. Using the values of impedance (calculated from V_L/I_L) in Table 42–1, calculate the phase angle θ and impedance using the phase-angle relationships. Complete Table 42–2 for the 47 mH and 100 mH inductor circuits.

11. In the space below Table 42–2 draw the impedance phasor diagrams for the respective circuits. If the triangle's individual sides are drawn somewhat to scale, the impedance angles will be more clear.

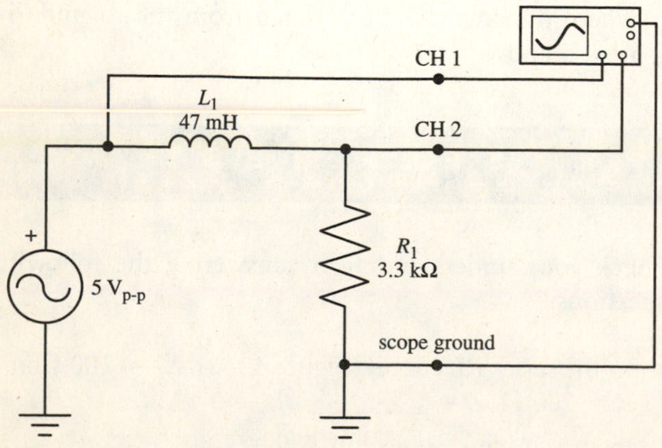

Optional Activity ·

This activity requires the use of electronic simulation software. Design and test a RL circuit that will produce a 45° phase shift when $L = 10$ mH and a frequency of 1 kHz. Draw the phasor diagram for your circuit.

ANSWERS TO SELF TEST

1. 223.6
2. 26.6
3. 223.6
4. impedance
5. 0.157
6. 5

Figure 42–5. Circuit for procedure step 2.

EXPERIMENT
42

TABLE 42–1. Verifying the Impedance Formula for an *RL* Circuit

Inductor Value, mH		V_{in}, V_{p-p}	Voltage Across Resistor V_R, V_{p-p}	Voltage Across Inductor V_L, V_{p-p}	Current Calculated V_R/R mA	Inductive Reactance (calculated) V_L/I_L, Ω	Circuit Impedance (calculated) Ohm's Law V_T/I_T, Ω	Circuit Impedance (calculated) R-X_L, Ω
Rated	Meas.							
47								
100								

TABLE 42–2. Determining Phase Angle and Impedence

Inductor Value, mH		Inductive Reactance (from Table 42-1) Ω	$\tan \theta = \dfrac{X_L}{R}$	Phase angle θ, degrees	Impedance $Z = \dfrac{R}{\cos \theta}$, Ω
Rated	Measured				
47					
100					

QUESTIONS

1. Explain, in your own words, the relationship between resistance, inductive reactance, and impedance in a series *RL* circuit.

_____

_____

_____

_____

_____

2. Explain, in your own words, how increases and decreases in inductance, with a constant resistance, affect the phase angle of a series *RL* circuit.

_____

_____

_____

_____

_____

3. If the series inductor, used in Figure 42–5, were to short, what would happen to the circuit's impedance angle and current.

_____

_____

_____

_____

4. Under what conditions will the impedance angle, θ, be 45°? Does your lab data support this?

_____

_____

_____

_____

5. Refer to your data in Tables 42–1 and 42–2. Have the results in this experiment verified Formula (42–2)? Cite specific data to support your answer.

_____

_____

_____

_____

_____

VOLTAGE RELATIONSHIPS IN A SERIES *RL* CIRCUIT

BASIC INFORMATION

Phasors

A *phasor* is a quantity that is identified by two characteristics, *amplitude* and *direction*. In electricity we refer to sinusoidal voltages and currents as *phasor* quantities. Because it is related to sinusoidal voltages and currents, impedance will also be referred to as a phasor quantity.

Sinusoidal voltages or currents of the same frequency may be added and subtracted when they are represented by phasors.

Phase Angle Between Applied Voltage and Current in a Series *RL* Circuit

Because current is the same in every part of a series circuit, current I is shown as the reference phasor in considering the phase relations among V_R, V_L, and V in the series *RL* circuit [Figure 43–1(a) (p. 330)]. Current in an inductance lags the voltage across the inductance by 90°. The phasor V_L in Figure 43–1(b) leads I by 90°; therefore, V_L leads V_R by 90°.

Phasor Sum of V_R and V_L Equals Applied Voltage V

Voltages V_R and V_L are, respectively, the voltages across R and L in the series *RL* circuit [Figure 43–1(a)]. Does the arithmetic sum of V_R and V_L equal the applied voltage V? No, because V_R and V_L are 90° out of phase. But V is the phasor sum of V_R and V_L. This fact is shown in Figure 43–1(b), where V_R and V_L are the legs of a right triangle and V is their phasor sum. Applying the pythagorean theorem, we have

$$V = \sqrt{V_R^2 + V_L^2} \qquad (43\text{--}1)$$

What is the phase relationship between the applied voltage V and the current I in the inductive circuit? In Figure 43–1(b), I is seen to lag the applied voltage V by an angle θ. This angle θ is the same as the angle θ between Z and R in the impedance phasor diagram (Figure 43–2, p. 330).

Figure 43–2(a) is Figure 43–1(b) redrawn, with the composition of each of the voltages as shown. Thus, V_R is the product of I and R, V_L is the product of I and X_L, and V is the product of I and Z. Since I is a common factor in each of these products, it may be canceled, leaving the

OBJECTIVES

1 To measure the phase angle θ between the applied voltage V and the current I in a series *RL* circuit

2 To verify experimentally that the relationships among the applied voltage V, the voltage V_R across R, and the voltage V_L across L are described by the formulas

$$V = \sqrt{V_R^2 + V_L^2}$$

$$V_R = V \times \frac{R}{Z}$$

$$V_L = V \times \frac{X_L}{Z}$$

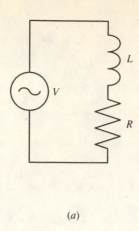

(a)

(b)

Figure 43–1. Phase relations in a series *RL* circuit.

impedance diagram [Figure 43–2(b)]. It is evident, therefore, that the angle θ is the same in Figures 43–1(b) and 43–2(b).

Further study of Figure 43–1(b) shows the relationship between V, V_R, V_L, and the phase angle θ. From the voltage phasor diagram

$$\frac{V_R}{V} = \cos \theta \qquad (43-2)$$

But from the impedance triangle we know that

$$\cos \theta = \frac{R}{Z}$$

Therefore,

$$\frac{V_R}{V} = \frac{R}{Z}$$

or the voltage across the resistor is

$$V_R = V \times \frac{R}{Z} \qquad (43-3)$$

From the voltage triangle [Figure 43–1(b)],

$$\frac{V_L}{V_R} = \tan \theta = \frac{X_L}{R}$$

Therefore,

$$V_L = V_R \times \frac{X_L}{R}$$

Substituting V_R from formula (43–3), we obtain

$$V_L = V \times \frac{R}{Z} \times \frac{X_L}{R}$$

$$V_L = V \times \frac{X_L}{Z} \qquad (43-4)$$

Formulas (43–3) and (43–4) may be used to calculate the voltages V_R and V_L in a series *RL* circuit where the applied voltage V, the resistance R, and the inductive reactance X_L are known.

Problem. If 15 V is applied across a circuit consisting of a 40-Ω resistor in series with an inductance whose re-

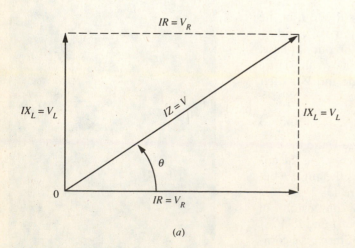

(a)

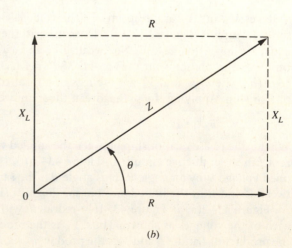

(b)

Figure 43–2. The phase angle of the voltage phasor diagram is the same as the phase angle of the impedance phasor diagram.

actance is 25 Ω, what are the values of the phase angle θ between V and I, the voltage V_R across R, and the voltage V_L across L?

Solution. Using formulas (43–3) and (43–4), we obtain the solutions using a scientific calculator.

$$\theta = \tan^{-1}\frac{X_L}{R} = \tan^{-1}\frac{25}{40} = 32.005°$$

$$Z = \frac{R}{\cos\theta} = \frac{40}{\cos 32.005°} = 47.170\ \Omega$$

(Answers have been rounded to three decimal places.)

$$V_R = V \times \frac{R}{Z}$$

$$= 15 \times \frac{40}{47.170} = 12.720\ \text{V}$$

$$V_L = V \times \frac{X_L}{Z}$$

$$= 15 \times \frac{25}{47.170} = 7.950\ \text{V}$$

We can check the solutions using formula (43–1). By substituting the calculated values of V_R and V_L in the formula, we find that

$$V = \sqrt{12.720^2 + 7.950^2}$$

$$= 15.000\ \text{V}$$

The calculated value of V is the same as the given applied voltage V. Our solutions are therefore verified.

SUMMARY

1. In a series RL circuit the current I lags the applied voltage V by an angle θ, called the phase angle.

2. The phase angle θ between V and I is the same as the angle θ between Z and R in the impedance phasor diagram of the RL circuit. The angle θ is also the same as the angle between V and V_R in Figure 43–1(b).

3. The value of θ depends on the relative values of X_L, R, and Z and may be calculated from the following formula:

$$\theta = \tan^{-1}\left(\frac{X_L}{R}\right)$$

4. In a series RL circuit the voltage drop V_L across the inductance leads the voltage drop V_R across the resistor by 90°.

5. Voltages V_R and V_L are added using their phasors to obtain the applied voltage V in the circuit. That is, V is the phasor sum of V_R and V_L.

6. The relationship between the applied voltage V, the voltage drop V_R across R, and the voltage drop V_L across L is given by the formula

$$V = \sqrt{V_R^2 + V_L^2}$$

7. If the applied voltage V and R, X_L, and Z are known in a series RL circuit, then V_R and V_L may be calculated from the formulas

$$V_R = V \times \frac{R}{Z}$$

$$V_L = V \times \frac{X_L}{Z}$$

S E L F   T E S T

Check your understanding by answering the following questions:

1. In the circuit of Figure 43–1 the measured value of $V_R = 30$ V, and $V_L = 20$ V. The applied voltage V must then equal _____ V.

2. (True/False) The phase angle between V and V_R is the same as the phase angle between the applied voltage V and the current I through the resistor in a series RL circuit. _____

3. The phase angle θ between V and I in question 1 is _____°.

4. In the circuit of Figure 43–1, $X_L = 68\ \Omega$, $R = 82\ \Omega$, and $V = 12$ V. In this circuit
   (a) $Z =$ _____ Ω
   (b) $\theta =$ _____°
   (c) $V_R =$ _____ V
   (d) $V_L =$ _____ V

MATERIALS REQUIRED

Instruments:
- Dual-trace oscilloscope
- DMM
- Function generator

Resistors (½-W, 5%):
- 1   1 kΩ
- 1   3.3 kΩ

Inductor:
- 100 mH coil

PROCEDURE

. . . .

1. Using an ohmmeter, measure the resistance of the 3.3 kΩ and 1 kΩ resistors. Record the values in Table 43–1.

2. With the function generator **off,** connect the circuit shown in Figure 43–3.

3. Turn **on** the function generator and, using Channel No. 1 of your oscilloscope, adjust its output to 10 V_{p-p} at a frequency of 5 kHz. Adjust your oscilloscope's controls to display 1 complete cycle that fills the graticule horizontally.

4. Notice that the trigger input should be set to Channel No. 2. In a series circuit, the current is the same throughout the circuit. Therefore, in a *series* circuit, the circuit current will be used as the reference point or 0° when taking measurements and drawing phasor diagrams. The voltage drop across R_1 is the result of the current flowing through it.

5. Adjust the oscilloscope's LEVEL and SLOPE controls so that V_{R1} fills the scope's graticule with 1 complete cycle. Horizontally most scopes are 10 divisions wide and 1 complete cycle occurs in 360°. If the display is adjusted for 10 divisions, this will result in a 36°/Div. setting on the oscilloscope.

6. With the Vertical MODE switch set to DUAL-ALT, measure the resulting phase shift between the circuit's current (represented by the V_{R1} sine wave) and the input voltage (V_{in}). Record your results in Table 43–1 for the 3.3 kΩ row.

7. Repeat steps 2 through 6 using the 1 kΩ resistor in place of the 3.3 kΩ.

8. Measure the voltage drop across the 1 kΩ resistor, (V_R) and across the inductor (V_L). Record these values in Table 43–2 for the 1 kΩ row. Turn **off** the oscilloscope and function generator.

9. Calculate the current in the circuit using Ohm's law with the measured values of V_R and R. Record your answer in Table 43–2 for the 1 kΩ resistor.

10. Calculate the inductive reactance X_L of the inductor using Ohm's law for inductors with the measured value of V_L and the calculated value of I. Record your answer in Table 43–2.

11. Using the calculated value of X_L from step 10 and the measured value of R, calculate the phase angle θ

$$\theta = \tan^{-1}\left(\frac{X_L}{R}\right)$$

Record your answer in Table 43–2 for the 1 kΩ resistor.

12. Repeat steps 8 through 11 for the 3.3 kΩ resistor.

13. Using the measured values of V_R and V_L for the 1 kΩ resistor, calculate V_{p-p} with the square root formula

$$V = \sqrt{V_R^2 + V_L^2}$$

Record your answer in the "Applied Voltage (Calculated)" column in Table 43–2. Repeat the calculation for V_R and V_L with the 3.3 kΩ resistor. Record your answer in Table 43–2.

14. In the space below Table 43–2, draw the respective impedance and voltage phasor diagrams for both the 3.3 kΩ and 1 kΩ circuits.

ANSWERS TO SELF TEST

1. 36.1

2. true

3. 33.7

4. (a) 106.5; (b) 39.7; (c) 9.24; (d) 7.66

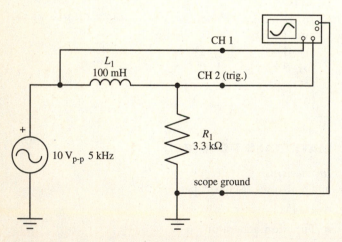

Figure 43–3. Circuit for procedure step 2.

EXPERIMENT 43

TABLE 43–1. Using the Oscilloscope to Find the Phase Angle θ in a Series *RL* Circuit

Resistance R, Ω		Width of Sine Wave D, divisions	Distance Between Zero Points d, divisions	Phase Angle θ, degrees
Rated Value	Measured Value			
3.3 k				
1 k				

TABLE 43–2. Phase Angle θ and Voltage Relationships in a Series *RL* Circuit

Resistor Rated Value, Ω	Applied Voltage V_{pp}, V	Voltage across Resistor V_R, $V_{p\text{-}p}$	Voltage across Inductor V_L, $V_{p\text{-}p}$	Current (Calculated) I, mA	Inductive Reactance X_L, (Calculated), Ω	Phase Angle θ (Cal. from X_L and R), degrees	Applied Voltage (Calculated) V_{pp}, V
3.3 k							
1 k							

QUESTIONS

1. Explain, in your own words, the relationship between the voltage across a resistance, the voltage across an inductance, and the applied voltage in a series RL circuit.

_____

_____

_____

_____

_____

_____

_____

2. Refer to your data in Tables 43–1 and 43–2. Is the phase angle θ found using the oscilloscope the same (for comparable resistors) as the phase angle found using resistance/inductive reactance formulas? Should the two angles be the same? Explain your answer.

_____

_____

_____

_____

_____

_____

_____

_____

_____

3. If L_1's inductance value decreased from 100 mH to 50 mH, what would the resulting voltage drop across R_1 do?

_____

_____

_____

_____

_____

_____

_____

_____

4. Do the results of this experiment verify Formula (43–1)? Refer to specific data to support your answer.

_____

_____

_____

_____

IMPEDANCE OF A SERIES *RC* CIRCUIT

BASIC INFORMATION

The impedance of a series *RC* circuit can be found the same way as the impedance of a series *RL* circuit. The same formulas can be used except that X_C replaces X_L. Thus,

$$Z = \sqrt{R^2 + X_C^2} \qquad (44\text{--}1)$$

Both R and X_C are phasor quantities and must be added using phasors to obtain Z.

Figure 44–1(*a*) (p. 336) is a series *RC* circuit whose impedance diagram is shown in Figure 44–1(*b*). Note that the phasor representing X_C is a vertical line in the downward direction (recall that the X_L phasor was a vertical line in the upward direction).

Problem 1. If in Figure 44–1(*a*), $R = 300 \ \Omega$, $X_C = 400 \ \Omega$, and $V = 25$ V, find Z and I.

Solution. We can find the impedance Z using formula (44–1):

$$
\begin{aligned}
Z &= \sqrt{R^2 + X_C^2} \\
&= \sqrt{300^2 + 400^2} = \sqrt{250{,}000} \\
Z &= 500 \ \Omega
\end{aligned}
$$

From Ohm's law

$$I = \frac{V}{Z} = \frac{25 \text{ V}}{500 \ \Omega}$$

$$I = 0.05 \text{ A or } 50 \text{ mA}$$

Alternative Method for Calculating Impedance · · · · · ·

As with series *RL* circuits, the $\tan^{-1}$ function on a scientific calculator may be used to find the phase angle θ between R and Z when R and X_C are known.

$$\theta = \tan^{-1}\left(\frac{-X_C}{R}\right) \qquad (44\text{--}2)$$

The impedance can then be found using this value of θ and the formula

$$Z = \frac{R}{\cos \theta} \qquad (44\text{--}3)$$

OBJECTIVES

1. To verify experimentally that the impedance Z of a series *RC* circuit is given by the formula
$$Z = \sqrt{R^2 + X_C^2}$$

2. To study the relationships among impedance, resistance, capacitive reactance, and phase angle

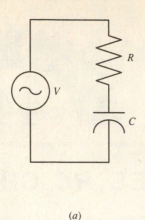
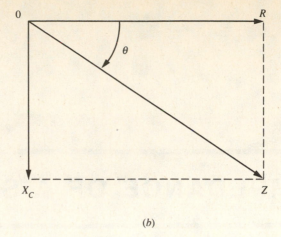

(a)   (b)

Figure 44–1. Phase relations in a series *RC* circuit.

Again, we can use the scientific calculator to produce the answer.

Problem 2. The values of problem 1 will be used. In addition to finding Z and I, the phase angle θ is also required.

Solution.

$$\theta = \tan^{-1}\left(\frac{-X_C}{R}\right)$$

$$= \tan^{-1}\left(\frac{-400}{300}\right)$$

$$= -53.130°$$

$$Z = \frac{R}{\cos \theta} = \frac{300 \ \Omega}{\cos -53.130}$$

$$Z = 500 \ \Omega$$

which is the same as the answer in problem 1. To find I use the formula $I = V/Z$.

$$I = \frac{V}{Z} = \frac{25 \ V}{500 \ \Omega} = 50 \ mA$$

Using Ohm's Law to Verify *I*, *V*, and *Z* · · · ·

One form of Ohm's law for ac circuits is

$$Z = \frac{V}{I} \qquad\qquad (44–4)$$

This formula can be used to verify the relationships among R, X_C, and Z. In the series circuit of Figure 44–1(*a*), the voltage V across the combination of R and C can be measured as can the current I in the circuit. The impedance Z can then be calculated using formula (44–4) and the measured value of V and I. If the value calculated in this way is the same as that calculated using formulas (44–1) and (44–3), the relationships among θ, X_C, and R have been verified.

Problem 3. In the series circuit of Figure 44–1(*a*), $R = 50 \ \Omega$, $X_C = 120 \ \Omega$, and $V = 10 \ V$. An ac ammeter connected into the circuit measures 77 mA. Verify the relationships among R, X_C, and θ.

Solution. The impedance can be found using formula (44–1)

$$Z = \sqrt{R^2 + X_C^2}$$

$$= \sqrt{50^2 + 120^2}$$

$$Z = 130 \ \Omega$$

Using Ohm's law to find Z, we have

$$Z = \frac{V}{I} = \frac{10 \ V}{77 \ mA}$$

$$Z = 129.870, \text{ or } Z = 130 \ \Omega$$

This verifies the value found using formula (44–1). The phase angle θ can be found from formula (44–2):

$$\theta = \tan^{-1}\left(\frac{-120}{50}\right)$$

$$\theta = -67.380°$$

Also,

$$Z = \frac{R}{\cos \theta} = \frac{50 \ \Omega}{\cos -67.380} = 130 \ \Omega$$

Again, the relationships among R, X_C, and θ lead to the same answer.

SUMMARY

1. In a series-connected *RC* circuit, the impedance Z is the phasor sum of R and X_C, where X_C lags R by 90° [Figure 44–1(*b*)].

2. From the impedance right triangle in Figure 44–1(*b*), the numerical value of Z can be found from the formula

$$Z = \sqrt{R^2 + X_C^2}$$

3. In Figure 44–1(b), if θ is the phase angle between Z and R, then

$$\theta = \tan^{-1}\left(\frac{-X_C}{R}\right)$$

4. The impedance Z may also be calculated if θ and R are known. Thus,

$$Z = \frac{R}{\cos\theta}$$

where θ is the phase angle calculated from the formula in item 3 above.

5. The impedance Z of an RC circuit may be found experimentally if the applied voltage V and the current I in the circuit are known. Z is calculated using Ohm's law:

$$Z = \frac{V}{I}$$

6. If the value of Z is substantially the same for each of the methods used, the right-triangle relationship among Z, X_C, R, and θ has been verified.

SELF TEST

Check your understanding by answering the following questions.

1. In the series RC circuit [Figure 44–1(a)], $R = 300\ \Omega$, $X_C = 120\ \Omega$, and $Z =$ _____ Ω.

2. In the circuit of question 1, $\theta =$ _____ °.

3. In the circuit of question 1, $R/\cos\theta =$ _____ Ω.

4. The ratio $R/\cos\theta$, where θ is the phase angle between Z and R, gives the _____ of the circuit of Figure 44–1(a).

5. In a series RC circuit, $R = 120\ \Omega$, $X_C = 150\ \Omega$, and the applied voltage $V = 5\ V$, the current $I =$ _____ mA.

6. In the circuit of question 1, $V_R = 45\ V$. The current I in the capacitor C is _____ mA.

MATERIALS REQUIRED

Instruments:
- DMM
- Function Generator
- Capacitor/inductor analyzer of LCR meter

Resistors:
- 1   2 k-Ω, ½-W, 5%

Capacitors:
- 1   0.033 μF
- 1   0.1 μF

PROCEDURE

1. Using a capacitor/inductor analyzer or LCR meter, measure the 0.033 μF and 0.1 μF capacitors to verify their values. Record the measured values in Table 44–1 (p. 339).

2. With the function generator's power switch in the **off** position, connect the circuit shown in Figure 44–2.

3. Turn the function generator **on** and adjust its output for a value of 10 $V_{p\text{-}p}$ at a frequency of 1 kHz using the oscilloscope. Record this input value in Table 44–1 in the V_{in} column.

4. Measure the $V_{p\text{-}p}$ values across the resistor and capacitor. Remember to use the ADD Mode and INVERT button of the oscilloscope to measure across C_1. Record these values in the Table 44–1.

5. Using the measured voltage across R_1 and its resistance value, calculate and record the series circuit cur-

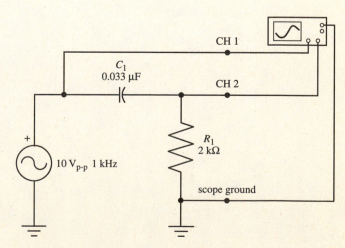

Figure 44–2. Circuit for procedure step 2.

rent. Since the resistor and capacitor are in series, this calculated current for R_1 is the same for C_1.

6. Calculate and record C_1's capacitive reactance value using the formula $X_C = \dfrac{1}{2\pi f C}$. Also, using the measured capacitor voltage drop and its series current value, calculate and record C_1's capacitive reactance.

7. Next use Ohm's law and the series reactance equation (44–1) to compute the circuit's impedance. Record both values in Table 44–1.

8. Replace the 0.033 μF capacitor with the 0.1 μF capacitor measured in step 1.

9. Repeat steps 3 through 7, recording all values in the respective 0.1 μF row of Table 44–1.

10. Using the values of impedance (calculated from V_C/I_C) in Table 44–1, calculate the phase angle θ and

impedance using the phase-angle relationships. Complete Table 44–2 for the 0.033 μF and 0.1 μF capacitor circuits.

11. In the space below Table 44–2, draw the impedance phasor diagrams for the respective circuits. If the triangle's individual sides are drawn somewhat to scale, the impedance angles will be more clear.

ANSWERS TO SELF TEST

1. 323
2. −21.8
3. 323
4. impedance
5. 26
6. 150

TABLE 44–1.   Determining the Impedance of a Series *RC* Circuit

Capacitor Value, μF		V_{in}, V_{p-p}	Voltage Across Resistor V_{Rp-p}	Voltage Across Capacitor Vc_{p-p}	Current Calculated V_R/R mA_{p-p}	Capacitive Reactance (calculated) Xc, Ω	Capacitive Reactance (calculated) Vc/Ic, Ω	Circuit Impedance (calculated) Ohm's Law V_T/I_T, Ω	Circuit Impedance (calculated) R-X_C, Ω
Rated	Measured								
.033									
0.1									

TABLE 44–2.   Determining Phase Angle and Impedance of a Series *RC* Circuit

Capacitor Value, μF		Capacitive Reactance (from Table 44-1) Ω	$\tan \theta = \dfrac{X_C}{R}$	Phase Angle θ, degrees	Impedance $Z = \dfrac{R}{\cos \theta}$ Ω
Rated	Measured				
.033					
0.1					

QUESTIONS

1. Explain, in your own words, the relationship between resistance, capacitive reactance, and impedance in a series *RC* circuit.

_____

_____

_____

_____

_____

2. Explain, in your own words, how increases and decreases in capacitive reactance affect the phase angle of a series *RC* circuit with a fixed resistance.

_____

_____

_____

_____

_____

3. Referring to the data in Tables 44–1 and 44–2, when using the 0.033 μF capacitor was the circuit resistive dominant or capacitive dominant?

_____

_____

_____

_____

4. If the 2 kΩ resistor increased to 4 kΩ, what would happen to the circuit's impedance and phase angle?

_____

_____

_____

_____

_____

5. What factors might introduce errors in this experiment? Include both human and mechanical factors.

_____

_____

_____

_____

VOLTAGE RELATIONSHIPS IN A SERIES *RC* CIRCUIT

BASIC INFORMATION

The relationships among R, X, Z, and θ in RC and RL circuits are very similar. The difference is that in an RL circuit, current I lags the applied voltage V, whereas in an RC circuit, current I *leads* the applied voltage V.

Phase Relationships Between the Applied Voltage and Current in a Series *RC* Circuit

The current I is common in every part of the RC series circuit of Figure 45–1(*a*) (p. 342). Therefore, current is used as the reference phasor in the phasor diagram showing V_R and V_C. Because current and voltage in a resistor are in phase, the voltage phasor V_R is shown in Figure 45–1(*b*) on the same line as the current phasor. But *current* in a *capacitor leads the voltage* across the capacitor by 90°. Therefore, the phasor V_C is shown lagging the current I and V_R by 90°.

Phasor V_R is the voltage across R, and V_C is the voltage across C in the series RC voltage divider [Figure 45–1(*a*)]. As in the case of an RL circuit, the applied voltage V is the phasor sum of V_R and V_C, as shown in Figure 45–1(*b*). We also see that V is the hypotenuse of a right triangle of which V_R and V_C are the legs. Therefore, applying the pythagorean formula, we obtain

$$V = \sqrt{V_R^2 + V_C^2} \qquad \text{(45–1)}$$

Figure 45–1(*b*) also shows the phase relationship between the applied voltage V and the current I in the series RC circuit. The current I is seen to lead the applied voltage V by the angle θ.

The angle θ by which current leads the applied voltage in a series RC circuit is the same as the angle θ between the impedance phasor Z and the resistance phasor R. Figure 45–2 (p. 342) is Figure 45–1(*b*) redrawn showing the phase relationships among V, V_R, and V_C. Voltage V_R is the product of I and Z. Because I is a common factor in each of these products, it may be canceled, leaving the impedance diagram [Figure 45–2(*b*)]. This diagram shows that the angle θ is the same in Figures 45–1(*b*) and 45–2(*b*).

Further study of Figure 45–1(*b*) shows the relationship among V, V_R, V_C, and the phase angle θ. From the voltage phasor diagram

$$\frac{V_R}{V} = \cos \theta \qquad \text{(45–2)}$$

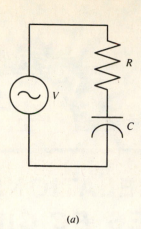

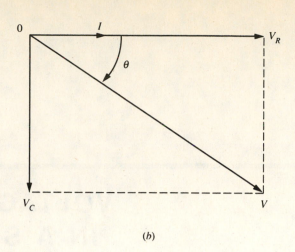

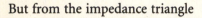

Figure 45–1. Phase relations in a series RC circuit.

But from the impedance triangle

$$\cos \theta = \frac{R}{Z}$$

Therefore,

$$\frac{V_R}{V} = \frac{R}{Z}$$

or the voltage across the resistor is

$$V_R = V \times \frac{R}{Z} \qquad (45\text{–}3)$$

From the voltage triangle [Figure 45–1(b)]

$$\frac{V_C}{V_R} = \tan \theta = \frac{-X_C}{R}$$

Therefore,

$$V_C = V_R \times \frac{V_C}{R}$$

Substituting V_R from formula (45–3), we have

$$V_C = V \times \frac{R}{Z} \times \frac{X_C}{R}$$

$$V_C = V \times \frac{X_C}{Z} \qquad (45\text{–}4)$$

Formulas (45–3) and (45–4) may be used to calculate the voltages V_R and V_C in a series RC circuit when the applied voltage V, the resistance R, and the capacitive reactance X_C are known.

Problem. If 12 V is applied across a circuit consisting of a 47-Ω resistor in series with a capacitor whose reactance is 100 Ω, what is the phase angle θ between V and I. What is the phase angle between the voltage V_R across R, and the voltage V_C across C?

Solution. Using formulas (45–3) and (45–4), we obtain the solutions using a scientific calculator.

$$\theta = -64.826°$$

This is also the angle between V_R and V_C.

$$Z = \frac{47}{\cos -64.826} = 110.494 \ \Omega$$

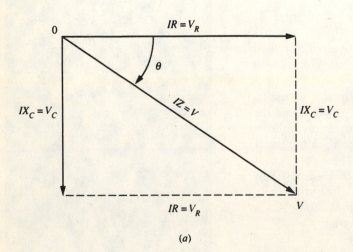

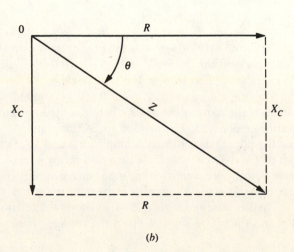

Figure 45–2. The phase angle of the voltage phasor diagram is the same as the phase angle of the impedance phasor diagram.

(Answers have been rounded off to three decimal places.)

$$V_R = V \times \frac{R}{Z} = 12 \text{ V} \times \frac{47 \text{ }\Omega}{110.494 \text{ }\Omega}$$

$$= 5.104 \text{ V}$$

$$V_C = V \times \frac{X_C}{Z} = 12 \text{ V} \times \frac{100 \text{ }\Omega}{110.494 \text{ }\Omega}$$

$$= 10.860 \text{ V}$$

We can check the solutions using formula (45–1). By substituting the calculated values of V_R and V_C into formula (45–1), we find that

$$V = \sqrt{5.104^2 + 10.860^2}$$

$$= 12.000 \text{ V}$$

The calculated value of V is the same as the given applied voltage V and our solutions are therefore verified.

SUMMARY

1. In a series RC (capacitive) circuit, the current I leads the applied voltage V by an angle θ, called the phase angle.

2. The phase angle θ between V and I is the same as the angle θ between Z and R in the impedance phasor diagram of the RC circuit. The angle θ is also the same as the angle between V and V_R [Figure 45–1(b)].

3. The value of θ depends on the relative values of X_C, R, and Z and may be calculated from the formula

$$\theta = \tan^{-1}\left(\frac{-X_C}{R}\right)$$

4. In a series RC circuit the voltage drop V_C across the capacitor lags the voltage drop V_R across the resistor by 90°.

5. Voltages V_C and V_R are phasors and are added using their phasors to obtain the applied voltage V in the circuit. That is, V is the phasor sum of V_R and V_C.

6. The relationship between the applied voltage V and the voltage drop V_R across R and the voltage drop V_C across C is given by the formula

$$V = \sqrt{V_R^2 + V_C^2}$$

7. If the applied voltage V and R, X_C, and Z are known in a series RC circuit, then V_R and V_C may be calculated from the formulas

$$V_R = V \times \frac{R}{Z}$$

$$V_C = V \times \frac{X_C}{Z}$$

SELF TEST

Check your understanding by answering the following questions:

1. In the circuit of Figure 45–1(a), the measured values of V_R and V_C are, respectively, 5 V and 12 V. The applied voltage V must then equal _____ V.

2. The phase angle between voltages V and _____ is the same as the phase angle between V and I in the circuit of Figure 45–1(a).

3. The phase angle θ between V and I in question 1 is _____°.

4. In the circuit of Figure 45–1(a), $X_C = 200$ Ω, $R = 300$ Ω, and $V = 6$ V, In this circuit,
   (a) $Z =$ _____ Ω;
   (b) $\theta =$ _____°;
   (c) $V_R =$ _____ V;
   (d) $V_C =$ _____ V.

5. (True/False) In a series RC circuit, $\dfrac{X_C}{R} = \dfrac{R}{Z}$. _____

MATERIALS REQUIRED

Instruments:
- Dual-trace oscilloscope
- DMM
- Function generator

Resistors (½-W, 5%):
- 1 1 kΩ
- 1 6.8 kΩ

Capacitor:
- 1 0.033 μF

1. Using an ohmmeter, measure the resistance of the 1 kΩ and 6.8 kΩ resistors. Record the values in Table 45–1.

2. With the function generator **off**, connect the circuit shown in Figure 45–3.

3. Turn on the function generator and, using Channel 1 of the oscilloscope, adjust its output to 10 V_{p-p} at a frequency of 1 kHz. Adjust the oscilloscope's controls to display one complete cycle that fills the graticule horizontally.

4. The trigger input should be set to Channel 2. In a series circuit, the current is the same throughout the circuit. Therefore, in a *series* circuit, the circuit current will be used as the reference or base line (0°) when taking measurements and drawing phasor diagrams. The voltage drop across R_1 is the result of the current flowing through it.

5. Adjust the oscilloscope's LEVEL and SLOPE controls so that V_{R1} fills the scope's graticule with one complete cycle. Horizontally most scopes are 10 divisions wide and one complete cycle occurs in 360°. If the display is adjusted for 10 divisions, this will result in a 36°/Div. setting on the oscilloscope.

6. With the Vertical MODE switch set to DUAL-ALT, measure the resulting phase shift between the circuit's current (as represented by the V_{R1} waveform) and the input voltage (V_{in}). Record your results in Table 45–1 for the 1 kΩ row. Turn **off** the oscilloscope and function generator.

7. Repeat steps 2 through 6 using the 6.8 kΩ resistor in place of the 1 kΩ. Do not turn off the function generator.

8. Measure the voltage drop across the 6.8 kΩ resistor, (V_R) and across the capacitor (V_C). Record these val-

ues in Table 45–2 for the 6.8 kΩ row. Turn **off** the function generator.

9. Calculate the current in the circuit for each value of V using Ohm's law with the measured values of V_R and R. Record your answers in Table 45–2 for the 6.8 kΩ resistor.

10. Calculate the capacitive resistance X_C of the capacitor using Ohm's law for capacitors with the measured value of V_C and the calculated value of I. Record your answers in Table 45–2 for the 6.8 kΩ resistor.

11. Using the calculated values of X_C from step 10 and the measured value of R, calculate the phase angle θ for each value of V_{p-p}.

$$\theta = \tan^{-1}\left(\frac{-X_C}{R}\right)$$

Record your answers in Table 45–2 for the 6.8 kΩ resistor.

12. Turn **on** the function generator and adjust the output as in step 3. Repeat steps 8 through 11 for the 1 kΩ resistor.

13. Using the measured values of V_R and V_C for the 1 kΩ resistor, calculate V_{p-p} with the square-root formula $V = \sqrt{V_R^2 + V_C^2}$. Record your answers in the "Applied Voltage (Calculated)" column in Table 45–2. Repeat the calculation for V_R and V_C with the 6.8 kΩ resistor. Record your answers in Table 45–2.

14. In the space below Table 45–2 draw the respective impedance and voltage phasor diagrams for both the 1 kΩ and 6.8 kΩ circuits.

ANSWERS TO SELF-TEST

1. 13
2. V_R
3. −67.4
4. (a) 361; (b) −33.7; (c) 5.0; (d) 3.32
5. false

Optional Activity ·

This activity requires the use of electronic simulation software.

A series *RC* coupling circuit has an input frequency of 1 kHz and a resistor value of 1 kΩ. Using electronic simulation software, design, build and test a circuit where $X_C = 1/10$ R for proper coupling.

Prepare a report showing your calculations, a fully labeled circuit diagram, and your test results.

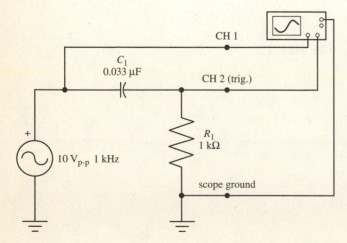

Figure 45–3. Circuit for procedure step 2.

EXPERIMENT 45

TABLE 45-1. Using the Oscilloscope to Find the Phase Angle θ in a Series *RC* Circuit

Resistance R, Ω		Capacitance C, μF	D, cm	Width of Sine Wave Points d, cm	Distance Between Zero Phase Angle θ, degrees
Rated Value	Measured Value				
1 k					
6.8 k					

TABLE 45-2. Phase Angle θ and Voltage Relationships in a Series *RC* Circuit

Resistance Rated Value, Ω	Capacitance (Rated Value) C, μF	Applied Voltage V_{pp}, V	Voltage across Resistor V_R, V_{p-p}	Voltage across Capacitor V_C, V_{p-p}	Current (Calculated) I, mA	Capacitive Reactance (Calculated) X_C, Ω	Phase Angle θ (Calculated from X_C and R), degrees	Applied Voltage (Calculated) V_{p-p}, V
1 k								
6.8 k								

Impedance and Voltage Vector Diagrams for $R = 1k\Omega$

Impedance and Voltage Vector Diagrams for R = 6.8 kΩ

QUESTIONS

1. Explain, in your own words, the relationship between the voltage across a resistance, the voltage across a capacitance, and the applied voltage in a series *RC* circuit.

_____

_____

_____

_____

_____

2. In reference to the series circuit using the 6.8 kΩ resistor, is this circuit resistive or capacitive dominant? Use lab data to support your answer.

_____

_____

_____

_____

_____

3. In a series circuit the impedance angle happens to be −30°. What is the relationship between the circuit current and the applied voltage?

_____

_____

_____

_____

_____

4. Do the results of this experiment verify formula (45–1)? Refer to specific data to support your answer.

_____

_____

_____

_____

_____

5. What factors might introduce errors in using the oscilloscope to determine the phase angle θ?

_____

_____

_____

_____

_____

_____

_____

POWER IN AC CIRCUITS

BASIC INFORMATION

Consumption of AC Power

Power dissipation in dc resistive circuits is defined as the product of the voltage and the current. That is,

$$P = V \times I \qquad (46\text{--}1)$$

where

P is power in watts (W),

V is voltage in volts (V),

I is current in amperes (A).

Power can also be calculated using the formulas

$$P = I^2R$$

$$P = \frac{V^2}{R} \qquad (46\text{--}2)$$

In dc circuits V and I are constant, not varying, values. In ac circuits V and I are continuously varying and may be in phase or out of phase with one another. An ac circuit may also contain reactive components in addition to resistive components. Resistive components dissipate power in ac circuits just as they do in dc circuits. However, pure reactive components do not dissipate net power. In one part of the cycle they take power from the circuit; in the next part of the cycle they return power to the circuit. The result is that there is no total power consumed by the reactive component.

Apparent Power

Apparent power P_A is the input power to the ac circuit. It is defined as the product of the voltage V and the current I in the ac circuit. Thus,

$$P_A = V \times I$$

for lack of a better unit and to avoid confusion with the unit watts, apparent power is usually measured in voltamperes (VA).

Current through a reactance produces reactive power

$$P_X = V_X \times I$$

which is usually abbreviated VAR.

OBJECTIVES

1. To differentiate between true power and apparent power in ac circuits

2. To measure power in an ac circuit

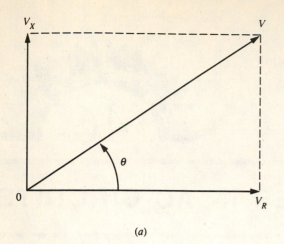

(a)

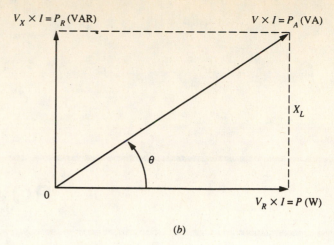

(b)

Figure 46–1. The voltage phasor diagram can be converted to a power phasor diagram.

True Power and Power Factor · · · · · · · · ·

The power in watts consumed by an electrical device with both resistive and reactive components is defined as true power. In a circuit containing both resistive and reactive components, true power is equal to or less than apparent power. The amount by which apparent power must be multiplied to obtain true power is called the *power factor* (PF) of the circuit. It is really the ratio of true power to apparent power. Thus,

$$\text{PF} = \frac{\text{true power}}{\text{apparent power}} = \frac{P(\text{W})}{P_A(\text{VA})}$$

and

$$P = P_A \times \text{PF} = V \times I \times \text{PF} \qquad (46\text{–}3)$$

The voltage phasor diagram in Figure 46–1(a) can be converted to a power diagram by multiplying each side of the voltage triangle by I, as in Figure 46–1(b). Thus,

$$\cos \theta = \frac{P(\text{W})}{P_A(\text{VA})} \qquad (46\text{–}4)$$

But from Figure 46–1(a) and Experiment 43,

$$\cos \theta = \frac{V_R}{V} = \frac{R}{Z} \qquad (46\text{–}5)$$

Therefore,

$$\text{PF} = \frac{R}{Z} \qquad (46\text{–}6)$$

If PF = 1, it means that the total impedance of the circuit equals the resistance. If PF = 0, it means that $R = 0$, and no true power is consumed in the circuit. Since R cannot be greater than Z, PF can never be greater than 1. And since R and Z are always positive, PF can never be a negative number.

True power in watts, represented by P, can be found using the same formulas as those used for dc circuits.

$$P = I^2R$$
$$P = V_RI \qquad (46\text{–}7)$$
$$P = \frac{V_R^2}{R}$$

Formulas (46–7) can be used in ac circuits because the voltage across a resistor is always in phase with the current through the resistor.

Two examples show how power can be calculated in an ac series circuit.

Problem 1. In the circuit of Figure 46–2, a coil has an inductive reactance X_L of 1 kΩ and resistance R of 250 Ω. The voltage source is 20 V ac. We wish to find the apparent power and true power delivered by the supply, the PF of the circuit, and the angle by which the current I leads or lags the voltage V.

Solution. The impedance of the circuit is

$$Z = \sqrt{R^2 + X_L^2} = \sqrt{(250^2) + (1\text{ k})^2}$$
$$= 1.03\text{ k}\Omega$$

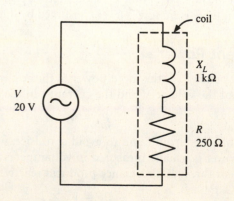

Figure 46–2. Circuit for problem 1.

The current in the circuit is

$$I = \frac{V}{Z} = \frac{20 \text{ V}}{1.03 \text{ k}\Omega}$$

$$= 19.4 \text{ mA}$$

Phase angle $\theta = \tan^{-1}\left(\frac{X_L}{R}\right) = \tan^{-1}\left(\frac{1 \text{ k}}{250}\right)$

$$\theta = 75.96°$$

Because the circuit is inductive, the current lags the voltage by 75.96°.

The apparent power P_A delivered by the supply is

$$P_A = VI = 20\text{V} \times 19.4 \text{ mA}$$

$$= 0.388 \text{ VA}$$

The true power P is

$$P = I^2R = (19.4 \text{ mA})^2 \times 250 \text{ }\Omega$$

$$= 94.1 \text{ mW}$$

The power factor PF is

$$\text{PF} = \frac{R}{Z} = \frac{250}{1.03 \text{ k}}$$

$$= 0.242$$

Power factor is often expressed as a percentage. To find percent PF, multiply by 100:

$$\% \text{ PF} = \text{PF} \times 100$$

Thus, in the preceding answer the percent power factor is

$$\% \text{ PF} = \text{PF} \times 100$$

$$= 0.242 \times 100$$

$$= 24.2\%$$

Expressed as a percentage, the power factor can never be greater than 100%.

Problem 2. The series circuit of Figure 46–3 contains a capacitor having a capacitive reactance $X_C = 300 \text{ }\Omega$ and $R = 500 \text{ }\Omega$. The supply voltage is 12 V ac. We wish to find P_A, P, PF, and θ.

Solution. We will find Z using the formulas

$$\theta = \tan^{-1}\left(\frac{-X_C}{R}\right)$$

$$= \tan^{-1}\left(\frac{-300}{500}\right)$$

$$= -30.96°$$

Because this circuit is capacitive, this negative impedance angle results in the current leading the voltage. Continuing the process to find Z,

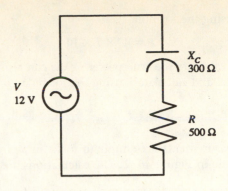

Figure 46–3. Circuit for problem 2.

$$Z = \frac{R}{\cos \theta} = \frac{500}{\cos 30.96}$$

$$Z = 583.10 \text{ }\Omega$$

$$I = \frac{V}{Z} = \frac{12 \text{ V}}{583.1 \text{ }\Omega}$$

$$= 20.6 \text{ mA}$$

$$P_A = VI = 12 \text{ V} \times 20.6 \text{ mA} = 0.247 \text{ VA}$$

$$P = P_A \times \frac{R}{Z} = 0.247 \times \frac{500 \text{ }\Omega}{583.1 \text{ }\Omega}$$

$$P = 0.212 \text{ W}$$

This can be verified using the formula

$$P = I^2R = (20.6 \text{ mA})^2 \times 500 \text{ }\Omega$$

$$= 0.212 \text{ W}$$

The power factor is

$$\text{PF} = \frac{R}{Z} = \frac{500 \text{ }\Omega}{583.1 \text{ }\Omega} = 0.857 \text{ or } 85.7\%$$

$$\text{PF} = \frac{P}{P_A} = \frac{0.212 \text{ W}}{0.247 \text{ VA}} = 0.858 \text{ or } 85.8\%$$

The difference of 0.1 is a result of the rounding process.

Measuring AC Power—Voltage-Current Method

Power in the ac circuit may be determined by a series of measurements using familiar instruments, namely, the voltmeter and the oscilloscope. With the voltmeter we can measure the voltage V_R across R and the applied voltage V. We can then determine the power factor by substituting the measured values in the formula

$$\text{PF} = \frac{V_R}{V}$$

We can measure the resistance of R and calculate the current I using the formula

$$I = \frac{V_R}{R}$$

Then, using the formula

$$P = V \times I \times \text{PF}$$

we can calculate the true power P. We can also measure V_R and R and calculate P using the formula

$$P = \frac{V_R^2}{R}$$

The measurements are simple to make in a circuit such as the one in Figure 46–2. The calculations are straightforward.

If we wish to measure the phase angle in the circuit directly, we can use an oscilloscope. By the method discussed in Experiment 43, we can determine θ by comparing the phase angle of the applied voltage with the phase angle of the voltage across R (which is the same as the phase angle of I).

Wattmeter

Power can be measured directly using a wattmeter. For low-frequency measurements, such as 50–60 Hz, an analog meter having two coils is usually used. Figure 46–4 is a simple schematic representation of such a meter. Wattmeters of this type can have three or four terminals. Terminals A and B in Figure 46–4 are connected across the load; terminals C and D are connected in series with the load. In effect, A and B are for the voltage connections, and C and D are for the current connections. The meter pointer will indicate true power directly; that is, $P = V \times I \times \text{PF}$.

SUMMARY

1. Power in ac circuits is consumed only by the resistive components.

2. Apparent power P_A in an ac circuit is the product of the source voltage and the line current. $P = V \times I$, where V is the applied voltage and I is the current drawn by the circuit.

3. The true power dissipated by the circuit is the product of V and I and the power factor PF. The power factor is equal to the cosine of the angle between the voltage and current in the circuit, that is,

$$P = V \times I \times \cos \theta$$

4. Other formulas for true power are

$$P = I^2 R = \frac{V_R^2}{R}$$

where

I = current in the circuit in amperes

R = total resistance of the circuit in ohms

V_R = voltage measured across the total resistance of the circuit

5. The power factor ($\cos \theta$) of an ac circuit may be determined by measuring θ, the phase angle between the applied circuit voltage V and the current I, and calculating $\cos \theta$.

6. Other power factor formulas are:

$$\text{PF} = \cos \theta = \frac{R}{Z} = \frac{V_R}{V} = \frac{P}{P_A}$$

7. Power in an ac circuit may be determined by measuring the applied voltage V, the current I, and the phase angle θ and substituting the measured values in the formula

$$P = V \times I \times \cos \theta$$

8. True power may be measured directly, using a wattmeter.

SELF TEST

Check your understanding by answering the following questions.

In the circuit of Figure 46–5, the coil has $L = 8$ H and $R = 1$ kΩ. The power source is $V = 12$ V at 60 Hz.

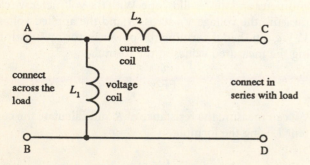

Figure 46–4. Simplified circuit diagram of a low-frequency analog wattmeter.

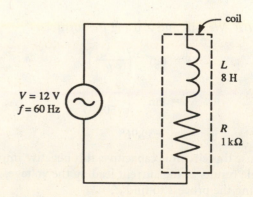

Figure 46–5. Circuit for Self-test.

1. The inductive reactance of $L = $ _____ Ω.
2. The impedance of the circuit is _____ Ω.
3. The current I drawn by the circuit is _____ A.
4. $P_A = $ _____ VA.
5. The power factor of the circuit is _____ %.
6. The phase angle of the circuit is _____ °.
7. $I^2R = $ _____ W.
8. The ratio of true power to apparent power is called _____.

MATERIALS REQUIRED

Power Supplies:
- Isolation transformer
- Variable-voltage autotransformer (Variac or equivalent)

Instruments:
- Dual-trace oscilloscope
- DMM or VOM
- Wattmeter
- 0–25-mA ac ammeter or second DMM with ac ammeter scales

Resistor:
- 1   100-Ω, 5-W

Capacitors:
- 1   5-μF or 4.7-μF, 100-V
- 1   10-μF, 100-V

Miscellaneous:
- SPST switch
- Polarized line cord with on-off switch and fuse

PROCEDURE

A. Determining Power Using the Voltage-Current Method

A1. Using an ohmmeter, measure the resistance of the 100-Ω resistor and record the value in Table 46–1 (p. 357).

A2. With the line cord unplugged, line switch **off,** and S_1 **open,** connect the circuit of Figure 46–6. Set the autotransformer to its lowest output voltage, and the ac ammeter to the 25-mA range.

A3. **Close** S_1. Increase the output voltage of the autotransformer until $V_{AB} = 50$ V. Measure the voltage across the resistor V_R and the current I. Record the

values in Table 46–1 in the 5-μF row. **Open** S_1; disconnect the 5-μF capacitor.

A4. Calculate the apparent power P_A, the true power P, the power factor, and the phase angle of the circuit. Use the measured values of V_{AB}, V_R, and I, as appropriate, in your calculations. Record your answers in Table 46–1 in the 5-μF row.

A5. With S_1 **open** and the autotransformer set at its lowest output voltage, connect a 10-μF capacitor in series with the 100-Ω resistor.

A6. **Close** S_1. Increase the output of the autotransformer until $V_{AB} = 25$ V. Measure V_R and I and record the

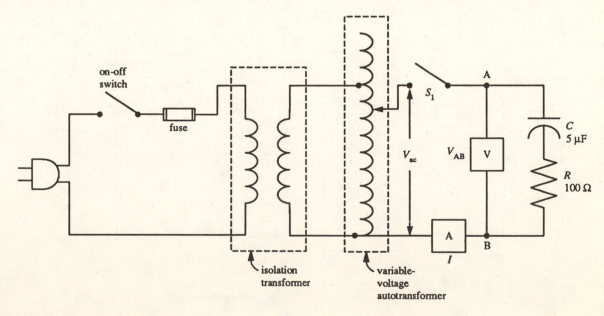

Figure 46–6. Circuit for procedure step A2.

values in Table 46–1 in the 10-μF row. After your last measurement, **open** S_1.

A7. Repeat step A4 for the 100-Ω/10-μF series circuit. Record your answers in Table 46–1 in the 10-μF row.

B. Measuring Power with a Wattmeter

B1. With S_1 **open** and the autotransformer set to its lowest output voltage, connect a wattmeter in the series circuit of Part A. The wattmeter connections are as shown in Figure 46–7. Although the connections to the meter are typical, check the instruction manual of your meter to verify the connections.

B2. **Close** S_1. Increase the output voltage until $V_{AB} = 25$ V. Measure the true power P, using the wattmeter, and the voltage across the resistor V_R. Record the values in Table 46–2 (p. 357) in the 10-μF row. **Open** S_1; disconnect the 10-μF capacitor.

B3. Calculate the true power using V_R and the measured value of R (from Table 46–1). Calculate the power factor using V_R and V_{AB}. Record your answers in the 10-μF row of Table 46–2.

B4. With S_1 **open,** connect the 5-μF capacitor in series with the 100-Ω resistor.

B5. **Close** S_1. Increase the autotransformer output until $V_{AB} = 50$ V. Measure the power of the circuit using the wattmeter, and the voltage across the resistor V_R. Record the values in Table 46–2 in the 5-μF row. **Open** S_1.

B6. Calculate the true power using V_R and the measured value of R (from Table 46–1). Calculate the PF using V_R and V_{AB}. Record your answers in Table 46–2 in the 5-μF row.

C. Determining Power Factor with an Oscilloscope

C1. Connect the dual-trace oscilloscope to the series RC circuit, as shown in Figure 46–8. The autotransformer should be set at its lowest output voltage. The trigger switch should be set to EXT.

C2. **Close** S_1. Increase the output of the autotransformer to 10 V rms. Channel 1 is the voltage reference channel. Turn on the oscilloscope. Adjust the controls of the scope so that a single sine wave, about 6 div. peak-to-peak, fills the width of the screen. Use the horizontal and vertical controls to center the waveform on the screen.

C3. Switch to Channel 2. This is the current channel. Adjust the controls so that a single sine wave, about 4 div. peak-to-peak, fills the width of the screen. Use the vertical control to center the waveform vertically. *Do not use the horizontal control.*

C4. Switch the oscilloscope to dual-channel mode. The Channel 1 and Channel 2 signals should appear together. Note where the curves cross the horizontal (x) axis. These are the zero points of the two sine waves. With a centimeter scale, accurately measure the horizontal distance d between the two positive or two negative peaks of the sine waves. Verify your measurement by measuring the distance between the corresponding zero points of the two waves (see Figure 46–9). Record the measurement in Table 46–3 (p. 357) in the 5-μF row. Also measure the distance D from 0 to 360° for the voltage sine wave. Record the value in Table 46–3 for the 100-Ω resistor. Turn **off** the scope; **open** S_1; disconnect the 5-μF capacitor.

C5. Using the formula in Figure 46–9, calculate the phase angle θ between the voltage and current in the cir-

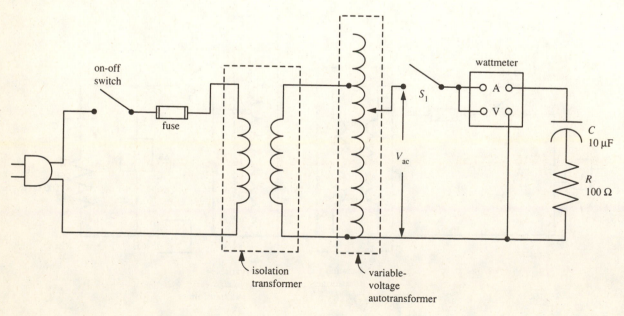

Figure 46–7. Circuit for procedure step B1.

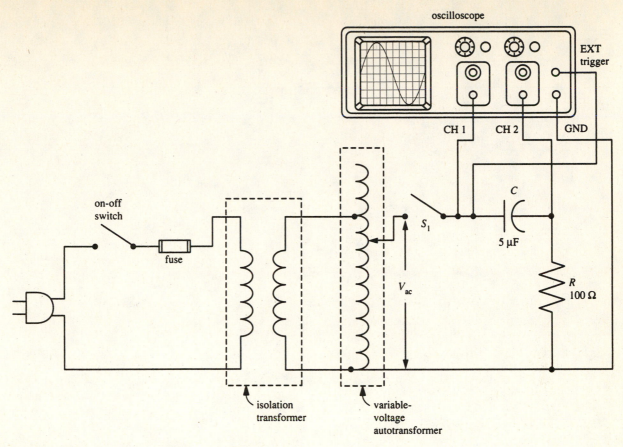

Figure 46–8. Circuit for procedure sep C1.

cuit of Figure 46–8. Using the calculated value of θ, calculate the power factor PF of the circuit. Record your answers in Table 46–3.

C6. Replace the 5-μF capacitor with a 10-μF capacitor in the circuit of Figure 46–8.

C7. **Close** S_1. Repeat steps C3 through C5 for the 10-μF capacitor. After taking the final measurement, turn **off** the scope; **open** S_1; and disconnect the scope from the circuit.

C8. Repeat step C5 for the 10-μF, 100-Ω series circuit.

ANSWERS TO SELF TEST

1. 3.02 k
2. 3.18 k
3. 3.78 mA
4. 0.0454
5. 31.5
6. 71.65
7. 14.3 mW
8. PF

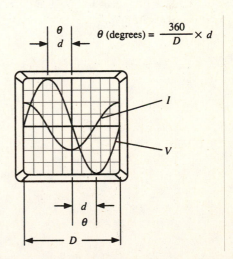

$$\theta \text{ (degrees)} = \frac{360}{D} \times d$$

Figure 46–9. Determining the power factor of an *RC* circuit using oscilloscope waveforms.

TABLE 46–1. Power Measurement—Voltage-Current Method

Resistance R, Ω		Capacitance (Rated Value) C, μF	Applied Voltage V_{AB}, V	Voltage across Resistor V_R, V	Current (Measured) I, mA	Apparent Power P_A, VA	True Power P, W	Power Factor PF	Phase Angle θ, degrees
Rated Value	Measured Value								
100		5							
100		10							

TABLE 46–2. Power Measurement—Wattmeter Method

Resistance R (Rated Value), Ω	Capacitance (Rated Value) C, μF	Applied Voltage V_{AB}, V	Voltage Across Resistor V_R, V	Power (Measured) P, W	Power (Calculated) P, W	Power Factor PF, %
100	5	50				
100	10	25				

TABLE 46–3. Determining Power Factor with an Oscilloscope

Resistance (Rated Value) R, Ω	Capacitance (Rated Value) C, μF	Distance Between Zero Points, d, cm	Width of Sine Wave D, cm	Phase Angle (Calculated) θ, degrees	Power Factor (Calculated) PF, %
100	5				
100	10				

QUESTIONS

1. Explain in your own words, the difference between true power and apparent power in an ac circuit.

_____

_____

_____

_____

2. Refer to your data in Tables 46–1 and 46–2. Compare the values of true power obtained by the wattmeter method with those obtained using the voltage-current method. Should they be equal? Explain.

_____

_____

_____

_____

3. A circuit in which the current leads the voltage has a leading power factor. If the current lags the voltage, the circuit has a lagging power factor, Refer to Tables 46–1 and 46–2. Is the power factor in this experiment leading or lagging? Explain why and support your answer with specific data.

_____

_____

_____

_____

_____

_____

_____

_____

4. Why is the power factor of an ac circuit significant? Can it be changed? If so, explain how.

_____

_____

_____

_____

_____

_____

_____

_____

5. Explain, in your own words, how phase angle and power factor are related.

_____

_____

_____

_____

_____

_____

_____

FREQUENCY RESPONSE OF A REACTIVE CIRCUIT

BASIC INFORMATION

Impedance of a Series *RL* Circuit

The impedance of a series *RL* circuit is given by the formula

$$Z = \sqrt{R^2 + X_L^2} \qquad (47\text{--}1)$$

If R remains constant, a change in X_L will affect Z. Thus, as X_L increases, Z also increases. As X_L decreases, Z decreases. Since

$$X_L = 2\pi f L \qquad (47\text{--}2)$$

we can change X_L by either increasing or decreasing the size of L, with f remaining constant. Or we can change X_L by increasing or decreasing f, with L remaining constant.

Let us consider the effect on X_L of varying f, with L constant. From formula (47–2) we see that as f increases, X_L increases; as f decreases, X_L decreases. The impedance Z will therefore also increase or decrease, respectively, as f increases or decreases. An example will illustrate this variation.

Problem. In the circuit of Figure 47–1 (p. 360), $R = 30\ \Omega$ and $L = 63.7$ mH. Show how the impedance will vary as the frequency is increased from 0 to 500 Hz.

Solution. Using formulas (47–1) and (47–2), we calculate X_L and Z for a range of frequencies and record the results in Table 47–1. Note that as X_L increases, the value of R has a decreasing effect on the value of Z. Figure 47–2 (p. 360) is a graph of the data in Table 47–1 showing how the impedance increases as the frequency increases.

Current Versus Frequency in an *RL* Circuit

The current in an ac circuit is given by the formula

$$I = \frac{V}{Z} \qquad (47\text{--}3)$$

Current varies inversely with Z. Since Z increases with f in a series *RL* circuit, current will decrease as f increases. If the voltage in the problem were 100 V, the circuit current would be as given in Table 47–1.

OBJECTIVES

1 To study the effect on impedance and current of a change in frequency in a series *RL* circuit

2 To study the effect on impedance and current of a change in frequency in a series *RC* circuit

TABLE 47–1. Variation of X_L and I with Frequency

f, Hz	X_L, Ω	$Z = \sqrt{R^2 + X_L^2}$, Ω	$I = \dfrac{V}{Z}$, A
0	0	30	3.33
100	40	50	2.00
150	60	67.1	1.49
200	80	85.4	1.171
250	100	104.4	0.958
300	120	123.7	0.808
400	160	162.8	0.614
500	200	202.2	0.495

In Figure 47–2 the graph of current I on the vertical scale versus the frequency f on the horizontal scale is known as the *frequency response curve* for the circuit.

Impedance of a Series *RC* Circuit

The impedance of a series *RC* circuit is given by the formula

$$Z = \sqrt{R^2 + X_C^2} \qquad (47\text{–}4)$$

Although formulas (47–4) and (47–1) are the same except for the substitution of X_C for X_L, the variation of Z with frequency is very different for the *RC* circuit. The reason is that X_C varies inversely with frequency. The formula for X_C is

$$X_C = \frac{1}{2\pi f C} \qquad (47\text{–}5)$$

As f increases, X_C decreases. As f decreases, X_C increases. Therefore, the impedance of a series *RC* circuit increases with a decrease in frequency but decreases with an increase in frequency.

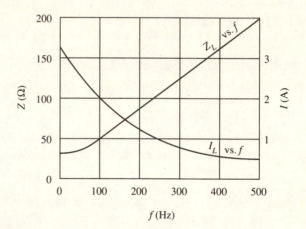

Figure 47–2. Graphs of impedance versus frequency and current versus frequency for a series *RL* circuit.

Current Versus Frequency in an *RC* Circuit

In a series *RC* circuit, as f decreases, X_C increases, Z increases, and I decreases. As f increases, X_C decreases, Z decreases, and I increases. Again, this relationship is opposite that of a series *RL* circuit.

The foregoing discussion shows that the effects of a capacitor and an inductor on current in series *RC* and *RL* circuits are opposite. This fact is related to the phase relationship between voltage and current in a capacitor and an inductor. In an inductor, current lags voltage, whereas in a capacitor, current leads the voltage.

SUMMARY

1. In a series *RL* circuit, with R and L constant, as X_L increases, Z increases.

2. As the frequency f increases or decreases, X_L increases or decreases.

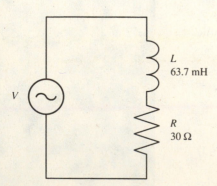

Figure 47–1. Circuit for illustrative problem.

3. As f increases in a series RL circuit, Z increases. As f decreases in a series RL circuit, Z decreases.

4. As f increases in a series RL circuit, I in the circuit decreases. As f decreases, I increases.

5. In a series RC circuit, with R and C constant, Z increases as f decreases.

6. In a series RC circuit, X_C decreases as the frequency f increases. The effect of f on X_C is opposite the effect of frequency on X_L.

7. In a series RC circuit I decreases as f decreases, and I increases when f increases.

8. The effects of a capacitor on impedance and current in a series-connected circuit are opposite the effects of an inductor in a series circuit.

9. The reactance of an inductor is

$$X_L = 2\pi fL$$

The impedance of a series RL circuit is

$$Z = \sqrt{R^2 + X_L^2}$$

10. The reactance of a capacitor is

$$X_C = \frac{1}{2\pi fC}$$

The impedance of a series RC circuit is

$$Z = \sqrt{R^2 + X_C^2}$$

11. The frequency response curve of a circuit is the graph of I on a vertical axis versus f on a horizontal axis.

SELF TEST

Check your understanding by answering the following questions:

1. The X_L of a 3.19-H choke at 100 Hz is _____ Ω.

2. The Z of a series RL circuit at 1 kHz where $R = 1$ kΩ and $L = 3.19$ H, is _____ Ω.

3. The Z of the circuit of question 2, at 2 kHz, is _____ Ω.

4. In a series RL circuit, Z _____ (increases/decreases) with an increase in frequency.

5. In the circuit of question 2, if the applied voltage $V = 15$ V, the current I in the circuit is _____ mA.

6. In the circuit of question 3, if the applied voltage $V = 12$ V, the current I in the circuit is _____ mA.

7. In a series RL circuit, I _____ (increases/decreases) with an increase in frequency.

8. In a series RC circuit, the X_C of the capacitor is 1 kΩ at 100 Hz. The X_C of the same capacitor at 200 Hz is _____ Ω, whereas at 50 Hz the X_C of the capacitor is _____ Ω.

9. The Z of a series RC circuit, with R and C constant, _____ (increases/decreases) with an increase in frequency.

10. If the Z of an RC circuit at 1 kHz = 1.41 kΩ, the Z of the same circuit at 2 kHz is _____ (greater/less) than 1.41 kΩ.

11. The current I in the RC circuit of question 10 _____ (increases/decreases) when f increases from 50 Hz to 100 Hz.

MATERIALS REQUIRED

Instrument:
- DMM
- Function generator

Resistor (½-W, 5%):
- 1   3.3 kΩ

Capacitor:
- 1   0.01 μF

Inductor:
- 100 mH coil

PROCEDURE

A. Frequency Response of an *RL* Circuit

A1. Using the DMM, measure the resistance of the 3.3 kΩ resistor and record the value in Table 47–2 (p. 363).

A2. With the function generator **off,** connect the circuit of Figure 47–3 (p. 362). Set the signal generator to its lowest output voltage and frequency.

A3. Turn on the function generator and set the output frequency to 1 kHz. Using channel 1 of the oscilloscope increase the output voltage until the voltage across the *RL* series circuit is $V = 10$ V p-p. Maintain this voltage throughout the experiment. Using channel 2 of the oscilloscope measure the voltage across the resistor V_R and record the value in the 1 kHz row of Table 47–2.

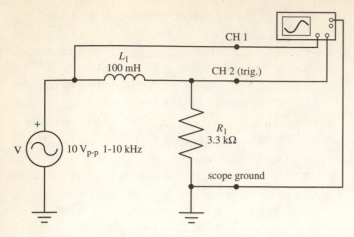

Figure 47–3. Circuit for procedure step A2.

A4. Increase the frequency to 2 kHz. Check to verify that $V = 10\ V_{p\text{-}p}$; adjust output voltage if necessary. Measure V_R and record the value in Table 47–2 in the 2 kHz row.

A5. Repeat step A4 by increasing the frequency in 1 kHz steps: 3 k, 4 k, 5 k, 6 k, 7 k, 8 k, 9 k, 10 kHz. At each frequency measure V_R and record the value in Table 47–2. At each frequency verify that $V = 10\ V_{p\text{-}p}$; adjust voltage if necessary. After all measurements have been taken, turn **off** the function generator.

A6. Using the measured values of V_R and R, calculate the current in the circuit for each frequency. Record your answers in Table 47–2.

A7. Using the calculated value of current I and voltage V, calculate the circuit impedance Z for each frequency. Record your answers in Table 47–2.

B. Frequency Response of an *RC* Circuit

B1. With the function generator **off,** connect the circuit of Figure 47–4. Set the function generator to its lowest voltage output and frequency.

B2. Turn **on** the function generator and set the output frequency to 1 kHz. Increase the output voltage of the function generator until the voltage across the *RC* series circuit $V = 10\ V_{p\text{-}p}$. Maintain this voltage throughout the experiment, checking periodically and adjusting the voltage if necessary.

B3. Measure the voltage across the resistor V_R, and record the value in Table 47–3 in the 1 kHz row.

B4. Increase the frequency to 2 kHz. Check to see that $V = 10\ V_{p\text{-}p}$; adjust if necessary. Measure V_R and record the value in the 2 kHz row of Table 47–3.

B5. Repeat step B4 in increasing 1 kHz steps: 3 k, 4 k, 5 k, 6 k, 7 k, 8 k, 9 k, 10 kHz. At each frequency, measure V_R and check $V = 10\ V_{p\text{-}p}$. Record the values at each frequency in Table 47–3. After all measurements have been taken, turn **off** the signal generator.

B6. Using the measured values of V_R (from Table 47–3) and R (from Table 47–2), calculate the current I in the circuit for each frequency. Record your answers in Table 47–3.

B7. Using the calculated values of current I and voltage V, calculate the circuit impedance for each value of frequency. Record your answers in Table 47–3.

ANSWERS TO SELF TEST

1. 2 k
2. 20.06 k
3. 40.08 k
4. increases
5. 0.748
6. 0.3
7. decreases
8. 500; 2000
9. decreases
10. less
11. increases

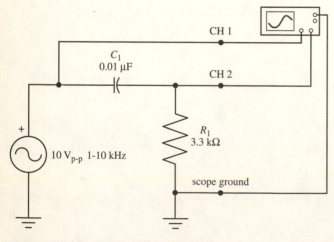

Figure 47–4. Current for procedure step B1.

TABLE 47–2. Frequency Response of an *RL* Series Circuit

Frequency f, Hz	Applied Voltage V, V_{p-p}	Voltage Across R V_R, V_{p-p}	Circuit Current (Calc.) I, mA	Circuit Imped. (Calc.) Z, Ω
1 k	10			
2 k	10			
3 k	10			
4 k	10			
5 k	10			
6 k	10			
7 k	10			
8 k	10			
9 k	10			
10 k	10			

R (rated) = 3.3 kΩ: R (measured)

TABLE 47–3. Frequency Response of an *RC* Series Circuit

Frequency f, Hz	Applied Voltage V, V_{p-p}	Voltage Across R V_R, V_{p-p}	Circuit Current (Calc.) I, mA	Circuit Imped. (Calc.) Z, Ω
1 k	10			
2 k	10			
3 k	10			
4 k	10			
5 k	10			
6 k	10			
7 k	10			
8 k	10			
9 k	10			
10 k	10			

1. Explain, in your own words, how changes in frequency affect impedance and current in a series *RL* circuit.

_____

_____

_____

_____

_____

_____

_____

2. Explain, in your own words, how changes in frequency affect impedance and current in a series *RC* circuit.

_____

_____

_____

_____

_____

_____

_____

3. On a separate sheet of 8½ × 11 graph paper plot a graph of impedance versus frequency using the data from Table 47–2. The horizontal (*x*) axis should be frequency, and the vertical (*y*) axis should be impedance. Label the axes as in Figure 47–2. On the same set of axes, plot current versus frequency using the data from Table 47–2. (See Figure 47–2 for the arrangement of axes.)

4. On a separate sheet of 8½ × 11 graph paper (not the sheet used in Question 3), plot a graph of impedance versus frequency using the data from Table 47–3. The horizontal (*x*) axis should be frequency, and the vertical (*y*) axis should be impedance. Label the axes as in Figure 47–2. On the same set of axes, plot current versus frequency using the data from Table 47–3. (See Figure 47–2 for the arrangement of axes.)

5. Refer to your graph of Questions 3 and 4. How does the impedance-versus-frequency graph for the series *RL* circuit differ from the impedance-versus-frequency graph for the series *RC* circuit? Explain.

_____

_____

_____

_____

_____

_____

_____

6. Refer to your graphs of Questions 3 and 4. How does the current-versus-frequency graph for the series *RL* circuit differ from the current-versus-frequency graph for the series *RL* circuit? Explain.

_____

_____

_____

_____

_____

7. Looking at the data in Tables 47–2 and 47–3, at approximately what input frequency are the circuit impedances equal? Calculate X_L and X_C at this frequency.

_____

_____

_____

_____

IMPEDANCE OF A SERIES *RLC* CIRCUIT

BASIC INFORMATION

Impedance of an *RLC* Circuit

To understand the effects on alternating current of series-connected *RLC* circuits, it is necessary to recall the individual effect of each of these components.

The effects of a resistor in an ac circuit are the same as those in a dc circuit, because alternating current and voltage are in phase in a resistor. In an ac voltage divider made up of resistors only, the voltage drop across each individual resistor may be added arithmetically to find the voltage across the entire combination of resistors.

Reactances in an ac circuit depend on the frequency of the source. The value of reactance varies with changes in frequency. In addition, the current through a reactance and the voltage across the reactance are not in phase. For a pure inductance (that is, $R = 0$) the current through the inductance lags the voltage across the inductance by 90°. For a pure capacitance, the current through the capacitance leads the voltage across the capacitance by 90°.

The effects of an inductor connected in series with a resistor in an ac circuit also depend on the frequency and the size of the inductor. In a series *RL* circuit, the current lags the voltage by an angle less than 90°.

When a capacitor is connected in series with a resistor, the resistance of the capacitor together with the resistance of the resistor determines the effect on alternating current. The effects of a capacitor are determined both by its size (capacitance) and by the frequency. In a series *RC* circuit, the alternating current leads the voltage by a phase angle less than 90°.

It may be seen from the characteristics of an inductance and a capacitance that they have opposite effects on current and voltage in an ac circuit. This fact is demonstrated in the phasor diagram of X_L and X_C in the series *RLC* circuit of Figure 48–1 (p. 368).

The phasor sum of X_L and X_C is the arithmetic difference of their numerical values. The resultant phasor, usually labeled X_T, is in the direction of the larger reactance. If X_L is larger than X_C, the resultant phasor will be a vertical line in the *upward* direction from the origin. In that case the circuit will be inductive and current will *lag* voltage. If X_C is larger than X_L, the resultant phasor will be a vertical line in the *downward* direction from the origin. In that case the circuit will be capacitive, and current will *lead* voltage, as shown in Figure 48–2(*b*) (p. 368).

OBJECTIVE

1. To verify experimentally that the impedance Z of a series *RLC* circuit is

$$Z = \sqrt{R^2 + (X_L - X_C)^2}$$

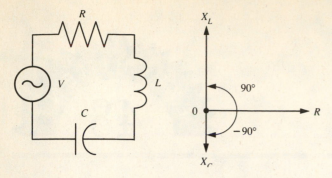

Figure 48–1. A series *RLC* circuit and its impedance phasor diagram. In this circuit the phasor diagram shows X_L is greater than X_C.

Figure 48–2 indicates that the impedance Z of the *RLC* circuit is given by the formula

$$Z = \sqrt{R^2 + (X_L - X_C)^2}$$

or

$$Z = \sqrt{R^2 + X_T^2} \qquad (48\text{–}1)$$

where X_T is the difference between X_L and X_C.

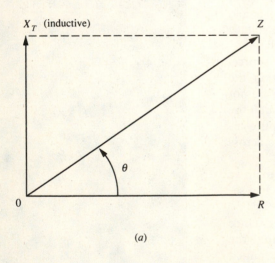

(a)

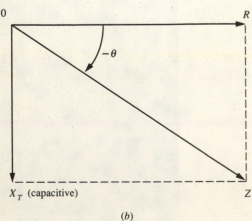

(b)

Figure 48–2. Impedance phasor diagram for an *RLC* series circuit.

If $X_T = 0$ in formula (48–1), the impedance of the circuit will be completely resistive. That is, $Z = R$. In an *RLC* series circuit, $X_T = 0$, only if $X_L = X_C$. In other words, if the inductive reactance X_L in an *RLC* circuit cancels the capacitive reactance X_C, the circuit will behave as if it contained only resistance. At this point the impedance of the circuit has its lowest value.

Alternate Method for Finding the Impedance of an *RLC* Circuit · · · · · · ·

In previous experiments the impedance of a reactive circuit was found using the phase angle θ. The solution involved the use of a scientific calculator and the $\tan^{-1}$ function. A similar process can be used to find Z in an *RLC* circuit. From the phasor diagram in Figure 48–2 the phase angle can be found using

$$\theta = \tan^{-1}\left(\frac{X_T}{R}\right) \qquad (48\text{–}2)$$

and

$$Z = \frac{R}{\cos \theta} \qquad (48\text{–}3)$$

Because X_T is the difference between X_L and X_C, the first step in the solution of θ is to find $X_L - X_C$. This form is used rather than $X_C - X_L$ so that the phase angle will agree with those in Figure 48–2. A negative θ indicates a capacitive circuit; a positive θ indicates an inductive circuit.

Illustrative Examples · · · · · · · · · · · · · · · · ·

As an illustrative example, we solve an *RLC* circuit using formulas (48–1), (48–2), and (48–3).

Problem 1. In the *RLC* circuit of Figure 48–3, $R = 82$ Ω, $X_L = 70$ Ω, and $X_C = 45$ Ω. The applied voltage is $V = 12$ V. Find the impedance of the circuit, whether the circuit is inductive or capacitive, the angle θ by which the current leads or lags the voltage, the power factor PF of the circuit, and power consumed in watts.

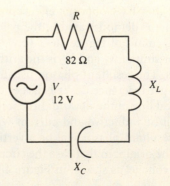

Figure 48–3. Circuit diagram for problems 1 and 2.

Solution. From formula (48–1),

$$Z = \sqrt{R^2 + X_T^2}$$

The difference X_T between X_L and X_C is

$$X_T = X_L - X_C = 70 - 45 = 25 \ \Omega$$

Substituting this value in formula (48–1) gives

$$Z = \sqrt{82^2 + 25^2}$$

$$= 85.73 \ \Omega$$

$$PF = \frac{R}{Z} = \frac{82}{85.73} = 0.9565 = 95.65\%$$

$$P = \frac{V^2}{Z} \times PF = \frac{12^2}{85.73} \times 0.9565$$

$$= 1.680 \times 0.9565$$

$$P = 1.607 \ W$$

The phase angle θ can be found by many methods, but since $PF = \cos\theta$, we can apply this formula and—using a calculator—solve for θ directly.

$$\theta = 16.96°$$

The positive angle indicates the direction of X is upward—that is, in the inductive area. The current therefore lags the voltage by 16.96°.

Problem 2. We will use the circuit of Figure 48–3 with the values of X_L and X_C reversed. That is, $X_L = 45 \ \Omega$ and $X_C = 70 \ \Omega$.

Solution. We solve this circuit using formulas (48–2) and (48–3).

$$\theta = \tan^{-1}\left(\frac{X_T}{R}\right)$$

where $X_T = X_L - X_C = 45 - 70 = -25 \ \Omega$

Therefore

$$\theta = \tan^{-1}\left(\frac{-25}{82}\right)$$

$$\theta = -16.96°$$

The negative sign indicates that the reactive phasor X is in the downward direction and the circuit is capacitive. This means the current will *lead* the voltage by 16.96°.

$$Z = \frac{R}{\cos\theta} = \frac{82}{\cos(-16.96)}$$

$$Z = 85.73 \ \Omega$$

which is exactly the same value found in problem 1. This was to be expected since X_T and R are the same. However, in problem 1 the reactance X_T was positive, indicating an inductive circuit, whereas in problem 2 the reactance X_T was negative, indicating a capacitive circuit. In both cases, of course, Z was a positive number.

Because the numerical values of R, X_T, and Z are the same in problems 2 and 1, the power factor and power consumption of the circuit will be exactly the same. In problem 1 we say that the circuit has a *lagging* power factor; in problem 2 the circuit has a *leading* power factor.

SUMMARY

1. An inductor L and a capacitor C have opposite effects on current in an ac circuit. Thus, in an inductor, current lags, whereas in a capacitor, current leads the applied voltage.

2. The total reactance X_T of a series RLC circuit is the arithmetic difference of X_L and X_C.

3. If X_L is greater than X_C in a series RLC circuit, then the circuit is inductive and the current lags the source voltage by the phase angle θ.

4. If X_C is greater than X_L in a series RLC circuit, then the circuit is capacitive, and the current leads the source voltage by the phase angle θ.

5. If X_L is greater than X_C, the phase angle θ is positive; if X_C is greater than X_L, the phase angle θ is negative.

6. The impedance of an RLC series circuit can be found using the formula

$$Z = \sqrt{R^2 + X_T^2}$$

7. Impedance can also be found using the phase angle in the following formulas:

$$\theta = \tan^{-1}\left(\frac{X_T}{R}\right)$$

$$Z = \frac{R}{\cos\theta}$$

SELF TEST

Check your understanding by answering the following questions:

1. In a series RLC circuit $X_L = 40 \ \Omega$, $X_C = 70 \ \Omega$, and $R = 40 \ \Omega$. The net reactance X is _____ Ω. This reactance is _____ (inductive/capacitive).

2. The impedance Z of the circuit in question 1 is _____ Ω.

3. The circuit in question 1 acts like an _____ (RC/RL) circuit.

4. In a series RLC circuit $X_L = 10 \ \Omega$, $X_C = 15 \ \Omega$, and $R = 12 \ \Omega$. In this circuit the net reactance is _____ (inductive/capacitive) and equal to _____ Ω. The impedance phasor (leads/lags) the resistance R.

5. In the circuit of question 4, the phase angle $\theta =$ _____ ° and is _____ (negative/positive).

6. In the circuit of question 4, $Z =$ _____ Ω.
7. The reason θ is negative is that $X_L - X_C$ is _____ (positive/negative).

MATERIALS REQUIRED

Instrument:
- DMM
- Function generator

Resistor:
- 1   2 kΩ, ½-W, 5%

Capacitor:
- 1   0.022 μF

Inductor:
- 100 mH coil

PROCEDURE

. . . .

1. With the function generator turned **off,** connect the circuit of Figure 48–4(a). Set the generator to its lowest output voltage.

2. Turn **on** the function generator. Increase the output voltage of the generator until $V_{AB} = 10$ V$_{p\text{-}p}$. Maintain this voltage throughout the experiment. Check it from time to time and adjust it if necessary.

3. Measure the voltage across the resistor V_R and the inductor, V_L. Record the values in Table 48–1 (p. 373) for the *RL* circuit. Turn off the generator.

4. Calculate the current in the circuit using the measured value of V_R and the rated value of R. Record your answer in Table 48–1 for the *RL* circuit.

5. Using the calculated value of I and the measured value of V_L, calculate X_L. Record your answer in the "*RL*" row of Table 48–1.

6. Calculate the total impedance of the circuit by two methods: Ohm's law (using the calculated value of I and the applied voltage V_{AB}) and the square-root formula (using R and X_L). Record your answers in the "*RL*" row of Table 48–1.

7. Add a 0.022 μF capacitor in series with the resistor and coil as in the circuit of Figure 48–4(b).

8. Turn **on** the generator. Check to see that $V_{AB} = 10$ V. Measure the voltage across the resistor V_R, the coil V_L, and the capacitor V_C. Record the values in the "*RLC*" row of Table 48–1. After taking all measurements, turn **off** the function generator.

9. Calculate I and X_L as in steps 4 and 5. Similarly, using the measured value of V_C and the calculated value of I, calculate the capacitive reactance of the circuit. Record your answer in the "*RLC*" row of Table 48–1.

10. Calculate the impedance Z of the circuit by two methods: Ohm's law (using V_{AB} and I) and the square-root formula (using R, X_C, and X_L). Record your answers in the "*RLC*" row of Table 48–1.

11. Remove the coil from the circuit leaving only the resistor in series with the capacitor as in Figure 48–4(c).

12. Turn **on** the function generator. Check V_{AB}; adjust if necessary. Measure V_R and V_C. Record the values in the "*RC*" row of Table 48–1. After taking all measurements, turn **off** the generator.

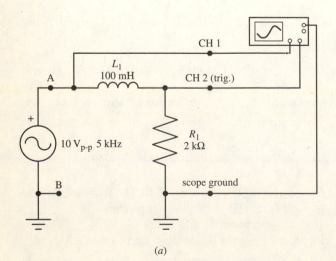

(a)

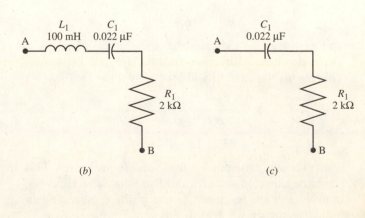

(b)   (c)

Figure 48–4. (a) Circuit for procedure step 1. (b) Circuit for procedure step 7. (c) Circuit for procedure step 11.

13. Using measured values of V_R and V_C and the rated value of R, calculate the current I in the circuit. Then using the calculated value of I, calculate X_C. Record your answers in Table 48–1 in the "RC" row.

14. Calculate the total impedance of the circuit by two methods: Ohm's law (using V_{AB} and I) and the square-root formula (using R and X_C). Record your answers in the "RC" row of Table 48–1.

ANSWERS TO SELF TEST

1. 30; capacitive
2. 50
3. RC
4. capacitive; 5 Ω; lags
5. 22.6; negative
6. 13
7. negative

TABLE 48–1.   Determining Impedance of an *RLC* Circuit

Circuit	Component			Applied Voltage V_{AB}, V$_{p-p}$	Voltage Across Resistor V_R, V$_{p-p}$	Voltage Across Inductor V_L, V$_{p-p}$	Voltage Across Capacitor V_C, V$_{p-p}$	Current I, mA	Reactance, Ω		Impedance Z, Ω	
	R, Ω	L, mH	C, μF						Ind. X_L	Cap. X_C	Ohm's Law	Square-Root Formula
RL	2 k	100	✕	10			✕			✕		
RLC	2 k	100	0.022	10								
RC	2 k	✕	0.022	10		✕			✕			

QUESTIONS

1. Explain, in your own words, the relationship between resistance, capacitance, inductance, and impedance in a series *RLC* circuit.

_____

_____

_____

_____

_____

2. Refer to your data in Table 48–1. How was the impedance of the series *RL* circuit affected when the series capacitor was added?

_____

_____

_____

_____

3. Refer to Figure 48–4(*b*). Under what conditions will the impedance have its lowest value? What is the lowest possible value of impedance for this circuit? (The values of *R*, *L*, and *C* are fixed.)

_____

_____

_____

_____

4. Refer to Figure 48–4(b). Under what conditions will the series current have its maximum value? What is the maximum current possible in this circuit? (The values of R, L, and C are fixed.)

_____

_____

_____

_____

_____

5. Refer to your data in Table 48–1. Is the RLC circuit inductive, capacitive, or resistive? Cite specific data to support your answer.

_____

_____

_____

_____

_____

6. On a separate sheet of $8\frac{1}{2} \times 11$ paper, draw the impedance phasor diagrams for the circuits of Figures 48–4 (a), (b), and (c).

EFFECTS OF CHANGES IN FREQUENCY ON IMPEDANCE AND CURRENT IN A SERIES *RLC* CIRCUIT

BASIC INFORMATION

In Experiment 48 we verified that the impedance Z of a series *RLC* circuit (Figure 49–1) is given by the formula

$$Z = \sqrt{R^2 + X_T^2} \qquad (49\text{–}1)$$

where X is the difference between X_L and X_C. In this experiment we will observe how a change in frequency of the voltage source affects circuit impedance and current.

Effect of Frequency on the Impedance of an *RLC* Circuit

Formula (49–1) shows that the minimum value of impedance in an *RLC* circuit, for a given value of R, occurs when $X_L = X_C$, or when $X_L - X_C = 0$. At that point $Z = R$, and the circuit acts like a pure resistance. The line current I is limited only by R. Therefore, I is maximum when $X_L = X_C$.

Because inductive reactance X_L and capacitive reactance X_C are dependent on frequency, there must be a frequency f_R at which X_L and X_C are equal. In this experiment we study the effect on impedance of frequencies greater and less than f_R. In a later experiment we study the effects on impedance when $f = f_R$.

OBJECTIVE

1 To study experimentally the effect on impedance and current in a series *RLC* circuit of changes in frequency.

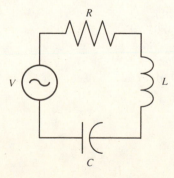

Figure 49–1. An *RLC* circuit.

In the series *RLC* circuit we have been studying, as the frequency *f* of the sinusoidal voltage source increases beyond f_R, X_L increases and X_C decreases. Hence, the circuit acts like an inductance whose reactance X_T increases as *f* increases. As the frequency *f* decreases below f_R, X_C increases and X_L decreases. The circuit now acts like a capacitance whose reactance X_T increases as *f* decreases.

A graph of impedance versus frequency for a series *RLC* circuit is shown in Figure 49–2. This curve shows minimum *Z* at f_R. It shows also that *Z* is capacitive (Z_C) for frequencies less than f_R and inductive (Z_L) for frequencies greater than f_R.

The variation of current versus frequency is also shown graphically in Figure 49–2.

SUMMARY

1. In a series *RLC* circuit there is a frequency f_R of the voltage source *V* at which $X_L = X_C$. At this frequency the impedance *Z* of the *RLC* circuit is minimum, and $Z = R$, the resistance in the circuit.

2. For frequencies higher than f_R, the X_L of the inductor increases while X_C decreases, and the circuit acts like an *RL* circuit, with

$$Z = \sqrt{R^2 + X_T^2}$$

As *f* increases beyond f_R, *Z* increases beyond *R*.

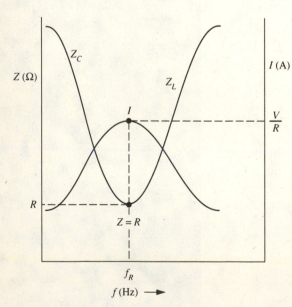

Figure 49–2. Variation of impedance (*Z*) and current (*I*) with changes in frequency (*f*) in an *RLC* circuit.

3. For frequencies lower than f_R, the X_C of the capacitor increases and X_L decreases. The circuit acts like an *RC* circuit. As *f* decreases below f_R, *Z* increases beyond *R*.

4. In a series *RLC* circuit the current can be calculated using Ohm's law,

$$I = \frac{V}{Z}$$

where *V* is the applied voltage and *Z* the circuit impedance.

SELF TEST

Check your understanding by answering the following questions:

1. In the circuit of Figure 49–1 for a particular frequency f_R, $X_L = X_C = 1$ kΩ and $R = 4.7$ kΩ. The impedance *Z* of the circuit is _____ Ω.

2. The circuit in question 1 acts like a pure _____.

3. For the frequency $f_R + 100$ Hz the impedance of the circuit in question 1 (Figure 49–1) is _____ (greater/less) than 4.7 kΩ, and the circuit is _____ (inductive/capacitive).

4. For the frequency $f_R - 100$ Hz the impedance of the circuit in question 1 is _____ (greater/less) than 4.7 kΩ, and the circuit is _____ (capacitive/inductive).

5. As *f* increases above f_R or decreases below f_R, the current *I* in the circuit _____ (increases/decreases/remains the same).

MATERIALS REQUIRED

Instruments:
- Function generator
- Oscilloscope

Resistor:
- 1   1 kΩ, ½-W, 5%

Capacitor:
- 1   0.01 μF

Inductor:
- 100 mH coil

PROCEDURE

1. With the function generator **off** and set at its lowest output voltage, connect the circuit of Figure 49–3. The dual-trace oscilloscope is triggered on channel 1.

2. Turn **on** the function generator. Set the frequency of the generator to 4 kHz. Increase the voltage output of the generator to 10 $V_{p\text{-}p}$. Adjust the oscilloscope to display two cycles of a sine wave with an amplitude approximately 4 units peak-to-peak.

3. Increase the frequency output of the generator slowly while observing the waveforms on the oscilloscope. If the amplitude of the V_R waveform increases, continue increasing the frequency until the amplitude begins to decrease. Determine the frequency at which the amplitude is at a maximum. This is f_R. Also, note the phase shift is 0° at f_R. If the amplitude decreased with an increase in frequency, decrease the frequency, again observing the amplitude of the sine wave on the scope. Continue to decrease the frequency until you are able to determine the frequency f_R at which the amplitude of the V_R wave is at a maximum. Measure the output voltage V of the generator at frequency f_R. Set and maintain this voltage to 10 $V_{p\text{-}p}$ throughout the experiment. Check it from time to time to verify its value; adjust the output voltage if necessary.

4. With the generator output frequency set at f_R, measure the voltage across the resistor V_R, the capacitor V_C, the coil V_L, and across the capacitor-inductor combination V_{LC}. All measurements should be made by moving the channel 1 and channel #2 connections as required. Record the values in Table 49–1 (p. 379) in the "f_R" row.

5. Add 500 Hz to the value of f_R and set the function generator to this frequency. Record the value in Table 49–1. Check V (it must be the same as step 3; adjust if necessary). Measure V_R, V_C, V_L, and V_{LC}. Record the values in Table 49–1 in the "f_R + 500" row.

6. Continue to increase the frequency in 500 Hz steps while measuring and recording V_R, V_C, V_L, and V_{LC} until the frequency is f_R + 2.5 kHz. Be sure to keep the input voltage amplitude constant.

7. Reduce the frequency of the generator to f_R-500 Hz. Record this value in Table 49–1. Check V again, then measure V_R, V_C, V_L, and V_{LC}. Record the values in Table 49–1.

8. Continue to reduce the frequency in 500-Hz steps until the final setting is equal to f_R − 2.5 kHz. In each step verify and record V (and adjust if necessary to maintain the constant voltage of this experiment); also measure V_R, V_C, V_L, and V_{LC}. Record all values in Table 49–1. After all measurements have been taken, turn **off** the function generator.

9. For each frequency in Table 49–1 calculate the difference between the V_L and V_C measurements. Record your answer as a positive number in Table 49–1.

10. For each frequency in Table 49–1 calculate the current in the circuit from the measured value of V_R and the rated value of R. Using the calculated value of I, calculate the impedance Z at each frequency by applying Ohm's law, $Z = V/I$.

11. Transfer the frequency steps from Table 49–1 to Table 49–2. Calculate X_C and X_L for each step, using the measured values of V_C and V_L from Table 49–1. Record

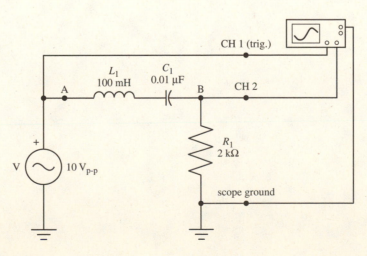

Figure 49–3. Circuit for procedure step 1.

your answers in Table 49–2. Calculate the circuit impedance at each step, using the square-root formula,

$$Z = \sqrt{R^2 + (X_L - X_C)^2}$$

and your calculated values of X_C and X_L and the rated value of R. Record your answers in Table 49–2.

Optional Activity ·

Plot a graph of frequency versus impedance using the data from Table 49–1. On the same graph, plot the frequency versus circuit current.

ANSWERS TO SELF TEST

1. 4.7 k
2. resistance
3. greater; inductive
4. greater; capacitive
5. decreases

TABLE 49–1. Effect of Frequency on Impedance of a Series *RLC* Circuit

Step	Frequency Hz	Voltage Across Res. V_R, $V_{p\text{-}p}$	Voltage Across Ind. V_L, $V_{p\text{-}p}$	Voltage Across Cap. V_C, $V_{p\text{-}p}$	Voltage Across AB V_{LC}, $V_{p\text{-}p}$	Voltage Difference $V_L - V_C$, $V_{p\text{-}p}$	Current (Calculated) I, mA	Impedance Z (Ohm's Law Cal.) Ω
$f_R + 2.5$ k								
$f_R + 2$ k								
$f_R + 1.5$ k								
$f_R + 1$ k								
$f_R + 500$								
f_R								
$f_R - 500$								
$f_R - 1$ k								
$f_R - 1.5$ k								
$f_R - 2$ k								
$f_R - 2.5$ k								

TABLE 49–2. Comparison of Impedance Calculations for a Series *RLC* Circuit

Step	Frequency, Hz	Inductive Reactance (Calculated) X_L, Ω	Capacitive Reactance (Calculated) X_C, Ω	Impedance (Calculated—Square-Root Formula) Z, Ω
$f_R + 2.5$ k				
$f_R + 2$ k				
$f_R + 1.5$ k				
$f_R + 1$ k				
$f_R + 500$				
f_R				
$f_R - 500$				
$f_R - 1$ k				
$f_R - 1.5$ k				
$f_R - 2$ k				
$f_R - 2.5$ k				

QUESTIONS

1. Explain, in your own words, the effect that changes in frequency have on the impedance and current in a series *RLC* circuit.

_____

_____

_____

_____

_____

_____

2. Explain the phase shift across V_R, as the input frequency was adjusted above and below f_R.

_____

_____

_____

_____

_____

_____

3. Refer to your data in Table 49–1. Compare the voltage across the inductance, capacitance, and the voltage difference $V_L - V_C$ for each frequency in the table.

_____

_____

_____

_____

_____

_____

4. Why didn't V_L and V_C totally cancel out each other at resonance?

_____

_____

_____

_____

_____

5. Why might a series *RLC* circuit load down a generator at resonance?

_____

_____

_____

_____

_____

IMPEDANCE OF PARALLEL
RL AND *RC* CIRCUITS

BASIC INFORMATION

Impedance of a Parallel *RL* Circuit

The circuit of Figure 50–1 consists of two parallel branches, a branch consisting of a resistor *R* only, and a branch consisting of an inductor *L* only. Since *R* and *L* are connected in parallel, the same voltage *V* appears across each branch. By Ohm's law the current through each branch is

$$\text{Current in } R \text{ branch:} \quad I_R = \frac{V}{R}$$

$$\text{Current in } L \text{ branch:} \quad I_L = \frac{V}{X_L}$$

By Kirchhoff's current law, the total current I_T delivered by *V* is the sum of I_R and I_L. But I_R and I_L are not in phase; I_R is in phase with the voltage, and I_L is out of phase and lagging the voltage by 90°. The currents must therefore be added using phasors.

Because the voltage, *V*, is the same across each branch, *V* will be used as the reference.

With *V* as the reference phasor, Figure 50–2 (p. 382) is the current phasor diagram for the *RL* parallel circuit of Figure 50–1.

OBJECTIVES

1 ▸ To determine experimentally the impedance of a parallel *RL* circuit

2 ▸ To determine experimentally the impedance of a parallel *RC* circuit

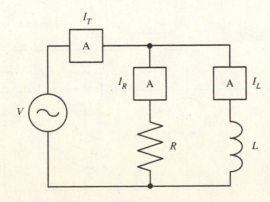

Figure 50–1. Parallel *RL* circuit.

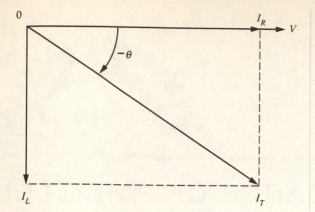

Figure 50–2. Total current I_T in a parallel RL circuit is the phasor sum of I_R and I_L.

From this diagram we can derive the formula for I_T using the right-triangle relationship

$$I_T = \sqrt{I_R^2 + I_L^2} \qquad (50\text{–}1)$$

The current phasor diagram (Figure 50–2) also shows that the current delivered by the source lags the voltage by the phase angle θ. We can solve for this angle using the formula

$$\theta = \tan^{-1}\left(\frac{-I_L}{I_R}\right) \qquad (50\text{–}2)$$

Again the scientific calculator provides a convenient method for finding θ.

The total impedance of the parallel RL circuit can be found from the total current I_T and the applied voltage V using Ohm's law

$$Z = \frac{V}{I_T} \qquad (50\text{–}3)$$

Problem 1. In the circuit of Figure 50–1, $R = 20\ \Omega$, $X_L = 45\ \Omega$, and $V = 15$ V. Find the total current I_T, the impedance Z, and the angle θ by which the line current lags the voltage V.

Solution. The current in each branch is found using Ohm's law:

$$I_R = \frac{V}{R} = \frac{15\text{ V}}{20\ \Omega} = 0.75\text{ A}$$

$$I_L = \frac{V}{X_L} = \frac{15\text{ V}}{45\ \Omega} = 0.333\text{ A}$$

The total circuit current is found using formula (50–1):

$$I_T = \sqrt{I_R^2 + I_L^2} = \sqrt{0.75^2 + 0.333^2}$$

$$= 0.821\text{ A}$$

From formula (50–3) the impedance of the circuit is

$$Z = \frac{V}{I_T} = \frac{15\text{ V}}{0.821\text{ A}}$$

$$= 18.270\ \Omega$$

The phase angle θ is found using formula (50–2) and the scientific calculator:

$$\theta = -23.94°$$

Since the circuit is inductive, the current lags the voltage by $-23.94°$.

The power factor of a parallel circuit can be found using the phasor diagram of Figure 50–2. Since PF = $\cos\theta$, Figure 50–2 shows that

$$\cos\theta = \frac{I_R}{I_T}$$

The power factor PF of the circuit in problem 1 is

$$\text{PF} = \frac{I_R}{I_T} = \frac{0.75}{0.821}$$

$$= 0.914 = 91.4\%\text{ lagging}$$

Impedance of a Parallel RC Circuit

The circuit of Figure 50–3 consists of a capacitor C in parallel with a resistor R. The supply voltage V is applied across both C and R. By Ohm's law the current in each branch is

$$\text{Current in } R \text{ Branch:} \quad I_R = \frac{V}{R}$$

$$\text{Current in } C \text{ Branch:} \quad I_C = \frac{V}{X_C}$$

By Kirchhoff's current law, the sum of I_R and I_C is equal to the total current I_T delivered by the supply. But I_R and I_C cannot be added arithmetically, because they are not in phase; I_R is in phase with the applied voltage, and I_C is out of phase and leading the applied voltage by 90°. The two branch currents I_R and I_C must be combined by phasor addition. With the applied voltage V as the reference phasor, Figure 50–4 is the current phasor diagram for the RC parallel circuit. From this diagram we can derive the formula for the total current I_T delivered by the voltage source based on the right-triangle relationship

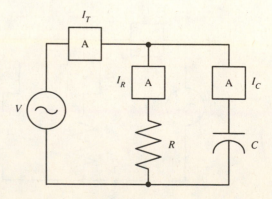

Figure 50–3. Parallel RC circuit.

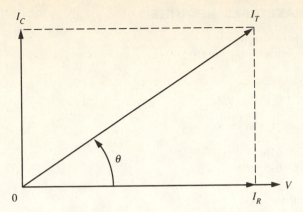

Figure 50–4. Total current I_T in a parallel RC circuit is the phasor sum of I_R and I_C.

$$I_T = \sqrt{I_R^2 + I_C^2} \qquad (50\text{–}4)$$

The phasor diagram (Figure 50–4) also shows that the current delivered by the supply leads the voltage by the phase angle θ. We can solve for this angle using the formula

$$\theta = \tan^{-1}\left(\frac{I_C}{I_R}\right) \qquad (50\text{–}5)$$

The total impedance of the parallel RC circuit can be calculated using the total current I_T and the supply voltage V. By Ohm's law,

$$Z = \frac{V}{I_T} \qquad (50\text{–}6)$$

Problem 2. The circuit of Figure 50–3 consists of a resistor $R = 470\ \Omega$ in parallel with a capacitive reactance $X_C = 200\ \Omega$. The applied voltage is $V = 20$ V. Find the total current I_T, the total impedance Z, and the angle θ by which I_T leads V.

Solution. The current in each branch is found using Ohm's law:

$$I_R = \frac{V}{R} = \frac{20\ \text{V}}{470\ \Omega}$$

$$= 42.6\ \text{mA}$$

$$I_C = \frac{V}{X_C} = \frac{20\ \text{V}}{200\ \Omega}$$

$$= 100\ \text{mA}$$

The total current is found using formula (50–4)

$$I_T = \sqrt{I_R^2 + I_C^2} = \sqrt{42.6 + 100}$$

$$= 109\ \text{mA}$$

From formula (50–6)

$$Z = \frac{V}{I_T} = \frac{20\ \text{V}}{109\ \text{mA}}$$

$$= 183\ \Omega$$

The phase angle θ is calculated with a scientific calculator

$$\theta = 67°$$

Since the circuit is capacitive, the current leads the voltage by 67°.

The power factor of the circuit is

$$\text{PF} = \cos\theta = \frac{I_R}{I_T} = \frac{42.6\ \text{mA}}{109\ \text{mA}}$$

$$= 0.391 = 39.1\%\ \text{leading}$$

SUMMARY

1. In a parallel RL or RC circuit, the voltage across each branch of the circuit is the same.

2. In a parallel RL circuit, branch currents I_R and I_L are phasors. The total (line) current I_T is a phasor equal to the phasor sum of I_R and I_L.

3. In a parallel RC circuit, branch currents I_R and I_C are phasors. The total (line) current I_T is a phasor equal to the phasor sum of I_R and I_C.

4. In a parallel RL or RC circuit the applied voltage V is used as the reference phasor.

5. In a parallel RL circuit the current in L lags the current in R by 90°. The value of I_T can be calculated from the formula

$$I_T = \sqrt{I_R^2 + I_L^2}$$

6. In a parallel RL circuit the total current lags the applied voltage V by some angle θ, where

$$\theta = \tan^{-1}\left(\frac{-I_L}{I_R}\right)$$

7. In a parallel RC circuit, the current in C leads the current in R by 90°. The value of I_T can be calculated from the formula

$$I_T = \sqrt{I_R^2 + I_C^2}$$

8. In a parallel RC circuit, the total current leads the applied voltage V by some angle θ, where

$$\theta = \tan^{-1}\left(\frac{I_C}{I_R}\right)$$

9. The impedance Z of an RL or an RC circuit may be calculated from the formula

$$Z = \frac{V}{I_T}$$

where V is the applied voltage, and I_T is the total, or line, current.

Check your understanding by answering the following questions:

1. In a parallel RL circuit the applied voltage is $V = 10$ V; the measured values are $I_L = 0.2$ A and $I_R = 0.5$ A. The total current is $I_T =$ _____ A.

2. The angle θ by which total current I_T (in question 1) _____ (leads/lags) the applied voltage V is _____°.

3. The impedance of the circuit in question 1 is _____ Ω.

4. In a parallel RC circuit the applied voltage is $V = 6$ V, and the measured values are $I_C = 0.1$ A and $I_R = 0.2$ A. The total current is $I_T =$ _____ A.

5. The angle θ by which the total current I_T (in question 4) _____ (leads/lags) the applied voltage V is _____°.

6. The impedance of the circuit in question 4 is _____ Ω.

MATERIALS REQUIRED

Instruments:
- Function generator

Resistor:
- 1 2 kΩ, ½ W
- 3 10 Ω, ½ W

Capacitor:
- 1 0.1 μF

Inductor:
- 100 mH coil

PROCEDURE

A. Impedance of a Parallel RL Circuit

A1. With the function generator turned **off,** connect the parallel RL circuit shown in Figure 50–5(a). A 10 Ω sense resistor has been added to each branch to enable current and phase angle measurements to be made with the dual-trace oscilloscope. Use channel 1, connected to the input voltage, as the trigger source.

A2. Turn **on** the function generator and adjust its sine-wave output voltage to 10 V$_{\text{p-p}}$ at 2 kHz.

A3. Using channel 1 as the reference, measure the voltage drop and phase shift occurring across R_1's series sensing resistor. Calculate the branch current using Ohm's law. Record both the current value and angle for the resistive branch in Table 50–1 (p. 387).

A4. Using the oscilloscope and series sensing resistor, measure the current through L_1 and its phase shift angle. Record the current value and its phase shift angle for the inductor branch in Table 50–1. Turn **off** the generator.

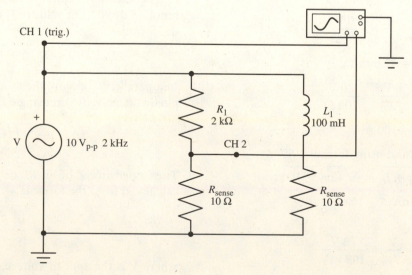

Figure 50–5(a). Circuit for procedure step A1.

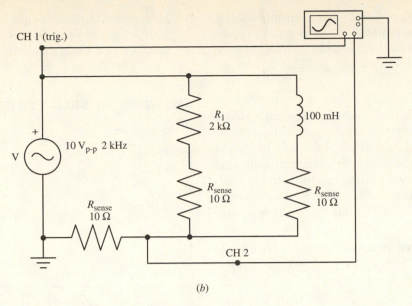

R_1
2 kΩ

100 mH

R_{sense}
10 Ω

R_{sense}
10 Ω

10 V$_{\text{p-p}}$ 2 kHz

V

R_{sense}
10 Ω

CH 2

(b)

Figure 50–5(b). Circuit for procedure step A5.

A5. Add a 10 Ω series sensing resistor to the parallel circuit as shown in Figure 50–5(b). This resistor will have both branch currents flowing through it. Turn **on** the generator, making sure its output is still 10 V$_{\text{p-p}}$ at 2 kHz. Once again, measure the current value and phase shift of the new sensing resistor with the oscilloscope. Record these values in the total line current column of Table 50–1. Turn **off** the function generator.

A6. Using the measured values of I_R and I_L, calculate I_T by applying the square-root formula (50–1). Record your answer in Table 50–1. Calculate the circuit impedance with Ohm's law using V and the measured value of I_T. Record your answer in Table 50–1.

B. Impedance of a Parallel *RC* Circuit

B1. With the function generator turned **off**, connect the parallel *RC* circuit shown in Figure 50–6(a). A 10 Ω sense resistor has been added to each branch to enable current and phase angle measurements to be made with the oscilloscope. Use Channel No. 1, connected to the input voltage, as the trigger source.

B2. Turn **on** the function generator and adjust its sine-wave output voltage to 10 V$_{\text{p-p}}$ at 2 kHz.

B3. Using Channel 1 as the reference, measure the voltage drop and phase shift occurring across R_1's series sensing resistor. Calculate the branch current using Ohm's law. Record both the current value and angle for the resistive branch in Table 50–2 (p. 387).

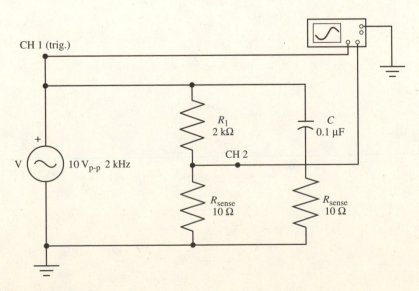

CH 1 (trig.)

R_1
2 kΩ

C
0.1 µF

CH 2

10 V$_{\text{p-p}}$ 2 kHz

V

R_{sense}
10 Ω

R_{sense}
10 Ω

Figure 50–6(a). Circuit for procedure step B1.

B4. Using the oscilloscope and series sensing resistor, measure the current through C and its phase shift angle. Record the current value and its phase shift angle for the capacitor branch in Table 50–2. Turn **off** the generator.

B5. Add a 10 Ω series sensing resistor to the parallel circuit as shown in Figure 50–6(b). This resistor will have both branch currents flowing through it. Turn on the generator, making sure its output is still 10 V_{p-p} at 2 kHz. Once again, measure the current value and phase shift of the new sensing resistor with the oscilloscope. Record these values in the total line current column of Table 50–2. Turn **off** the function generator.

B6. Using the measured values of I_R and I_C, calculate I_T

by applying the square-root formula (50–1). Record your answer in Table 50–2. Calculate the circuit impedance with Ohm's law using V and the measured value of I_T. Record your answer in Table 50–2.

ANSWERS TO SELF TEST

1. 0.539
2. lags; 21.8
3. 18.6
4. 0.224
5. leads; 26.6
6. 26.8

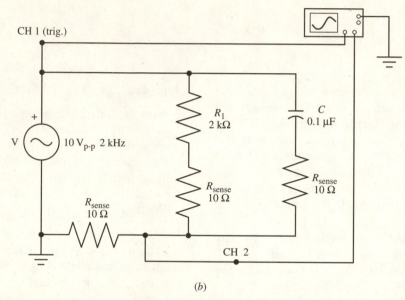

(b)

Figure 50–6(b). Circuit for procedure step B5.

TABLE 50–1. Impedance of a Parallel *RL* Circuit

Applied Voltage V, V_{p-p}	Current and Angle in Resistor Branch I_R, mA_{p-p}	Current and Angle in Inductor Branch I_L, mA	Total Line Current and Angle (Measured) I_T, mA_{p-p}	Total Line Current (Calculated Using Square-Root Formula) I_T, mA_{p-p}	Circuit Impedance (Calculated Using Ohm's Law) Z, Ω
10 V					

TABLE 50–2. Impedance of a Parallel *RC* Circuit

Applied Voltage V, V_{p-p}	Current and Angle in Resistor Branch I_R, mA_{p-p}	Current and Angle in Inductor Branch I_C, mA	Total Line Current and Angle (Measured) I_T, mA_{p-p}	Total Line Current (Calculated Using Square-Root Formula) I_T, mA_{p-p}	Circuit Impedance (Calculated Using Ohm's Law) Z, Ω
10 V					

QUESTIONS

1. Refer to your own data in Table 50–1 and Table 50–2. Do the measured values of I_R, I_L, and I_T verify formula (50–1)? Explain your answer with reference to specific data in the table.

_____

_____

_____

_____

_____

_____

_____

2. If the inductor in Figure 50–5(*a*) doubled in value, what would happen to Z_T and I_T?

_____

_____

_____

_____

_____

_____

_____

3. Calculate the phase angle of the *RL* circuit from Figure 50–5(*a*)'s rated values. Ignore the 10 Ω resistors. State whether it is a leading or lagging angle.

_____

_____

_____

_____

_____

_____

_____

4. Calculate the phase angle of the *RC* circuit from Figure 50–6(*a*)'s rated values. Ignore the 10 Ω resistors. State whether it is a leading or lagging angle.

_____

_____

_____

_____

_____

_____

_____

5. On a separate sheet of $8^{1}/_{2} \times 11$ paper draw the phasor diagrams for current in the *RL* and *RC* parallel circuits. Use your data from Tables 50–1 and 50–2.

IMPEDANCE OF A PARALLEL *RLC* CIRCUIT

BASIC INFORMATION

The simplest type of parallel *RLC* circuit is illustrated in Figure 51–1. The ac voltage *V* is common to each leg in the circuit. The current I_R in *R* may be calculated using Ohm's law,

$$I_R = \frac{V}{R} \tag{51–1}$$

where *I* is in amperes, *V* is in volts, and *R* is in ohms. The currents I_C in *C* and I_L in *L* may be calculated similarly.

$$I_C = \frac{V}{X_C} \tag{51–2}$$

$$I_L = \frac{V}{X_L} \tag{51–3}$$

Because *R*, *L*, and *C* are in parallel, the voltage across each leg is the applied voltage *V*.

The total line current I_T is equal to the phasor sum of I_R, I_C, and I_L. This is consistent with the results of Experiment 50, where parallel *RL* and *RC* circuits were studied.

Figure 51–2 (p. 390) is a phasor diagram showing the phase relationship between the applied voltage *V* and the currents in each leg. Thus, I_R is in phase with *V*, I_C leads *V* by 90°, and I_L lags *V* by 90°. Since I_C and I_L are 180° out of phase, they may be subtracted arithmetically to give the reactive current in the circuit, as in Figure 51–3 (p. 390). If I_L is greater than I_C, the circuit is inductive. If I_C is greater than I_L, the circuit is capacitive, as in Figure 51–3. A special case exists when $I_L = I_C$. This is dis-

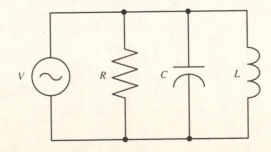

Figure 51–1. Parallel *RLC* circuit.

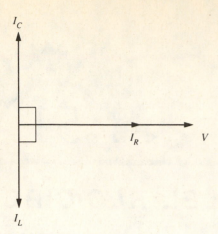

Figure 51–2. Current phasor diagram for a parallel *RLC* circuit. The applied voltage V is used as the reference phasor.

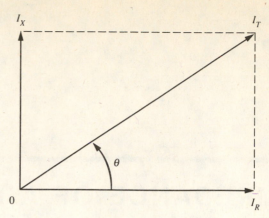

Figure 51–3. The reactive current I_x is the difference between I_c and I_L. In this case I_c is greater than I_L, so θ is positive and the *RLC* circuit is capacitive.

cussed in a later experiment. The line current may be obtained by finding the phasor sum of I_R and the reactive current I_X, where I_X is the difference between I_L and I_C, as in Figure 51–3.

The line current may also be calculated by substituting the values of I_R and I_X in the formula

$$I_T = \sqrt{I_R^2 + I_X^2} \qquad (51\text{–}4)$$

This formula can be used as long as there is no resistance in series with L or C in the circuit of Figure 51–1.

The line current I_T is the phasor sum of the resistive and reactive currents and is not in phase with either current. It differs from I_R, the resistive current, by the angle θ, with the applied voltage V as the reference phasor. This angle may be calculated from the formula

$$\theta = \tan^{-1}\left(\frac{I_X}{I_R}\right) \qquad (51\text{–}5)$$

Having found the line current I_T, we can now calculate the value of impedance Z in ohms using Ohm's law.

$$Z = \frac{V}{I_T} \qquad (51\text{–}6)$$

Differences Between AC and DC Parallel Circuits ·····················

There is an important difference between dc and ac parallel circuits. In dc parallel circuits the total line current I_T increases with every parallel branch added. In ac parallel circuits the line current does not always increase as more branches are added. For example, if a circuit initially consists of a resistive branch in parallel with a capacitive branch, the effect of adding an inductive branch *may* be to *reduce* the line current, a fact that is evident from formula (51–4).

The line current in the circuit of Figure 51–1 would be less than the line current of an *RC* parallel circuit if the difference between I_L and I_C were less than I_C. This would happen if I_L were less than twice I_C. The situation

would be similar if the original circuit were a parallel *RL* circuit and *C* were added in parallel. The line current I_T would drop if the inductive current I_L were greater than the difference between I_L and I_C. Such a condition would arise if I_C were less than twice I_L.

It follows that the total impedance of ac parallel circuits is not always *less* than the smallest branch impedance but is sometimes *greater* than the smallest branch impedance.

Parallel Circuits with *RL* and *RC* Series Branches ·····················

A parallel *RLC* circuit may contain branches with R and L or R and C in series, as in Figure 51–4. Branch AB consists of a resistor R_1 in series with a capacitor C. Branch CD consists of a resistor R_2 in series with an inductor L.

The impedance of the series combinations in each branch can be found using formulas studied in previous experiments.

For branch AB

$$Z_{AB} = \sqrt{R_1^2 + X_C^2} \qquad (51\text{–}7)$$

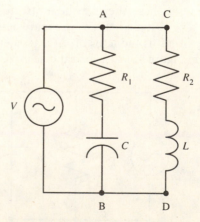

Figure 51–4. Parallel *RLC* circuit consisting of series *RC* and *RL* branches.

The current I_{AB} in branch AB is

$$I_{AB} = \frac{V}{Z_{AB}} \qquad (51\text{–}8)$$

Because branch AB is capacitive, the current leads the voltage across AB. The phase angle θ can be found using the formula

$$\theta_1 = \tan^{-1}\left(\frac{X_C}{R_1}\right) \qquad (51\text{–}9)$$

For branch CD

$$Z_{CD} = \sqrt{R_2^2 + X_L^2} \qquad (51\text{–}10)$$

The current I_{CD} in branch CD is

$$I_{CD} = \frac{V}{Z_{CD}} \qquad (51\text{–}11)$$

Because branch CD is inductive, the current lags the voltage across CD. The phase angle θ_2 can be found using the formula

$$\theta_2 = \tan^{-1}\left(\frac{X_L}{R}\right) \qquad (51\text{–}12)$$

The total, or line, current delivered by the voltage source is the phasor sum of I_{AB} and I_{CD}. The phasor diagram of Figure 51–5 shows the reference voltage V at 0°. Current I_{AB} leads the voltage at an angle θ_1. Current I_{CD} lags the voltage at an angle θ_2.

To find I_T, the parallelogram OPQR is completed by drawing PQ parallel to phasor OR and QR parallel to phasor OP. The phasor OQ is the total current I_T,

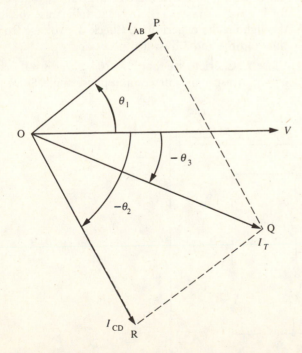

Figure 51–5. Current phasor diagram for an *RLC* circuit with *RL* and *RC* branches. The applied voltage *V* is used as the reference phasor.

and θ_3 is the angle by which the total current is out of phase with the applied voltage. In Figure 51–5 angle θ_3 is a negative angle, indicating that I_T lags the applied voltage, and the parallel circuit is behaving inductively. This is indicated by the fact that phasor I_{CD} is larger than phasor I_{AB}. If the reverse were true, θ_3 would be positive, and the total circuit would be capacitive.

SUMMARY

A method for finding the impedance of a parallel *RLC* circuit requires the following steps:

1. Calculate the impedance of each branch, or leg, of the parallel circuit by the methods already discussed in previous experiments. Calculate the value of Z using V and I and Ohm's law. Calculate the phase angle θ using $\tan^{-1}$.

2. Find the current in each branch and the phase angle between the applied voltage V and the branch current. Branch current is computed using the formula

$$I = \frac{V}{Z}$$

Note that the phase angle (in degrees) is the same as the angle θ in statement 1, but it is opposite in sign. Thus, the phase angle of a capacitive current is positive $(+)$ and that of an inductive current is negative $(-)$.

3. Obtain the phasor sum of all branch currents. The resultant is the total, or line, current. The value of current and phase angle can be found from the phasor diagram.

4. If the parallel *RLC* circuit has no resistance in the capacitive or inductive branch—that is, if the circuit is exactly as in Figure 51–1—then the net reactive current I_X is the difference between I_L and I_C.

5. Find the total current I_T for the circuit in statement 4 by using the formula

$$I_T = \sqrt{I_R^2 + I_X^2}$$

and the phase angle θ between the applied voltage V and the total current is

$$\theta = \tan^{-1}\left(\frac{I_X}{I_R}\right)$$

6. Find the impedance Z of the parallel *RLC* circuit by using the formula

$$Z = \frac{V}{I_T}$$

where V is the voltage across the parallel circuit and I_T is the total, or line, current.

Check your understanding by answering the following questions:

1. In the circuit of Figure 51–1 $R = 12\ \Omega$, $X_C = 15\ \Omega$, and $X_L = 18\ \Omega$. The applied voltage $V = 6$ V.
   (a) The current I_R in $R =$ _____ A.
   (b) The phase angle between V and I_R is _____°.

2. For the circuit in question 1, (a) the current I_C in $C =$ _____ A, and (b) the phase angle between V and $I_C =$ _____°.

3. For the circuit in question 1, (a) the current I_L in $L =$ _____ A, and (b) the phase angle between V and I_L is _____°.

4. The net reactive current in the circuit in question 1 is _____ A, and it is _____ (capacitive/inductive).

5. The total current I_T in the circuit of question 1 is _____ A.

6. The phase angle between the applied voltage V and I_T in the circuit of question 1 is _____° and its sign is _____ (+/−).

7. The total impedance in the circuit of question 1 is _____ Ω.

MATERIALS REQUIRED

Instruments:
- Function generator
- Oscilloscope

Resistor:
- 1 2 kΩ, ½-W
- 1 10 Ω, ½-W

Capacitor:
- 1 0.022 µF

Inductor:
- 100 mH coil

Miscellaneous:
- 3 SPST switches

PROCEDURE

1. With the function generator turned **off,** and switches S_1 through S_3 **open,** connect the circuit of Figure 51–6. Channel 2, of the oscilloscope, is placed across the sensing resistor. By measuring the voltage drop across R_{sense} and using Ohm's law, the circuit current can be indirectly measured.

2. Turn the generator **on.** Increase the output voltage V to $V = 10$ V$_{\text{p-p}}$ at 5 kHz. Maintain this voltage throughout the experiment. Check the voltage from time to time and adjust it if necessary.

3. **Close S_1.** Check to see that $V = 10$ V$_{\text{p-p}}$, adjust if necessary. Measure the current and phase angle. Since S_2

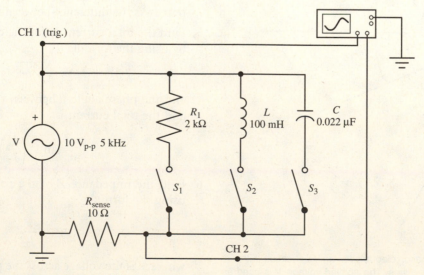

Figure 51–6. Circuit for procedure step 1.

and S_3 are **open,** the only current in the circuit is the current in the resistor I_R. Record the value in Table 51–1 (p. 395). **Open** S_1.

4. **Close** S_2. Check that $V = 10\,V_{\text{p-p}}$. Measure the current and phase angle. Since S_1 and S_3 are **open,** the only current in the circuit is the current in the inductor I_L. Record the value in Table 51–1. **Open** S_2.

5. **Close** S_3. Check V; adjust if necessary. Measure the current and phase angle. Since S_1 and S_2 are **open,** the only current in the circuit is the current in the capacitor branch I_C. Record the value in Table 51–1.

6. **Close** S_1 (S_3 is still **closed**). Verify that $V = 10\,V_{\text{p-p}}$. Measure the current and phase angle in the circuit. With S_1 and S_3 **closed** and S_2 **open,** the current in the circuit is the total of I_R and I_C, or I_{RC}. Record the value in Table 51–1. **Open** S_3.

7. **Close** S_2 (S_1 is still **closed**). $V = 10\,V_{\text{p-p}}$. Measure the current in the circuit. With S_1 and S_2 **closed** and S_3 open, the current in the circuit is the total of I_R and I_L, or I_{RL}. Record the value in Table 51–1.

8. **Close** S_3. Now $S_1, S_2,$ and S_3 are **closed.** Check V. Measure the current and angle in the circuit. Since all branch circuit switches are **closed,** the ammeter will measure the total current I_T of the parallel RLC circuit. Record the value in Table 51–1. **Open** all switches; turn **off** the function generator.

9. Calculate the line current I_T using the measured values of I_R, I_L, and I_C and the square-root formula (51–4). Record your answer in Table 51–1.

10. Using the measured value of V (it should be $10\,V_{\text{p-p}}$) and the measured value of I_T, calculate the circuit impedance and indicate whether the circuit is inductive, capacitive, or resistive. Record your answers in Table 51–1.

11. Calculate the phase angle and power factor for the parallel RLC circuit and indicate whether the circuit has a leading or lagging power factor. Record your answers in Table 51–1.

ANSWERS TO SELF TEST

1. (a) 0.5; (b) 0
2. (a) 0.4; (b) 90
3. (a) 0.333; (b) −90
4. 0.0667; capacitive
5. 0.504
6. 7.6; +
7. 11.9

Name _____ Date _____

TABLE 51-1.   Determining the Impedance of a Parallel *RLC* Circuit

Applied Voltage V, V_{p-p}	Resistor Current and Phase I_R, mA$_{p-p}$	Inductor Current and Phase I_L, mA$_{p-p}$	Capacitor Current and Phase I_C, mA$_{p-p}$	Resistor and Capacitor Current and Phase I_{RC}, mA$_{p-p}$	Resistor and Inductor Current and Phase I_{RL}, mA$_{p-p}$	Total Current in RLC Circuit and Phase (Measured) I_T, mA$_{p-p}$	Total Current (Calculated Using Square Root Formula) I_T, mA$_{p-p}$	Circuit Impedance Z (R, L, or C), Ω
10 V								

Power factor _____% Leading/lagging? _____ Phase Angle (degrees) _____

QUESTIONS

1. Refer to your data in Table 51–1. Do the measured values of I_R, I_L, I_C, and I_T verify Formula (51–4)? Explain your answer with reference to specific data in the table.

_____

_____

_____

_____

_____

_____

_____

_____

_____

2. Refer to your data in Table 51–1. Compare I_{RC} with I_T. Which is greater? Explain.

_____

_____

_____

_____

_____

_____

_____

_____

_____

_____

_____

_____

3. Using your data in Table 51–1, calculate the impedance of the *RL* section alone. Also, calculate the impedance of the *RC* section alone. Compare the two impedances with the total impedance of the *RLC* circuit. Are the results as expected? Explain.

_____

_____

_____

_____

_____

_____

_____

_____

_____

_____

_____

_____

_____

_____

_____

_____

4. On a separate sheet of 8½ × 11 cross-section paper, draw a phasor diagram to scale of the currents in the parallel *RLC* circuit. Label all phasors and the phase angle of the circuit. On the same sheet of cross-section paper, draw the series circuit equivalent to the *RLC* circuit. Label the values of the series components. Show all calculations.

RESONANT FREQUENCY AND FREQUENCY RESPONSE OF A SERIES *RLC* CIRCUIT

BASIC INFORMATION

Resonant Frequency of Series *RLC* Circuit

In the circuit of Figure 52–1 the voltage V is produced by an ac generator whose frequency and output voltage are manually adjustable. For a particular frequency f and output voltage V there will be a current I such that $I = V/Z$, where Z is the impedance of the circuit. The voltage drops across R, L, and C will be given by IR, IX_L, and IX_C, respectively.

If the generator frequency is changed but V remains constant, the current I and the voltage drops across R, L, and C will change.

There is a frequency f_R, called the *resonant frequency*, at which

$$X_L = X_C \tag{52-1}$$

The resonant frequency can be found as follows:

$$X_L = 2\pi fL \tag{52-2}$$

and

$$X_C = \frac{1}{2\pi fC}$$

At f_R, when $X_L = X_C$,

$$2\pi f_R L = \frac{1}{2\pi f_R C} \tag{52-3}$$

Formula (52–3) can be solved for f_R:

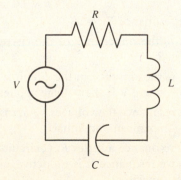

Figure 52–1. Series *RLC* circuit.

OBJECTIVES

1 To determine experimentally the resonant frequency f_R of a series *LC* circuit

2 To verify that the resonant frequency of a series *LC* circuit is given by the formula

$$f_R = \frac{1}{2\pi\sqrt{LC}}$$

3 To develop experimentally the frequency-response curve of a series *LC* circuit

$$f_R^2 = \frac{1}{(2\pi)^2 LC} \qquad (52\text{--}4)$$

or

$$f_R = \frac{1}{2\pi\sqrt{LC}} \qquad (52\text{--}5)$$

where f_R is given in hertz, L in henrys, and C in farads.

This is the formula for the resonant frequency f_R of a resonant series RLC circuit. If the value of L and C are known, the resonant frequency can be calculated directly using formula (52–5).

Problem. Find the resonant frequency of a series RLC circuit where $L = 8$ H, $C = 0.01$ μF, and $R = 47$ Ω.

Solution. Formula (52–5) will be used to find f_R.

$$f_R = \frac{1}{2\pi\sqrt{LC}} = \frac{1}{2\pi\sqrt{8 \times 0.01 \times 10^{-6}}}$$

$$f_R = \frac{10^4}{2\pi\sqrt{8}}$$

$$f_R = 563 \text{ Hz}$$

Note that the value of R did not enter into the calculation of f_R.

Characteristics of a Series Resonant Circuit ·

Some characteristics of a series RLC circuit were investigated in a previous experiment. The following facts were observed:

1. The voltage drop across a reactive component is equal to the product of the current I in the circuit and the reactance X of the component.

2. The total reactive effect on a circuit is the difference between the capacitive reactance X_C and the inductive reactance X_L.

3. The impedance Z of a series RLC circuit is
$$Z = \sqrt{R^2 + X_T^2}$$

4. The impedance Z of the circuit is minimum when $X_L = X_C$, and circuit current I is maximum at this pint.

When $X_L = X_C$ the circuit is said to be resonant. *At resonance, the impedance of a series RLC circuit is minimum and the current I in the circuit is maximum.* This fact, which was established experimentally in a previous experiment, can also be found using the formula
$$Z = \sqrt{R^2 + X_T^2}$$

At f_R, $X_T = 0$,
$$Z = \sqrt{R^2 + (0)^2} = R$$

Therefore, a *minimum impedance $Z = R$ exists at f_R.* Also,

since
$$I = \frac{V}{Z} = \frac{V}{R}$$

and V is assumed to be fixed, there will be *maximum current I* in the circuit at resonance. Note that at f_R the applied voltage V appears across R, and resonant current I may be calculated from

$$I = \frac{V}{R} \qquad (52\text{--}6)$$

Because I is limited at resonance only by the value of R, the circuit is said to be *resistive*. For all frequencies higher than f_R, X_L is greater than X_C, and the circuit is *inductive*. For all frequencies lower than f_R, X_C is greater than X_L, and the circuit is *capacitive*.

At resonance the voltage V_L across L and the voltage V_C across C are *maximum* and equal. Theoretically, at the resonant frequency if R becomes zero, the current I becomes infinite and the voltages V_L and V_C become infinitely large. Practically, this condition is never realized because L has some resistance R_L that may be considered in series with it.

Frequency Response of a Series Resonant Circuit ·

The frequency-response characteristics of a series RLC circuit can be found experimentally by applying a constant-amplitude signal voltage to the circuit at the resonant frequency and at frequencies on both sides of resonance. The voltage across L or C is measured, and a graph of V_L or V_C versus f is plotted. This is one form of the frequency-response curve of the circuit.

The circuit current I can also be measured at f_R and at frequencies on both sides of f_R. A graph of I versus f is another form of the frequency-response curve of the circuit.

SUMMARY

1. In a series RLC circuit there is a frequency f_R, called the resonant frequency, which may be calculated from the formula

$$f_R = \frac{1}{2\pi\sqrt{LC}}$$

At resonance $X_L = X_C$.

2. At f_R the impedance of the circuit is minimum, and $Z = R$.

3. At f_R the current I in the circuit is maximum, and $I = V/R$ where V is the voltage applied across the circuit.

4. At resonance the voltages V_L across L and V_C across C are maximum and equal.

5. At resonance a series RLC circuit acts as a resistance; above resonance it is inductive, and below resonance it is capacitive.

SELF TEST

Check your understanding by answering the following questions:

1. In a series resonance circuit, X_L = 120 Ω at f_R. The value of X_C at resonance is _____ Ω.

2. When the frequency f of the source applied to a series *RLC* circuit is higher than the resonant frequency f_R, then X_L _____ (equals/is greater than/is less than) X_C and the circuit is _____.

3. In a series *RLC* circuit R = 1 kΩ, L = 100 μH, C = 0.001 μF, and the applied voltage V = 12 V. The resonant frequency f_R of this circuit is _____ Hz.

4. In the circuit of question 3, the circuit current I = _____ mA at resonance.

5. In the circuit of question 3, the voltage V_L across L is _____ V.

MATERIALS REQUIRED

Instruments:
- Function generator
- Oscilloscope

Resistor:
- 1   1 kΩ, ½-W, 5%

Capacitors:
- 1   0.001-μF
- 1   0.01-μF
- 1   0.0033-μF

Inductor:
- 10-mH coil

PROCEDURE

• •

A. Determining the Resonant Frequency of a Series *RLC* Circuit

A1. Calculate the resonant frequencies for 10-mH–0.01-μF; 10-mH 0.0033-μF; and 10-mH–0.001-μF series *LC* combinations. Use formula (52–5) and the rated values of L and C. Record your answers in Table 52–1 (p. 401).

A2. With the function generator and oscilloscope **off**, connect the circuit of Figure 52–2.

A3. Turn **on** the function generator and set the frequency for 15 kHz. Turn **on** the oscilloscope and calibrate it for voltage measurements. Adjust the scope to view

the sine-wave output of the generator. Increase the output of the generator until the scope indicates a 5-V$_{p-p}$ voltage. Maintain this voltage throughout the experiment.

A4. Observe the p-p voltage across the resistor V_R as the frequency of the generator is varied above and below 15 kHz. Note the frequency at which V_R is a maximum. In a series *RLC* circuit, V_R is a maximum at the resonant frequency, f_R. Also, note that the phase shift is 0° at resonance. Make sure to view this on your scope. Record the value of f_R in Table 52–1 (p. 401) in the 0.01-μF row. Turn **off** the function generator.

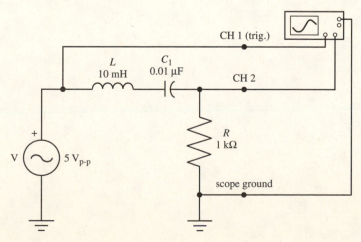

Figure 52–2. Circuit for procedure step A2.

A5. Replace the 0.01-μF capacitor with a 0.0033-μF capacitor. Turn **on** the function generator. Check the generator output voltage to verify that it is 5 V$_{p-p}$; adjust the voltage if necessary.

A6. Set the frequency of the generator to 27 kHz. Observe the voltage across the resistor V_R as the frequency is varied above and below 27 kHz. At the point of maximum V_R the frequency is f_R. Record this value in Table 52–1 in the 0.0033-μF row. Turn **off** the function generator.

A7. Replace the 0.0033-μF capacitor with a 0.001-μF capacitor. Turn **on** the function generator. Check the generator output voltage and adjust to maintain 5 V$_{p-p}$ if necessary.

A8. Set the frequency of the generator to 50 kHz. Observe the voltage across the resistor V_R as the frequency is varied above and below 5 kHz. At the resonant frequency f_R the voltage across the resistor will be maximum. Record the value of f_R in Table 52–1 in the 0.001-μF row.

B. Plotting a Frequency-Response Curve

B1. With the circuit of Figure 52–2 still connected and the 0.001-μF capacitor in the circuit, check the oscilloscope to verify that the output voltage is still 5 V p-p. Also verify the value of f_R for the 10-mH and

0.001-μF circuit. (It should be the same as that obtained in step A8.)

B2. Examine Table 52–2. In this part of the experiment you will take a series of readings at frequencies above and below the resonant frequency. For each frequency setting you will measure and record the voltage across the 1 kΩ resistor. Since f_R may not be a round number, you may not be able to set the exact frequencies on the generator. Therefore, choose values of frequency as close to the step values as possible. For example, if $f_R = 9227$, $f_R + 3000 = 12{,}227$; in that case, choose the closest frequency for which an exact setting can be made. It is important to continue monitoring the output voltage of the generator and adjusting it to 5 V$_{p-p}$ if necessary. Complete Table 52–2. After completing all measurements, turn **off** the oscilloscope and function generator.

ANSWERS TO SELF TEST

1. 120

2. is greater than; inductive

3. 503,292

4. 12

5. 3.79

6. 1

TABLE 52–1. Resonant Frequency of a Series _RLC_ Circuit

Inductor LmH	Capacitor C, μF	Resonant Frequency f_R, Hz	
		Calculated	Measured
10	0.01		
10	0.0033		
10	0.001		

TABLE 52–2. Frequency Response of a Series _RLC_ Circuit

Step	Frequency f, Hz	Voltage Across Resistor V_R, V_{p-p}
$f_R - 21$ kHz		
$f_R - 18$ kHz		
$f_R - 15$ kHz		
$f_R - 12$ kHz		
$f_R - 9$ kHz		
$f_R - 6$ kHz		
$f_R - 3$ kHz		
f_R		
$f_R + 3$ kHz		
$f_R + 6$ kHz		
$f_R + 9$ kHz		
$f_R + 12$ kHz		
$f_R + 15$ kHz		
$f_R + 18$ kHz		
$f_R + 21$ kHz		

QUESTIONS

1. Explain, in your own words, what the resonant frequency of a series _RLC_ circuit is.

_____

_____

_____

_____

2. Refer to your data in Table 52–2. Why wasn't the output voltage across the resistor equal to V (5 V_{p-p}) when the input frequency was at resonance?

_____

_____

_____

_____

_____

3. What was the impedance of your series *RLC* circuit at resonance? Explain.

_____

_____

_____

_____

_____

4. Refer to your data in Table 52–1. Compare the calculated value of resonant frequency with the measured value. Explain any unexpected results.

_____

_____

_____

_____

_____

_____

5. Discuss the effect on the resonant frequency of changes in capacitance with fixed resistance and inductance in a series *RLC* circuit. Refer to your data in Table 52–1.

_____

_____

_____

_____

_____

6. On a separate sheet of 8½ × 11 graph paper, plot a graph of frequency versus current through the resistor using your data in Table 52–2. The horizontal (x) axis should be frequency, and the vertical (y) axis should be current through the resistor. Draw a broken vertical line at the resonant frequency extending from the frequency axis to the frequency response curve. Label all axes; label the resonant frequency line f_R.

7. Why was it important to constantly monitor the output voltage of the function generator during the frequency response measurements?

_____

_____

_____

_____

_____

_____

_____

EFFECTS OF Q ON FREQUENCY RESPONSE AND BANDWIDTH OF A SERIES RESONANT CIRCUIT

BASIC INFORMATION

Circuit Q and Frequency Response

In Experiment 52 we studied the frequency-response curve of a series resonant RLC circuit. Theoretically, at resonance $X_L = X_C$ and if the only resistance was that of the coil the impedance would be $Z = R_L$, where R_L equals the resistance of the coil.

The amount of coil resistance R_L, therefore, determines the current flow through the circuit at resonance if there is no resistance other than the coil resistance. The R_L and X_L of the coil determine the *quality*, or Q, of the circuit, which is given by the formula

$$Q = \frac{X_L}{R_L} \qquad (53\text{--}1)$$

The Q of the circuit also determines the rise in voltage across L and C at the resonant frequency f_R. The voltage developed across L is given by the formulas

$$V_L = IX_L$$

$$= \frac{V}{R} \times X_L \qquad (53\text{--}2)$$

If the circuit resistance R is the coil resistance R_L, then

$$V_L = V \times \frac{X_L}{R_L}$$

$$V_L = VQ \qquad (53\text{--}3)$$

Also, since $X_L = X_C$ at resonance,

$$IX_L = IX_C$$

and

$$V_L = V_C$$

Therefore,

$$V_C = VQ \qquad (53\text{--}4)$$

OBJECTIVES

1. To measure the effect of circuit Q on frequency response

2. To measure the effect of circuit Q on bandwidth at the half-power points

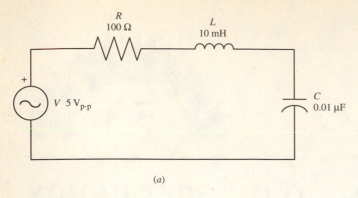

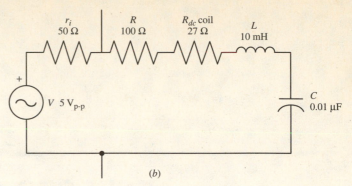

Figure 53–1(a). Basic *RLC* circuit.

Figure 53–1(b). Practical *RLC* circuit.

Formulas (53–3) and (53–4) become significant for values of $Q > 1$. For such values, V_C and V_L are greater than the applied voltage V. Also, the higher the value of Q, the greater the voltage gain in the circuit. This is the *first example* of voltage gain.

In a practical sense, other resistance values need to be taken into account when calculating a circuit's Q. Figure 53–1(a) shows a basic *RLC* resonant circuit. The resonant frequency and Q for this circuit calculated is

$$f_R = \frac{1}{2\pi\sqrt{LC}} = \frac{1}{6.28\sqrt{(10 \text{ mH})(0.01 \text{ }\mu\text{F})}} = 15.9 \text{ kHz}$$

$$X_L = 2\pi fL = (6.28)(15.9 \text{ kHz})(10 \text{ mH}) = 999 \text{ }\Omega$$

$$Q = \frac{X_L}{R} = \frac{999 \text{ }\Omega}{100 \text{ }\Omega} = 9.9$$

Actually the circuit appears more like Figure 53–1(b). Notice the internal resistance, r_i, of the generator is 50 Ω and the dc resistance of the inductor is 27 Ω. Therefore, the actual circuit Q, when considering all resistance, will be

$$Q = \frac{X_L}{(r_i + R_{dc} + R)}$$

$$= \frac{999 \text{ }\Omega}{(50 \text{ }\Omega + 27 \text{ }\Omega + 100 \text{ }\Omega)} = 5.64$$

Circuit Q is also significant when we consider the frequency response of a series resonant circuit. The frequency-response characteristic can be determined by applying a constant-amplitude signal voltage V into the circuit at the resonant frequency and at frequencies on either side of resonance. The voltage across L or C is measured, and a graph of V_L or V_C versus f is plotted. This is one form of the frequency-response curve of the circuit.

The circuit current I can also be determined. A graph of I versus f is another form of the frequency-response curve of the circuit.

Circuit *Q* and Bandwidth · · · · · · · · · · · · ·

Figure 53–2 is a graph of the response of a series resonant circuit. Three significant points have been marked

on the curve. These are f_R, the resonant frequency, and f_1 and f_2. Points f_1 and f_2 are located at 70.7 percent of maximum (maximum is at f_R) on the curve. They are called the *halfpower points*, and the frequency separation between them is $f_2 - f_1$. This frequency separation is called the *bandwidth (BW)* of the circuit. Bandwidth, therefore, is given by the formula

$$BW = f_2 - f_1 \qquad (53–5)$$

Bandwidth is related to Q. It can be shown that BW is given by the formula

$$BW = \frac{f_R}{Q} \qquad (53–6)$$

As was noted in the preceding experiment, the resonant frequency f_R of a series *RLC* circuit can be calculated from the formula

$$f_R = \frac{1}{2\pi\sqrt{LC}} \qquad (53–7)$$

where f_R is in hertz, L is in henrys, and C is in farads. Since the formula does not include R, it is apparent that the resonant frequency of the series *RLC* circuit is not affected by the size of the resistance R. However, R does affect the bandwidth and amplitude of the response curve. The higher the value of R, the lower is the value of Q, as shown

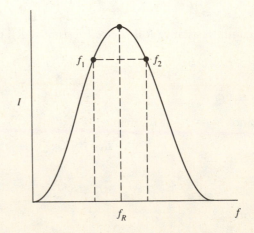

Figure 53–2. Frequency-response curve of a series resonant circuit.

in formula (53–1). The higher the resistance R, the wider is the bandwidth, as shown by formulas (53–1) and (53–6). Moreover, the lower the value of Q, the lower is the value of current I in the circuit, and the lower are the voltages V_L across L and V_C across C.

Series resonant circuits are used in communication, video, and industrial electronics as frequency-selective circuits and as traps to eliminate unwanted signals. Normally such circuits require a highly peaked response with narrow bandwidth. To achieve such a response, the Q of the circuit must be high. Hence high-Q coils are employed. In such circuits the Q of the circuit is mainly determined by the Q of the coil.

However, there are some applications in electronics where wideband frequency-selective circuits are required. In such cases coil "loading" is achieved by the use of external resistors.

SUMMARY

1. The quality, or Q, of a coil is defined by the formula

$$Q = \frac{X_L}{R}$$

2. In a series resonant circuit, where the only resistance in the circuit is coil resistance R_L the Q of the circuit is determined by the Q of the coil.

3. In a series resonant circuit the voltages across L and C are equal and are affected by the Q of the circuit. These voltages may be calculated from the formula

$$V_L = V_C = VQ$$

where V is the applied voltage and V_L and V_C are the voltages across L and C, respectively.

4. If circuit Q is greater than one, then the voltages V_L and V_C are higher than the applied voltages. The voltages in a series resonant circuit provide the first example of amplification.

5. Bandwidth (BW) of a response curve is defined as the difference between the two frequencies f_2 and f_1 at the half-power points (Figure 53–2). The half-power points are those points that appear at 70.7 percent of maximum on the frequency-response curve. Maximum output occurs at the resonant frequency f_R.

6. Bandwidth is related to circuit Q and is given by the formula

$$BW = \frac{f_R}{Q}$$

7. The lower the Q of the circuit, the wider the bandwidth and the flatter the response curve.

8. The lower the Q of the circuit, the lower the amplitude of the response curve and the lower the circuit gain.

SELF TEST

Check your understanding by answering the following questions.

In the circuit of Figure 53–3, $L = 50$ mH, $C = 0.01$ μF, and $R = 20$ Ω. Assume that coil resistance $R_L = 0$. The applied voltage $V = 5.0$ V. The output voltage is V_C.

1. The resonant frequency is $f_R =$ _____ kHz.
2. At resonance $X_L =$ _____ Ω.
3. The circuit Q is _____ at resonance.
4. The voltage V_L across L is _____ V at resonance.
5. The bandwidth of the circuit is _____ Hz.
6. The voltage V_C across C at the half-power points is _____ V.

MATERIALS REQUIRED

Instruments:
- Function generator
- Oscilloscope

Resistors (½-W, 5%):
- 1 1 kΩ
- 1 220 Ω
- 1 100 Ω

Capacitor:
- 1 0.001 μF

Inductor:
- 10-mH coil

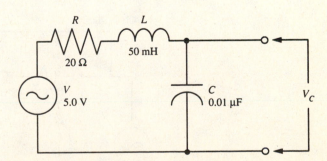

Figure 53–3. Circuit diagram for the self test.

PROCEDURE

A. Circuit Q and Frequency Response of a Series Resonant Circuit

A1. With the function generator and oscilloscope turned **off,** connect the circuit of Figure 53–4. The oscilloscope should be calibrated to measure the output voltage of the generator.

A2. Turn **on** the generator and scope. Adjust the output of the generator V to 2 $V_{p\text{-}p}$ as measured by the oscilloscope. Maintain this voltage throughout the experiment, checking it whenever the frequency of the generator is changed; adjust to 2 $V_{p\text{-}p}$ if necessary.

A3. Set the function generator to 50 kHz. Vary the frequency above and below 50 kHz until the maximum voltage across the capacitor V_C is determined. The maximum V_C is reached at the resonant frequency, f_R. Record f_R and V_C in Table 53–1 (p. 409).

A4. Examine Table 53–1. You will measure the voltage across the capacitor V_C as you vary the frequency from 21 kHz below the resonant frequency to 21 kHz above f_R in steps of 3 kHz. Choose the frequency on the generator as close as possible to the deviation shown. Record the actual frequency in the column shown. Record each voltage in the column marked "1 kΩ Resistor." After completing all measurements, turn **off** the function generator and disconnect the 1 kΩ resistor from the circuit.

A5. Replace the 1 kΩ resistor with a 220-Ω resistor. Turn **on** the generator and adjust the output voltage V to 2 $V_{p\text{-}p}$ as measured by the scope. Maintain this voltage throughout the experiment.

A6. Measure the voltage across the capacitor for each of the frequencies in Table 53–1. Record the values in Table 53–1 in the "220-Ω Resistor" column. After

completing all the measurements, turn **off** the generator and disconnect the 220-Ω resistor.

A7. Replace the 220-Ω resistor with a 100-Ω resistor. Turn **on** the generator and adjust the output V to 2 $V_{p\text{-}p}$ as measured by the oscilloscope. Maintain this voltage throughout the experiment.

A8. Measure the voltage across the capacitor V_C for each frequency in Table 53–1. Record the values in Table 53–1 in the "100-Ω Resistor" column. After completing all the measurements, turn **off** the generator and the oscilloscope; disconnect the 100-Ω resistor.

B. Effect of Resistance on Resonant Frequency; Determining the Phase Angle of a Resonant Circuit

B1. Reconnect the circuit of Figure 53–4 with the 1 kΩ resistor and the scope's probes across the resistor.

B2. Turn **on** the generator and the oscilloscope. Set the generator output V at 2 $V_{p\text{-}p}$ as measured by the scope. Maintain this voltage throughout the experiment; adjust if necessary.

B3. Vary the frequency until the voltage across the resistor V_R is at maximum. At maximum V_R the frequency is the resonant frequency of the circuit. Record f_R and V_R in Table 53–2 (p. 409) in the 1 kΩ row. Measure the voltage across the capacitor-inductor combination, V_{LC}. Record the value in Table 53–2 in the 1 kΩ row. Turn **off** the generator and disconnect the 1 kΩ resistor.

B4. Connect the 220-Ω resistor and repeat step B3. Record the frequency in the 220-Ω row. Measure the voltage across the capacitor-inductor combination, V_{LC}. Record the value in Table 53–2 in the 220-Ω row. Turn **off** the generator and disconnect the 220-Ω resistor.

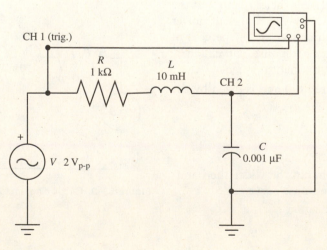

Figure 53–4. Circuit for procedure A1.

B5. Replace the 220-Ω resistor with a 100-Ω resistor and repeat step B3. Record the frequency in the 100-Ω row. Measure the voltage across the capacitor-inductor combination, V_{LC}. Record the value in Table 53–2 in the 100-Ω row. Turn **off** the generator and the oscilloscope. Disconnect the circuit.

B6. Measure the resistance of the inductor and record the value in Table 53–2.

B7. For each value of resistor, calculate the current in the circuit, using the measured value of V_R and the rated value of R. Record your answers in Table 53–2.

B8. Using your circuit's practical resistance values, calculate each circuit's Q. Then, using your measured V_C at resonance, determine the measured Q value. Record your answers in Table 53–2.

Optional Activity ·

This activity requires the use of electronic simulation software.

Simulate the circuit of Figure 53–4 and repeat steps B2 through B5 in the Procedure. Verify your experiment results from Table 53–2.

The simulation software contains a Bode plotter which is capable of producing a graph of your circuit's frequency response. Read the instructions for the use of the plotter. After you become familiar with the various settings of the plotter, connect the plotter to your simulated circuit. Measure and display the output response of your circuit for different values of R.

ANSWERS TO SELF TEST

1. 7.118
2. 2.24 k
3. 111.8
4. 559
5. 63.7
6. 395

53

TABLE 53–1. Circuit Q and Frequency Response of a Series Resonant Circuit

Frequency Deviation	Frequency f, Hz	1 k-Ω Resistor Voltage across Capacitor V_C, V_{p-p}	220-Ω Resistor Voltage across Capacitor V_C, V_{p-p}	100-Ω Resistor Voltage across Capacitor V_C, V_{p-p}
$f_R - 21$ k				
$f_R - 18$ k				
$f_R - 15$ k				
$f_R - 12$ k				
$f_R - 9$ k				
$f_R - 6$ k				
$f_R - 3$ k				
f_R				
$f_R + 3$ k				
$f_R + 6$ k				
$f_R + 9$ k				
$f_R + 12$ k				
$f_R + 15$ k				
$f_R + 18$ k				
$f_R + 21$ k				

TABLE 53–2. Effect of Resistance on a Series Resonant Circuit

Resistor R, Ω	Resonant Frequency f_R, Hz	Voltage across Resistor V_R, V_{p-p}	Voltage across Inductor/ Capacitor Combination V_{LC}, V_{p-p}	Circuit Current (Calculated) I, mA_{p-p}	Circuit Q Cal.	Circuit Q Meas.
1 k						
220						
100						

R_{dc} (resistance of 10–mH inductor) = _____ Ω

EFFECTS OF Q ON FREQUENCY RESPONSE AND BANDWIDTH   **409**

QUESTIONS

1. On a separate sheet of 8½ × 11 graph paper, plot a graph of frequency versus voltage across the capacitor using your data in Table 53–1. Plot a separate frequency response curve for each value of resistance. The horizontal (x) axis should be frequency, and the vertical (y) axis should be voltage. Label the curves with the appropriate resistor values. Indicate the resonant frequency on the frequency axis. Also indicate the half-power points for each of the curves. Label each curve with its bandwidth and Q.

2. Explain, in your own words, the relationship between the Q of a circuit and the frequency response of the circuit.

_____

_____

_____

_____

_____

_____

_____

3. Explain, in your own words, the relationship between the Q of a circuit and the bandwidth at the half-power points.

_____

_____

_____

_____

_____

_____

_____

4. Refer to your data in Table 53–2. What effect, if any, did changes in resistance have on the resonant frequency of the circuit (Figure 53–4)? Explain any unexpected results.

_____

_____

_____

_____

_____

_____

_____

CHARACTERISTICS OF PARALLEL RESONANT CIRCUITS

BASIC INFORMATION

Characteristics of a Parallel Resonant Circuit · · · · · · ·

Resonant Frequency of a High-Q Circuit.   The circuit of Figure 54–1 (p. 412) consists of C and L in parallel, with the coil resistance R_L shown in series with L. It is assumed here that the Q of this circuit is high (that is, the R_L is small compared with X_L) and that the resistance of C and the wiring resistance of the circuit are negligible and may be ignored.

There is a particular frequency at which $X_L = X_C$. This frequency may be defined as the condition for parallel resonance in a high-Q circuit and is similar to the condition for series resonance.

There are other definitions for parallel resonance. Thus, parallel resonance may be considered as the frequency at which the impedance of the parallel circuit is maximum. Also, parallel resonance may be considered as the frequency at which the parallel impedance of the circuit has unity power factor. These three definitions may lead to three different frequencies, each of which may be considered as the resonant frequency. In circuits whose Q is greater than 10, however, the three conditions lead to the same resonant frequency.

In a high-Q circuit, the formula for the resonant frequency f_R is the same as in the case of series resonance and is given by

$$f_R = \frac{1}{2\pi\sqrt{LC}} \tag{54–1}$$

Line Current.   If the resistance R_L of the inductance L in the circuit of Figure 54–1 is small, then at resonance the impedance X_C of the capacitive branch is practically equal to the impedance of the inductive branch. The current in each branch may therefore be considered equal. These currents, however, are practically 180° out of phase. Their phasor sum, which is the line, or total, current I_T is very small. Because the impedance that this parallel resonant circuit presents to the circuit is equal to V/I_T, it is apparent that the impedance is very high (since I_T is low). Although the line current is low, the circulating current in the parallel resonant circuit is high at resonance.

OBJECTIVES

1 To determine experimentally the resonant frequency of a parallel *RLC* circuit

2 To measure the line current and impedance of a parallel *RLC* circuit at the resonant frequency

3 To measure the effect of variations in frequency on the impedance of a parallel *RLC* circuit

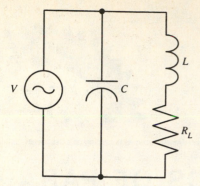

Figure 54–1. Parallel resonant circuit.

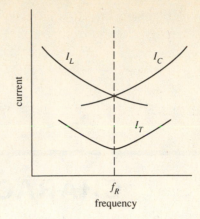

Figure 54–2. Current versus frequency in a parallel resonant circuit. The line current, I_T, is the phasor sum of I_C and I_L.

Frequency Response.

The characteristics of a parallel resonant circuit to frequencies on either side of the resonant frequency may now be studied. For a frequency f_a higher than f_R, the reactance X_C of the capacitive branch is lower than the reactance X_L of the inductive branch. There is, therefore, a greater capacitive than inductive current, and the circuit may be considered *capacitive*. Similarly, it may be shown that for a frequency f_b lower than f_R, the circuit is *inductive*. The graph of Figure 54–2 illustrates these relationships. Individual graphs for I_L, I_C, and I_T are shown. It is evident that at resonance, line current is minimum, that I_L equals I_C, and that I_L and I_C are each greater than the line, or total, current I_T.

Impedance.

Figure 54–3 is a graph of impedance versus frequency in a parallel resonant circuit. This graph resembles the current versus frequency-response characteristic of a series resonant circuit. It shows that circuit impedance is maximum at resonance and falls off on either side of resonance.

The resonant frequency f_R of a parallel LC circuit such as in Figure 54–4 can be found experimentally using a variable-frequency source. By measuring the voltage across the known resistor R in the circuit V_R we can calculate the line, or total, current of the circuit I_T using Ohm's law:

$$I_R = \frac{V_R}{R} = I_T \qquad (54\text{–}2)$$

We can observe the voltage across the resistor as the frequency of the voltage source is varied while making certain the output voltage is kept constant. An observation of minimum V_R indicates I_T is a minimum. The frequency at this point is f_R. By again varying the frequency above and below f_R while maintaining a constant voltage output, we can measure the value of V_R at different frequencies. We can calculate I_R from these values of V_R using formula (54–2). We can plot a graph of frequency versus I_R from these data.

We can also find the voltage across the parallel LC circuit (often called a *tank* circuit). Using this voltage V_t to-

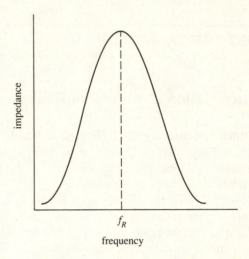

Figure 54–3. Impedance versus frequency in a parallel resonant circuit.

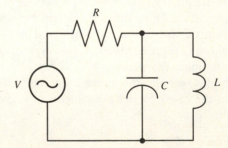

Figure 54–4. A parallel LC circuit used to determine f_R.

gether with I_T, we can calculate Z_t, the impedance of the tank circuit:

$$Z_t = \frac{V_t}{I_T} \qquad (54\text{–}3)$$

The voltage V_t across the tank circuit is directly proportional to the Q of the circuit. If we change the circuit by placing a resistor in parallel with the tank circuit, we are loading the circuit, and the voltage across the tank circuit will decrease. This loading effect increases as the size of the parallel resistor decreases.

SUMMARY

1. In a parallel *LC* circuit, if the *Q* of the circuit is greater than 10, the resonant frequency of the circuit may be calculated from the formula

$$f_R = \frac{1}{2\pi\sqrt{LC}}$$

At this frequency $X_L = X_C$. (The parallel *LC* circuit is often called a tank circuit.)

2. In a high-*Q* circuit the impedance of the inductive branch equals (approximately) the impedance of the capacitive branch at resonance.

3. Because the voltage across each of the parallel branches in a high-*Q* circuit is the same and the impedance of both branches are equal at f_R the currents in both branches are equal. That is, $I_L = I_C$.

4. The currents, however, are approximately 180° out of phase, since one is capacitive and the other is inductive. They therefore tend to cancel each other, and the line, or total, current, which is the phasor sum of I_C and I_L, is therefore very low. In a very high-*Q* circuit, the line current is close to zero at f_R.

5. The very low line current means that the impedance of the parallel *LC* circuit is very high, since

$$Z = \frac{V}{I_T}$$

where *V* is the applied voltage and I_T is the total, or line, current.

6. For frequencies higher than f_R, the current in the capacitive branch is higher than the current in the inductive branch, and the circuit acts like a capacitive circuit.

7. For frequencies lower than f_R, current in the inductive branch is higher than current in the capacitive branch, and the circuit acts like an inductance.

8. The frequency-response curve of *f* versus *Z* of a high-*Q* parallel resonant circuit resembles the current versus frequency-response curve of a series resonant circuit.

Check your understanding by answering the following questions:

1. The *Q* of a parallel resonant circuit in Figure 54–1 is 100. At resonance the current in the inductive branch is 50 mA, and the voltage across the tank circuit is 100 V. The current in the capacitive branch is _____ mA.

2. In the parallel resonant circuit of Figure 54–4, the line current at resonance is 0.05 mA. The voltage across the tank circuit is 2.5 V. The impedance of the tank circuit is _____ Ω.

3. For the circuit in question 2, at frequencies higher than f_R, the impedance of the parallel resonant circuit is _____ (higher/lower) than the impedance at f_R.

4. In the circuit of Figure 54–4, the coil resistance R_L is 12 Ω, *R* = 10 kΩ, *L* = 30 mH, and *C* = 0.01 μF. The resonant frequency of the circuit f_R = _____ Hz.

5. The *Q* of the coil in the circuit of question 4 is _____.

6. The voltage across *R*, in Figure 54–4, is 1.5 V at resonance. The line current in the circuit is _____ mA.

MATERIALS REQUIRED

Instruments:
- Oscilloscope
- Function generator

Resistors (½-W, 5%):
- 2   33-Ω
- 1   10 kΩ

Capacitor:
- 1   0.022 μF

Inductor:
- 10-mH coil

PROCEDURE

. . . .

A. Resonant Frequency and Impedance of a Parallel *LC* Resonant Circuit

A1. With the function generator and oscilloscope turned **off,** connect the circuit of Figure 54–5 (p. 414).

A2. Turn **on** the generator and oscilloscope. Adjust the scope to measure the voltage output of the generator. Increase the generator output voltage *V* to 4 V p-p. Maintain this voltage throughout the

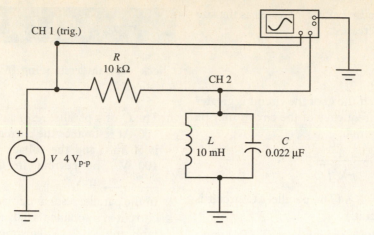

Figure 54–5. Circuit for procedure step A1.

experiment. Set the generator frequency to 10 kHz. Adjust the scope to display two or three cycles of the sine-wave.

A3. Vary the generator frequency above and below 10 kHz and observe the voltage across the resistor V_R using the differential (ADD/INVERT) mode of the oscilloscope. At the minimum V_R the frequency will equal the resonant frequency f_R. Verify that $V = 4$ V p-p; adjust if necessary. Record f_R and V_R in Table 54–1.

A4. Note in Table 54–1 a series of frequencies above and below the resonant frequency f_R. Set the generator as close as possible to each frequency. At each frequency, measure the p-p voltage across the resistor V_R and the parallel LC circuit (tank circuit) V_{LC}, checking periodically to make sure $V = 4$ V p-p. Record the frequency f, V_R, and V_{LC} in Table 54–1. After completing all measurements, turn **off** the generator and the oscilloscope and disconnect the circuit.

A5. Using the measured value of V_R for each frequency and the rated value of R, calculate the line current I at each frequency. Record your answers in Table 54–1.

A6. Using the calculated values of I from step A5 and the p-p value of V (4 V_{p-p}), calculate the impedance of the tank circuit at each frequency. Record your answers in Table 54–1.

B. Reactive Characteristics of a Parallel LC Circuit

B1. With the generator and oscilloscope **off,** connect the circuit of Figure 54–6. It may be assumed that the resonant frequency, f_R, of this circuit is the same as in Part A. Transfer the frequencies from Table 54–1 to Table 54–2 (p. 416).

B2. Turn **on** the generator and the oscilloscope. Set the generator voltage V to 4 V_{p-p} and maintain this voltage throughout the experiment. Check V from time to time and adjust as necessary.

B3. For each frequency in Table 54–2, measure the voltage across the resistor in the capacitive branch AB,

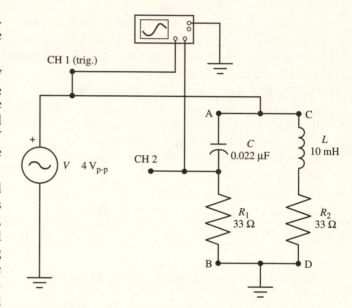

Figure 54–6. Circuit for procedure step B1.

V_{R1}, and the voltage across the resistor in the inductive branch CD, V_{R2}. Record the values in Table 54–2. After completing all measurements, turn **off** the generator and oscilloscope and disconnect the circuit.

B4. Using the measured values of V_{R1} and V_{R2} and the rated values of R_1 and R_2, calculate the currents in the capacitive branch I_C and the inductive branch I_L for each frequency. Record your answers in Table 54–2.

ANSWERS TO SELF TEST

1. 50

2. 50 k

3. lower

4. 9.19 k

5. 144

6. 0.15

TABLE 54–1. Frequency Response of a Parallel Resonant Circuit

Frequency Deviation	Frequency f, Hz	Voltage across Resistor V_R, V_{p-p}	Voltage across Tank Circuit V_{LC}, V_{p-p}	Line Current (Calculated) I, μA	Tank Circuit Impedance (Calculated) Z, Ω
$f_R - 6$ k					
$f_R - 5$ k					
$f_R - 4$ k					
$f_R - 3$ k					
$f_R - 2$ k					
$f_R - 1$ k					
$f_R - 500$ k					
f_R					
$f_R + 500$ k					
$f_R + 1$ k					
$f_R + 2$ k					
$f_R + 3$ k					
$f_R + 4$ k					
$f_R + 5$ k					
$f_R + 6$ k					

CHARACTERISTICS OF PARALLEL RESONANT CIRCUITS **415**

TABLE 54–2. Reactance Characteristics of a Parallel *LC* Circuit

Frequency f, Hz	Voltage across Resistor R_1 V_{R1}, mV$_{p-p}$	Voltage across Resistor R_2 V_{R2}, mV$_{p-p}$	Current in Capacitive Branch (Calculated) I_C, mA$_{p-p}$	Current in Inductive Branch (Calculated) I_L, mA$_{p-p}$

QUESTIONS

1. Explain, in your own words, the effects of changes in frequency on the impedance of a parallel *RLC* circuit.

_____

_____

_____

_____

2. Explain, in your own words, the effects of changes in frequency on the total current of a parallel *RLC* circuit.

_____

_____

_____

_____

3. What determines the value of *Q* in a parallel *RLC* circuit? What factor or condition determines if it is a high *Q*?

_____

_____

_____

4. On a separate sheet of 8½ × 11 graph paper, plot a graph of frequency versus tank circuit impedance using your data in Table 54–1. Frequency should be the horizontal (x) axis and impedance should be the vertical (y) axis. Label all axes. Indicate the resonant frequency with a dotted vertical line.

5. On a separate sheet of 8½ × 11 graph paper (not the sheet used for Question 4), plot a graph of frequency versus capacitive branch current I_C using your data in Table 54–2. On the same set of axes, plot frequency versus inductive branch current I_L from the data in the table. The horizontal (x) axis should be frequency, and the vertical (y) axis should be current. Label the axes and each curve. Indicate the resonant frequency with a dotted line.

6. Is the parallel *RLC* circuit of Part B resistive, capacitive, or inductive? Explain your answer.

_____

_____

_____

_____

LOW-PASS AND HIGH-PASS FILTERS

BASIC INFORMATION

Frequency Filters

Electronic signals are often made up of more than one frequency. For example, the ordinary AM radio signal consists of a radio-frequency component containing the voice, music, and sound frequencies we hear when we turn on a radio. The carrier is a unique frequency assigned to the radio station by the Federal Communications Commission (FCC). It is combined with the audio component by a process called *modulation.* The modulated signal is then transmitted over the air by the radio station. The radio receiver contains circuits that can select a particular carrier frequency and reject all others. Once the modulated signal is in the circuitry of the radio, the carrier is separated from the audio component and eliminated, and only the audio component is allowed to pass through to the speakers or earphones. The process of selecting and eliminating or rejecting particular frequencies or ranges of frequencies is called *filtering.* Filters are employed in such devices as FM and television receivers, security systems, computers, and motor controls.

There are different types of filters. Some are highly selective and permit a single frequency or a very narrow band of frequencies through, reducing or eliminating all others. These are called *narrowband filters.* Others, called *wideband filters,* pass a wide band of frequencies while rejecting all others. Filters are also classified as low-pass or high-pass. We examine the characteristics of the latter two types in this experiment.

High-Pass Filter

Filter circuits consist of capacitors, inductors, and resistors, in combination. An *LC* series resonant circuit is one example of a frequency filter, for these highly selective circuits favor one frequency, the resonant frequency, over all others.

A capacitor theoretically offers an infinite reactance to a zero-frequency signal—that is, to direct current. Therefore, if a capacitor is placed in series with a load resistance, it will pass the ac component of a complex signal but block the dc component of that signal. Figure 55–1(*a*) shows a 6 V_{P-P} ac signal combined with 5 V dc. The resulting combined signal varies between +8 and +2 V. Figure 55–1(*b*) shows how a coupling capacitor *C* is connected in the circuit to block the dc but permit

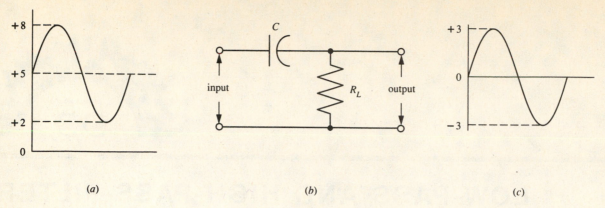

(a) (b) (c)

Figure 55–1. The dc blocking effect of a capacitor. (*a*) A 6 V$_{p-p}$ signal imposed on a +5 V dc axis. The resulting combined signal varies between +8 and +2 V. (*b*) The combined dc and ac signal is connected as shown to the input terminals. (*c*) The capacitor *C* blocks the dc component and permits only the ac component to pass through to the output. The signal across R_L is a pure 3 V$_{p-p}$ ac voltage.

the ac component through. Figure 55–1(*c*) shows the pure ac sine wave across R_L. Note that the +5 V dc has been filtered by capacitor *C*.

The reactance of a capacitor varies inversely with frequency. This characteristic is used to pass some frequencies but reject others. In the circuit of Figure 55–2(*a*), the level of ac voltage V_R coupled by capacitor *C* to resistor *R* is

$$V_R = V \cos \theta = V \times \frac{R}{Z}$$

and

$$\theta = \tan^{-1}\left(\frac{X_C}{R}\right) \qquad (55\text{–}1)$$

The angle θ depends on the relative values of *C* and *R* and on the frequency *f* of the applied signal source. Assume there are three frequencies in the complex signal *V*, namely, 159 Hz, 1590 Hz, and 15,900 Hz. The reactance X_C of *C* at each of these frequencies is, respectively, 10,000, 1000, and 100 Ω. The voltage V_R delivered to *R* at 159 Hz is less than 10 percent of *V*. At 1590 Hz (the frequency at which $X_C = R$) more than 70 percent of *V* is delivered to *R*, and at 15,900 Hz more than 99 percent of *V* is delivered to *R*. Thus it is evident *C* will permit the higher

frequencies to reach *R* with a minimum of reduction but will reduce, or attenuate, the lower frequencies. This is one type of high-pass filter.

The circuit of Figure 55–2(*a*) can be modified by adding an inductor *L* in parallel with the load resistor R_L to form another high-pass filter [Figure 55–2(*b*)]. The inductor has a low reactance at the low frequencies and effectively reduces the output voltage across R_L at frequencies for which X_L is less than $R_L/10$ (approximately). At higher frequencies the value of X_L increases. For frequencies at which X_L is greater than $10R_L$, the parallel impedance of X_L and R_L approaches R_L, stabilizing the load at the value R_L. The inclusion of inductor *L* modifies the response characteristics of the circuit, which is still a high-pass filter.

Low-Pass Filter

The reactance of an inductor varies directly with frequency. In Figure 55–3 this characteristic of *L* is used to pass the low frequencies through to R_L. If the input contains high and low frequencies, the high frequencies will be rejected, or blocked, from passing through to R_L As in the case of the *RC* circuit in Figure 55–2, the *RL* circuit in Figure 55–3 is a voltage divider. The amplitude of the voltage deliv-

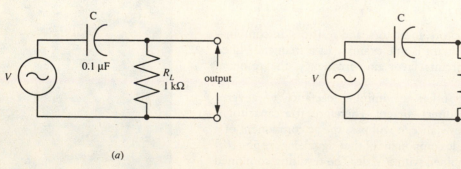

(a) (b)

Figure 55–2. High-pass filters. (*a*) capacitor *C* blocks the lower-frequency and passes the higher-frequency components of *V* through to R_L. (*b*) Adding inductor *L* modifies the filtering action of the circuit.

420 EXPERIMENT 55

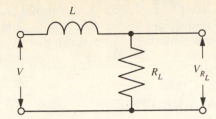

Figure 55–3. Low-pass filter. Inductor *L* blocks the higher frequencies and passes the lower frequencies through to R_L.

ered by *V* to R_L will depend on the inductance of *L*, the resistance of R_L, and the frequency of *V*. As frequency increases, X_L increases, and the voltage across R_L decreases. As frequency decreases, X_L decreases and the voltage across R_L increases. This is an example of a low-pass filter.

An example of another low-pass filter is shown in Figure 55–4(*a*). The addition of capacitor *C* in this circuit increases the filtering of higher frequencies. As the frequency *f* of the input signal increases, not only does X_L increase, reducing the output voltage V_{R_L}, but X_C decreases, thus increasing the amount of high-frequency signal bypassed by the capacitor. This, in effect, reduces the impedance of the output circuit, thus further reducing V_{R_L}.

If *L* is replaced by a resistor R_1, as in Figure 55–4(*b*), the result will still be a low-pass filter because of the bypass action of capacitor *C*.

Cutoff Frequency

Although a filter will pass or reject specific frequencies, it is not often necessary to eliminate or reduce the signal of the unwanted frequencies to zero. A value of frequency, called the *cutoff frequency*, is used to specify the point at which a frequency has been reduced, or attenuated, to a particular value. The cutoff frequency is defined here as the frequency whose output voltage is 70.7 percent of the maximum output signal. This, in effect, means the cutoff frequency has been attenuated by 29.3 percent of maximum.

In the case of a simple *RC* low-pass filter, the output signal will be attenuated to 70.7 percent of the input sig-

nal when the input frequency causes $X_C = R$. This cutoff frequency can be calculated by the equation

$$f_C = \frac{1}{2\pi RC} \tag{55–2}$$

This cutoff point is commonly used to describe the frequency response of electronic circuits and test equipment, such as the bandwidth of an oscilloscope. This cutoff point may also be referred to as the half-power point. When the output voltage is at 70.7 percent, the output current will normally also be at a 70.7 percent level. The product of $V \times I$ will result in a 50 percent output power level.

SUMMARY

1. Complex electronic signals contain many frequency components. Filters are used to separate these frequency components, attenuating those that are unwanted and passing on those that are required.

2. Filters consist of combinations of capacitors, inductors, and resistors.

3. The circuits of Figure 55–2(*a*) and (*b*) show examples of high-pass filters. These circuits are basically voltage dividers. With increasing frequency, X_C decreases, and the output voltage across R_L increases.

4. The circuits of Figures 55–3 and 55–4 are low-pass filters. Here the voltage-divider action is affected by the increase in X_L with an increase in frequency. Accordingly, the output voltage across R_L is greater for low frequencies than it is for high frequencies. High frequencies are attenuated, and low frequencies are passed on to the output.

5. The terms low and high frequencies are relative. The frequency-response characteristic of a circuit depends on the circuit parameters—that is, on the values of *L*, *C*, and *R*, and on the circuit configuration. In all cases the value of output voltage for each frequency component can be calculated using ac circuit formulas.

6. The cutoff frequency is defined as the frequency whose output voltage is 70.7 percent of the maximum output voltage.

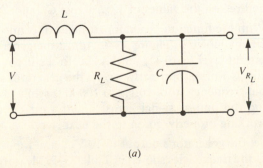

(*a*)

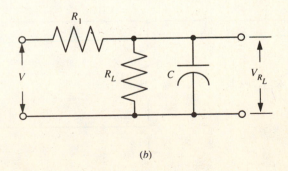

(*b*)

Figure 55–4. Other examples of low-pass filters.

Check your understanding by answering the following questions:

1. Circuits that attenuate some frequencies and pass others are called _____.

2. DC voltages can be separated from ac signals by means of a _____.

3. In the high-pass circuit of Figure 55–2(a), the frequency at which the output voltage across R_L is 70.7 percent of maximum is the frequency at which $X_C =$ _____ Ω.

4. The cutoff frequency in the filter of Figure 55–2(a) is _____ Hz.

5. In the low-pass filter of Figure 55–3, the output voltage across R_L, V_{R_L}, is $V \times$ _____ $= V \times$ _____.

6. In the circuit of Figure 55–3, if $R = 500 \, \Omega$ and $L = 2$ H, all frequencies higher than _____ Hz will be attenuated more than 29.3 percent; that is, the cutoff frequency is _____ Hz.

MATERIALS REQUIRED

Instruments:
- Oscilloscope
- Function generator

Resistors:
- 1 10 kΩ, ½-W, 5%
- 1 22 kΩ

Capacitor:
- 1 0.001 μF

PROCEDURE

• •

A. High-Pass Filter

A1. Examine the circuit shown in Figure 55–5. For the values shown, calculate the circuit's cutoff frequency, f_C. Record this value in the frequency column in Table 55–1 (p. 425).

A2. For each frequency shown in Table 55–1, calculate and record the values of V_{out}, across R_1, and X_C.

A3. With the function generator turned **off**, connect the circuit shown in Figure 55–5. Adjust the generator's output level to 10 V_{p-p} at 1 kHz. To see if your circuit is working properly, sweep the generator's output frequency from 10 Hz to 100 kHz

while observing the output signal across R_1. If your circuit is working properly, go on to the next step.

A4. Set the generator's output to 10 V_{p-p} and input each of the frequencies shown in Table 55–1. Measure and record the output signal across R_1, for each input.

A5. Turn off the function generator. For each output voltage measured in step A4, calculate the percent of V delivered to R and record these values in Table 55–1.

B. Low-Pass Filter

B1. Examine the circuit shown in Figure 55–6. For the values shown, calculate the circuit's cutoff frequency, f_C. Record this value in the frequency column in Table 55–2 (p. 425).

B2. For each frequency shown in Table 55–2, calculate and record the values of V_{out}, across C_1, and X_C.

B3. With the function generator turned **off**, connect the circuit shown in Figure 55–6. Adjust the generator's output level to 10 V_{p-p} at 1 kHz. To see if your circuit is working properly, sweep the generator's output frequency from 10 Hz to 100 kHz while observing the output signal across C_1. If your circuit is working properly, go on to the next step.

B4. Set the generator's output to 10 V_{p-p} and input each of the frequencies shown in Table 55–2. Measure and record the output signal across C_1, for each input.

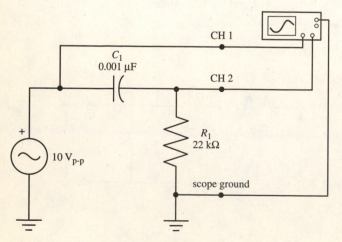

Figure 55–5. Circuit for procedure step A1.

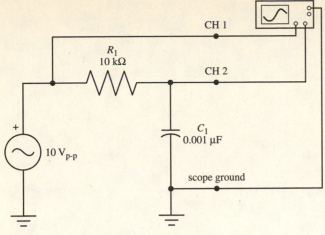

Figure 55–6. Circuit for procedure step B1.

B5. Turn **off** the function generator. For each output voltage measured in step B4, calculate the percent of V delivered to C and record these values in Table 55–2.

Copyright © by Glencoe/McGraw-Hill

Optional Activitiy ·

Design, construct, and test a filter circuit with the following criteria:

1. The component value must be available in your lab.

2. The circuit must "effectively" block frequencies above 20 kHz.

If available, first test your circuit using electronic simulation software.
After your instructor has verified your calculations and operation circuit, construct and test the circuit.

ANSWERS TO SELF TEST

1. filters
2. blocking capacitor
3. 1 k
4. 1.59 k
5. R_L/Z; $\cos \theta$
6. 39.8; 39.8

Name _____   Date _____

TABLE 55–1. High-Pass Filter

Frequency f, Hz	X_C Ω	V_{out} (Calculated) $V_{p\text{-}p}$	V_{out} (Measured) $V_{p\text{-}p}$	V_{out} Percent, % (Measured)
100				
500				
1 k				
2 k				
5 k				
$f_C =$				
10 k				
20 k				
50 k				
100 k				
200				

TABLE 55–2. Low-Pass Filter

Frequency f, Hz	X_C Ω	V_{out} (Calculated) $V_{p\text{-}p}$	V_{out} (Measured), $V_{p\text{-}p}$	V_{out} Percent, % (Measured)
100				
500				
1 k				
2 k				
5 k				
10 k				
$f_C =$				
20 k				
50 k				
100 k				
200 k				

1. Using semi-logarithmic graph paper, plot the frequency response curve for the *RC* high-pass filter. This graph paper permits recording the entire frequency range of your data on the X axis. Use your measured data from Table 55–1. Also, on your graph mark and label the half-power point.

2. Why is the 70.7 percent point sometimes referred to as the half-power point?

_____

_____

_____

_____

3. On the same sheet of graph paper used in question 1, plot the frequency response curve for the *RC* low-pass filter. Use your measured data from Table 55–2. Mark and label the half-power point.

4. If the capacitor in a high-pass *RC* filter circuit decreased in value, what would happen to this circuit's bandwidth?

_____

_____

_____

_____

5. In the space below, draw a circuit which could be used as a high-pass filter using only a resistor and inductor.

BANDPASS AND BANDSTOP FILTERS

BASIC INFORMATION

Bandpass Filters

In Experiment 55, "Low-Pass and High-Pass Filters," frequency response curves for RC low-pass and high-pass filters were discussed. For each of these circuits, it was determined that the output voltage would be at a 70.7 percent level of the input voltage when $X_C = R$. The frequency at which this occurred was called the cutoff frequency, f_C, and the output was at the half-power point.

Combining a low-pass filter with a high-pass filter, as shown in Figure 56–1, creates a special circuit call a *bandpass filter*. The components C_1 and R_1 form the high-pass filter section while C_2 and R_2 form the low-pass filter section of this bandpass filter. Figure 56–2 (p. 428) shows how the output voltage, across C_2, varies as the input frequency changes for each respective section. The frequency response for each section has been overlapped to demonstrate their individual affects. Notice that the low-pass filter determines the upper cutoff frequency, f_{C2}, while the high-pass filter determines the lower cutoff frequency, f_{C1}. The frequencies between f_{C1} and f_{C2} are in the circuit's effective passband. The bandwidth, BW, of this circuit is found by

$$BW = f_{C2} - f_{C1} \qquad (56\text{–}1)$$

The calculations for determining f_{C1} and f_{C2} are accomplished as before with simple series RC high-pass and low-pass filters. Because the low-pass filter section, as shown in Figure 56–1, is in parallel with the high-pass filter section, it is possible that circuit loading could occur. To pre-

OBJECTIVES

1. To determine experimentally the response of a bandpass filter

2. To determine experimentally the response of a bandstop filter

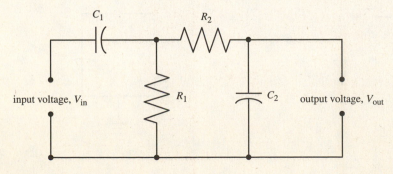

Figure 56–1. Bandpass filter.

427

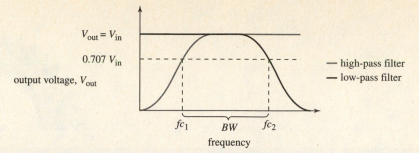

Figure 56–2. The *RC* bandpass filter V_{out} versus frequency curve.

vent this, R_2 should be 10 or more times larger than the resistance of R_1. An *RC* bandpass filter circuit will have the ability to attenuate the input frequencies outside of its passband.

Bandpass filters can also be constructed using the concepts of series and parallel resonant circuits. Circuits of this type are shown in Figures 56–3(*a*) and 56–3(*b*). When analyzing these circuits, the center of the respective bandwidth will occur at the coil-capacitor's resonant frequency. The bandwidth, $f_{C2} - f_{C1}$, will be determined by the circuit's Q.

Bandstop Filter · · · · · · · · · · · · · ·

Resistors, capacitors, and inductors can also be combined in a variety of ways to attenuate a narrow range of frequencies and passing all other frequencies. These circuits are called bandstop or notch filters. An *RC* bandstop filter is shown in Figure 56–4. This circuit is essentially a twin-T notch filter that works on the principle of a bridge circuit. When the component values are properly selected, at one frequency, f_N, the bridge will be balanced. When the bridge is balanced the output voltage will be nearly zero. If the components of this filter are precisely matched, at f_N the output voltage will be down to less than 3 percent of the input voltage. A typical frequency response curve is shown in Figure 56–5.

The notch frequency, f_N, can be calculated using the formula

$$f_N = \frac{1}{4\pi RC} \qquad (56\text{–}2)$$

Bandstop or notch filters can also be constructed using series and parallel resonant circuits. Figures 56–6(*a*) and 56–6(*b*) show examples of typical *LC* notch filters. *LC* bandstop filters can also be constructed using more elaborate schemes, such as bridges and twin-Ts. Sometimes these circuits have adjustable components and are used to tune out or trap out undesired frequencies.

SUMMARY

1. A bandpass filter passes a band of frequencies while attenuating all others.

2. A bandstop filter attenuates a narrow band of frequencies while passing all others.

3. A circuit's bandwidth is determined by finding the difference between its upper and lower cutoff frequencies.

4. Bandpass filters can be constructed using *RC* or *LC* circuits.

5. Bandstop filters are sometimes referred to as notch filters.

6. Besides using *RC* circuits, notch filters can be constructed using *LC* resonant components.

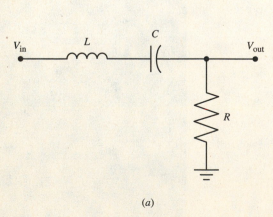

(*a*)

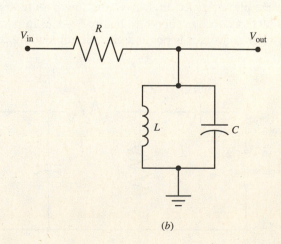

(*b*)

Figure 56–3. (*a*) An *LC* series bandpass filter. (*b*) An *LC* parallel bandpass filter.

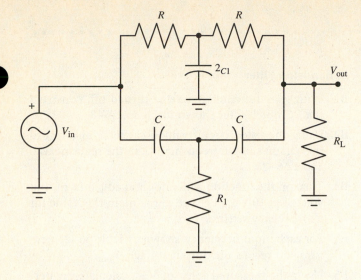

Figure 56–4. An *RC* bandstop filter.

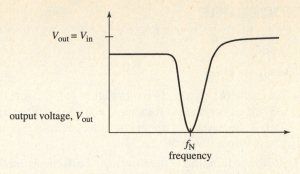

Figure 56–5. Notch filter V_{out} versus frequency response curve.

4. To prevent circuit loading of a *RC* bandpass filter, the low-pass resistor should be at least _____ times larger than the high-pass resistor.

5. The bandstop filter shown in Figure 56–4 has $R_1 = 10$ kΩ and $C_1 = 0.01$ μF. The notch frequency would be at _____ Hz.

MATERIALS REQUIRED

Instruments:
- Function generator
- Oscilloscope

Resistors (½-W, 5%)
- 1   3.3 kΩ
- 5   10 kΩ
- 1   100 kΩ

Capacitors:
- 4   0.001 μF
- 1   0.1 μF
- 1   500 pF

S E L F   T E S T

Check your understanding by answering the following questions:

1. A circuit has a bandwidth of 10 kHz. If the upper cut-off frequency is 30 kHz the lower cutoff frequency will be _____ Hz.

2. In a *RC* bandpass filter circuit, the low-pass section determines the _____ cutoff frequency while the high-pass section determines the _____ cutoff frequency.

3. In the circuit of Figure 56–1, $C_1 = 0.1$ μF, $R_1 = 1$ kΩ, $R_2 = 10$ kΩ, and $C_2 = 0.1$ μF. The upper cutoff frequency would be _____ Hz while the lower cutoff frequency would be _____ Hz.

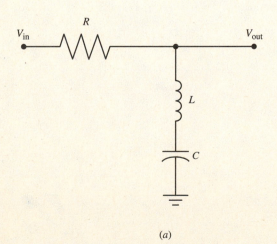

(a)

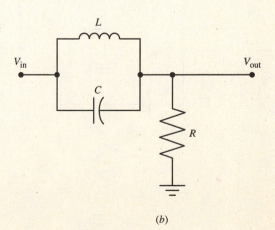

(b)

Figure 56–6. *LC* bandstop filters. (*a*) Series. (*b*) Parallel.

PROCEDURE

A. Bandpass Filter

A1. Examine the circuit shown in Figure 56–7. Using the circuit values shown for this *RC* bandpass filter, calculate the lower and upper cutoff frequencies. Record these values Table 56–1.

A2. With the function generator turned **off**, construct the *RC* bandpass circuit of Figure 56–7.

A3. Turn **on** the function generator and set its output voltage to 10 $V_{p\text{-}p}$. Be sure to maintain this output value for all frequency settings. For each input frequency listed in Table 56–1, measure and record the value of output voltage, V_{out}.

A4. At each frequency setting, calculate the amount of attenuation offered by the filter in percent output. Record the percentage values in Table 56–1.

B. Bandstop Filter

B1. With your function generator turned off, construct the *RC* notch filter shown in Figure 56–8.

B2. Using the listed circuit values, calculate and record this circuit's notch frequency, f_N, in the space next to Table 56–2.

B3. Turn on the function generator and adjust its output voltage to 10 $V_{p\text{-}p}$. Be sure to maintain this value for all frequency settings.

B4. For each input frequency, shown in Table 56–2, measure and record the output voltage, V_{out}.

B5. Using the measured value of V_{out}, calculate the percent output at each frequency.

Optional Activity

This activity requires the use of electronic simulation software. Design and construct a simulated parallel *LC* bandpass filter for a center frequency of 455 kHz. Demonstrate the proper operation of this circuit to your instructor.

ANSWERS TO SELF TEST

1. 20 kHz
2. upper; lower
3. 1.59 k; 159
4. 10
5. 796

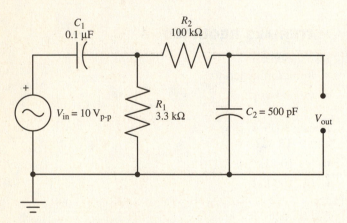

Figure 56–7. Bandpass filter for procedure step A1.

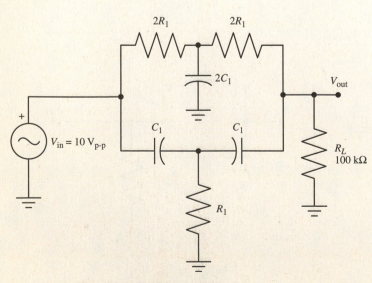

Figure 56–8. Circuit for procedure step B1.

TABLE 56–1 *RC* Bandpass Filter

$f_{C_1} =$ _____

$f_{C_2} =$ _____

Frequency	V_{out}	Percent Output, %
10 Hz		
50 Hz		
100 Hz		
200 Hz		
300 Hz		
400 Hz		
500 Hz		
600 Hz		
700 Hz		
800 Hz		
900 Hz		
1 kHz		
2 kHz		
3 kHz		
4 kHz		
5 kHz		
10 kHz		
20 kHz		
30 kHz		
40 kHz		
50 kHz		
60 kHz		
70 kHz		
80 kHz		
90 kHz		
100 kHz		
200 kHz		
1 MHz		

TABLE 56–2. *RC* Bandstop Filter

$f_N =$ _____

Frequency	$V_{out\ p\text{-}p}$	Percent Output, %
10 Hz		
100 Hz		
200 Hz		
500 Hz		
1 kHz		
2 kHz		
3 kHz		
4 kHz		
5 kHz		
6 kHz		
7 kHz		
8 kHz		
9 kHz		
10 kHz		
20 kHz		
30 kHz		
40 kHz		
50 kHz		
60 kHz		
70 kHz		
80 kHz		
90 kHz		
100 kHz		
200 kHz		
500 kHz		
1 MHz		

QUESTIONS

1. Using the graph paper provided with this lab, plot the frequency response curve for the *RC* bandpass filter in Figure 56–7. On the graph, label the respective cutoff points and identify the circuit's bandwidth.

2. When the *RC* bandpass filter input changed from 10 kHz to 100 kHz, what was the amount of attenuation in percent per decade? Cite measured values to support your answer.

_____

_____

_____

_____

_____

_____

_____

_____

_____

_____

_____

_____

_____

3. Using the graph paper provided with this lab, plot the measured frequency response for the *RC* notch filter. Be sure to label the notch frequency, f_N.

4. Explain, using the same frequencies as in question 2, how the *RC* notch filter's roll-off compared to the roll-off for the *RC* bandpass filter.

_____

_____

_____

_____

_____

_____

_____

_____

_____

_____

_____

_____

_____

NONLINEAR RESISTORS—THERMISTORS

BASIC INFORMATION

In all previous experiments the components that made up the circuits were assumed to have constant values. In particular, the values of resistors were assumed to remain unchanged despite changes in voltage and current. For the most part this assumption was acceptable for the purpose of the experiment.

However, there is a class of components known as *thermistors* whose resistance value changes with changes in operating temperature. As its name implies, a thermistor is a thermally sensitive resistor. It is part of a large family of materials called *semiconductors*. A semiconductor is a substance whose resistivity lies somewhere between that of a conductor and that of an insulator.

The thermistor is a two-terminal component that may be used in either ac or dc circuits. It is manufactured in a number of shapes, such as beads, rods, disks, washers, and surface mount. Thermistors are often part of sensor devices used in temperature control, flow measurement, voltage regulation, and such.

Temperature Characteristic

A fundamental characteristic of the thermistor is its change of resistance with changes in temperature. A number of formulas have been developed to approximate the manner in which the resistance varies with temperature. One such approximate formula is

$$R = R_0 \times e^k$$

and

$$k = B\left(\frac{1}{T} - \frac{1}{T_0}\right) \tag{57-1}$$

where

R = resistance at any temperature T, in degrees Kelvin (K)

R_0 = resistance at reference temperature T_0 (K)

B = a constant whose value depends on the thermistor material (determined from measurements at 0° and at +50°C)

e = 2.7183 (base of natural logarithms)

Formula (57-1) shows that the resistance variation is nonlinear. Thermistors are frequently referred to as *nonlinear resistors*.

OBJECTIVES

1 To observe the self-heating effect of current on the resistance of a thermistor

2 To determine experimentally the variation in resistance with time of current in a thermistor

Thermistors exhibit negative temperature coefficients (NTC) and positive temperature coefficients (PTC). The resistance of NTC thermistors *decreases* as their temperature *rises* and *increases* as their temperature *falls.* The resistance of PTC thermistors *increases* as their temperature *rises* and *decreases* as their temperature *falls.* Most materials used in circuits, such as copper conductors and the standard carbon composition and wirewound resistors, also exhibit the characteristic of a rising resistance when heated, although in most cases the effect is minimal. Therefore, their resistance can be considered relatively constant.

This discussion is concerned mainly with NTC thermistors. Figure 57–1 shows the variation in resistivity for three NTC thermistors, compared with the resistivity of platinum under similar temperature changes. Note the tremendous variation of specific resistance in thermistor 1, a change of 10,000,000 to 1 over a temperature range of 500°C, whereas that of platinum changes by a factor of less than 10 to 1.

Figure 57–2 shows the variation in resistance of another NTC thermistor. Here a range of 200°C causes a variation in resistance from 1 kΩ to approximately 2 Ω.

The resistance-temperature characteristic of thermistors has been utilized in electrical and electronic applications. In particular, thermistors are used as protective devices for electric-bulb filaments during the warmup period. They are used for temperature measurement, control, and compensation.

Static Volt-Ampere Characteristic · · · · · · ·

There are secondary characteristics that arise from the relationship between temperature and resistance that ther-

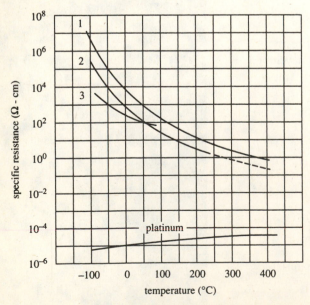

Figure 57–1. Temperature-specific resistance characteristics of three thermistors (labeled 1, 2, 3) compared with platinum.

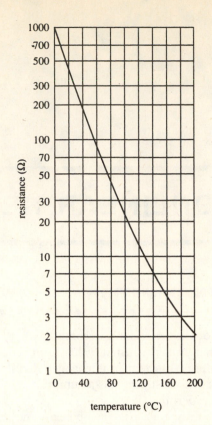

Figure 57–2. Temperature-resistance characteristics of a representative thermistor.

mistors exhibit. Among these is the static relationship between voltage and current. This secondary characteristic results from the self-heating that occurs when current flows through a thermistor. The static characteristic may be determined using the test circuit shown in Figure 57–3. In this circuit a rheostat R is used as a limiting resistor to vary the amount of current through the thermistor. As the resistance of R is decreased, the current I increases. As the resistance of R is increased, the current I decreases. A voltmeter is used to monitor the voltage across the thermistor for every value of current. Each time R is varied and a change in current is effected, enough time must be allowed for the voltage across the thermistor to reach a steady value. The resultant graph therefore reflects steady-state conditions of voltage versus current and is called the *static volt-ampere characteristic.*

The thermistor whose resistance-temperature characteristic is shown in Figure 57–2 has the static volt-ampere characteristic shown in Figure 57–4. A temperature of 25°C was the starting temperature for the tests from which the data for the graph were secured. The numbers on the graph—namely, 53, 73, 102, and so forth—indicate the thermistor temperature, measured after current had stabilized at the level shown on the graph. Thus, at 50 mA, the voltage across the thermistor was 11 V, and the temperature was 53°C.

The graph for this thermistor shows that up to a temperature of 53°C, there is a linear relationship between voltage and current. That is, Ohm's law holds during this

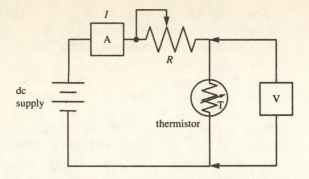

Figure 57–3. Circuit for determining the volt-ampere characteristics of a thermistor.

interval when only low current flow is involved. This is true because there is insufficient heat dissipated at this low level of current to affect the cold resistance of the thermistor. As current increases, however, the heat dissipated increases, thermistor temperature increases, and its resistance drops nonlinearly. The voltage across the thermistor now increases nonlinearly to a peak value V_M, designated the "self-heat" voltage. Beyond this value, the voltage across the thermistor drops nonlinearly with increasing current, as shown in Figure 57–4.

Applications based on the voltage-current characteristics of thermistors fall into four general categories, responding to changes in electrical parameters, in ambient temperatures, in dissipation (fluid and gaseous environments), and in radiation absorption. Devices falling into these categories include temperature alarm devices, pyrometers, flow meters, gas detectors, and microwave power regulators.

Dynamic Characteristic · · · · · · · · · · · ·

A thermistor requires a certain time interval to react to a change in the external circuit. A change in current through a thermistor from one value to another does not result in an instantaneous change in temperature; hence, there is no instantaneous change in resistance. These changes do occur, but they require a definite time. The thermal mass of the thermistor element will determine the time interval. A small element will change temperature more rapidly than an element with larger mass.

Figure 57–5 (p. 436) shows a test circuit for determining the dynamic characteristic of a thermistor. The applied voltage V is an ac power supply whose output is adjustable, and R_L is the load resistance. A voltmeter set on ac volts is connected across the load R_L to monitor the voltage across R_L. When switch S is open, there is no voltage applied to the circuit. At the instant S is closed, the cold resistance of the thermistor and the value of the load resistor R_L constitute the voltage divider that determines the voltage across R_L. As current starts to flow, the temperature of the NTC thermistor increases, its resistance decreases, and more of the source voltage appears across the load. Within a certain time interval, the circuit is stabilized. That is, the thermistor resistance is stabilized and the voltage across the load becomes constant.

Figure 57–6 (p. 436) is a graph showing the dynamic characteristic of a thermistor for two different values of supply voltage and load resistance. Note that with 48 V applied and a load resistance of 17 Ω, steady state is reached in approximately 75 s. With 115 V applied and a load of 50 Ω, steady state is reached in about 15 s.

The dynamic characteristic of a thermistor is utilized in audio devices, switching devices, voltage-surge protectors, and low-frequency negative-resistance devices.

SUMMARY

1. Thermistors are semiconductor devices whose resistance changes with changes in operating temperature.

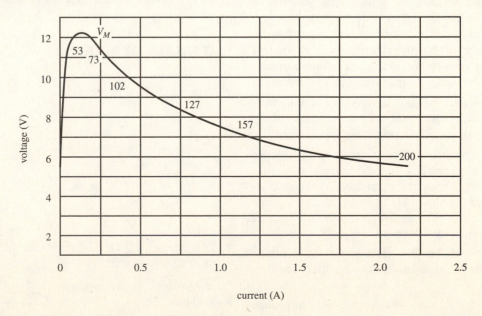

Figure 57–4. Static volt-ampere characteristics of a representative thermistor.

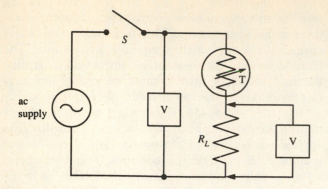

Figure 57–5. Circuit for determining the dynamic characteristic of a thermistor.

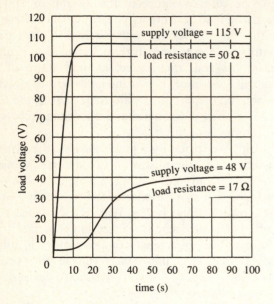

Figure 57–6. Dynamic characteristics of a representative thermistor.

2. The resistance of thermistors does not vary linearly with variations of temperature.

3. The resistance of negative temperature coefficient (NTC) thermistors decreases as their temperature increases and increases as their temperature decreases. The resistance of positive temperature coefficient (PTC) thermistors increases as their temperature increases and decreases as their temperature decreases.

4. Compared with the variation in resistance of a thermistor, the resistance of a carbon resistor remains relatively constant despite changes in operating temperature.

5. Thermistors exhibit unique temperature-resistance characteristics, depending on their size and construction. Thus, one particular thermistor changes in resistance from 1000 Ω to 2 Ω over a temperature change of 200°C. Another changes from 10,000,000 Ω to 1 Ω over a temperature variation from −100°C to +400°C.

6. Because of the self-heating effect of current in a thermistor, this device changes resistance with changes in current.

7. A secondary characteristic of a thermistor is the change in current it exhibits in an electric circuit. A graph of current versus voltage (Figure 57–4) of a thermistor, called a static volt-ampere characteristic, shows that current does *not* vary linearly with applied voltage.

8. A thermistor does not undergo instantaneous changes in resistance with changes in temperature. A certain time interval, determined by the thermal mass of the thermistor, is required to accomplish the resistance change. A thermistor with a smaller mass will change more rapidly than one with a larger mass.

9. The dynamic characteristic of a thermistor is a graph of the time it takes a thermistor to stabilize its resistance as a function of the applied voltage and current (Figure 57–6).

10. Thermistors are used in temperature alarm devices, pyrometers, switching devices, voltage-surge devices, and the like.

SELF TEST

Check your understanding by answering the following questions:

1. A thermistor is a _____ (linear/nonlinear) resistor.

2. The resistance of an NTC thermistor _____ (increases/decreases) with a decrease in temperature.

3. An NTC thermistor has a _____ (positive/negative/zero) temperature coefficient.

4. It takes _____ for a thermistor to undergo a change in resistance with a change in temperature.

5. The larger the thermal mass of a thermistor, the _____ (more/less) time it will take to change its resistance.

6. A(n) _____ (increase/decrease) in current in an NTC thermistor will cause a decrease in its resistance.

7. (True/False) Current in a thermistor varies linearly with voltage across it. _____

MATERIALS REQUIRED

Power Supply:
- Variable 0–15 V dc, regulated

Instruments:
- DMM

Resistors:
- 1   100-Ω, 5-W
- 1   200-Ω, 5-W

Miscellaneous:
- Thermistor, 100 Ω cold resistance (25°C). Resistance ratio $R_{25°}/R_{50°} \approx 3$
- SPST switch
- SPDT switch
- Optional electronic timer with alarm

PROCEDURE

. . . .

This experiment can be performed by two students working as a team. One student can be the timekeeper; the other student can read the voltmeter. If an electronic timer with an alarm is available, the entire experiment can be performed by one student.

A. Control Conditions

A1. Measure and record the resistance of the 200-Ω (R_1) and 100-Ω (R_2) resistors in Table 57–1 (p. 439).

A2. With power **off** and switch S_1 **open,** connect the circuit of Figure 57–7.

A3. Turn power **on; close** S_1. Switch S_2 should be in position 1. Adjust the power supply voltage to 15 V. Maintain this voltage throughout part A. Check the voltage from time to time and adjust if necessary.

A4. Once the power supply has been set at 15 V, measure the voltage across R_1 and record it in Table 57–1. Immediately throw S_2 to position 2 and measure the voltage across R_2. Record the value in Table 57–1. (Neglect changes in the polarity of the measurements.)

A5. Maintain power **on** for 5 min. After 5 min, measure the voltage across R_2 (S_2 should still be in position 2) and record the value in Table 57–1. Throw S_2 to position 1, measure the voltage across R_1, and record the value in Table 57–1. After measuring the voltages, **open** S_1 and turn **off** power. Disconnect R_1 from the circuit.

B. Dynamic Characteristic of a Thermistor

The timing element in this part is critical. Before performing the steps that follow, review them completely, including Table 57–2 (p. 441). Although the voltage measurements are important, changes in their polarity can be neglected and need not be recorded. Use a DMM for measuring voltages. An electronic timer with an alarm would be useful for accurate timing of the voltage measurement.

Switch S_2 makes it easy to measure the voltage across the thermistor and R_2, but it requires reading the voltmeter quickly and switching immediately from one switch position to another to read the second voltage.

B1. Measure and record the cold resistance of the thermistor in Table 57–2. Also record the room temperature.

B2. Connect the thermistor in the circuit of Part A as in Figure 57–8. Power should be **off** and S_1 **open.** Switch S_2 should be in position 1.

B3. Turn power **on; close** S_1. Adjust the power supply voltage to 15 V. Maintain this voltage for the following three steps (through step B6).

B4. As soon as the supply is set at 15 V, measure the voltage across the thermistor V_T, and record it in Table 57–2. Immediately set S_2 to position 2 and measure the voltage across resistor R_2, V_2, and record it in Table 57–2. These will be the voltage measurements at $t = 0$.

B5. Repeat step B4 15 s later (that is, at $t = 15$ s), then 15 s later at $t = 30$ s, followed by measurements at $t = 45$ s, $t = 1$ min, $t = 2$ min, $t = 3$ min, $t = 4$ min, and finally at $t = 5$ min.

B6. After making the voltage measurements at $t = 5$ min, turn **off** the DMM, set S_2 to position 1, and change the function switch of the DMM to the ohmmeter scales. Open S_1 and quickly turn **on** the DMM to read the hot resistance of the thermistor. Record the value in Table 57–2. After making the measurement, change the function switch of DMM back to the voltage scales.

B7. Adjust the voltage of the power supply to 9 V and maintain this voltage for the balance of Part B. Check periodically and adjust if necessary.

B8. After the thermistor has had a chance to return to room temperature (at least 5 min after power is

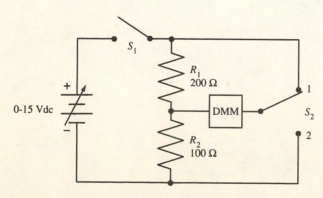

Figure 57–7. Circuit for control measurement; procedure step A2.

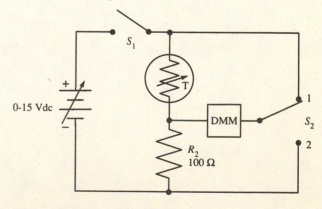

Figure 57–8. Circuit for procedure step B2.

turned off), repeat steps B4 through B6 for the time sequence $t = 0$, 15 s, 30 s, 45 s, 1 min, 2 min, 3 min, 4 min, and 5 min. After $t = 5$ min, measure the hot resistance of the thermistor using the procedure of step B6. Record all voltage and resistance measurements in Table 57–2.

ANSWERS TO SELF TEST

1. nonlinear
2. increases
3. negative
4. time
5. more
6. increase
7. false

EXPERIMENT 57

TABLE 57–1. Control-Circuit Measurements

| | Resistance, Ω | | | Voltage, V | |
	Rated Value	Measured Value		$t = 0$	$t = 5$ min
R_1	200				
R_2	100				

TABLE 57–2. Dynamic Characteristics of a Thermistor

	Thermistor Resistance R_T, Ω
Cold (room temp. = °C)	
After 5 min at 15 V	
After 5 min at 9 V	

Power Supply, V		Time t								
		0	15 s	30 s	45 s	1 min	2 min	3 min	4 min	5 min
15	V_T									
	V_2									
9	V_T									
	V_2									

QUESTIONS

1. Explain, in your own words, the effect that current has on a thermistor over a period of time.

———————————————————————————————————————

———————————————————————————————————————

———————————————————————————————————————

———————————————————————————————————————

2. Explain, in your own words, the difference between negative and positive temperature coefficients of a thermistor.

———————————————————————————————————————

———————————————————————————————————————

———————————————————————————————————————

3. Refer to your data in Table 57–1. Is the temperature coefficient of the resistor in this experiment negative, positive, or zero? Support your answer with specific references to your data.

_____

_____

_____

_____

_____

_____

4. Refer to your data in Table 57–2. Is the temperature coefficient of the thermistor in this experiment positive, negative, or zero? Support your answer with reference to specific data in the table.

_____

_____

_____

_____

_____

_____

_____

5. What specification of the thermistor is an indicator of the maximum current that the thermistor should carry?

_____

_____

_____

_____

6. Refer to your data in Table 57–2. Calculate the maximum current through the thermistor and the minimum current through the thermistor for the 15-V circuit.

_____

_____

_____

_____

_____

_____

NONLINEAR RESISTORS—VARISTORS (VDRs)

BASIC INFORMATION

. . . .

In the previous experiment you discovered that there is a class of resistors called thermistors, whose resistance is dependent on temperature. There is another group of nonlinear resistors whose resistance is voltage-dependent. These voltage-sensitive resistors are called *varistors,* or VDRs. The current in a varistor varies as a power of the applied voltage, and for a particular varistor it may increase by many orders of magnitude when the applied voltage is only doubled.

The older popular varistors were made of silicon carbide mixed with a ceramic binder, fired at a high temperature to produce a solid fused blank, and then covered with a metallic coating. Electrical contact is made to this coating. One drawback to the use of silicon carbide varistors is that they are limited to low-power circuits.

Modern varistors are made from metal oxides and are usually referred to as MOVs. The most common oxide used is zinc oxide; these varistors are called ZNR varistors. Metal-oxide varistors are fabricated in much the same manner as silicon carbide varistors. Varying the thickness of the MOV makes different operating voltages possible.

One important advantage of the MOV over the older silicon carbide varistor is its higher resistance. Since the MOV is typically connected across the line or across a load, the high resistance means that a much lower current is drawn by the MOV in its normal, or standby, condition.

The MOV provides an economical and reliable means of protecting against voltage surges, or "spikes." These spikes are transient conditions that occur rapidly and without warning. The rapid response time (as little as 50 ns) of the MOV makes them excellent circuit and device protectors.

Volt-Amperes Characteristics · · · · · · · · · · · · · · · · · · ·

Because the MOV is now the most widely used varistor, only its characteristics are discussed. The MOV has a bilateral and symmetrical volt-ampere characteristic. This means that current direction through the MOV does not affect its behavior. Figure 58–1 (p. 442) shows the symmetrical nature of its volt-ampere curve. The MOV is not polarized and can be used in either ac or dc circuits. It protects equally well against positive as well as negative spikes.

Varistors possess negative temperature coefficients (as did the NTC thermistors in Experiment 57). This means the resistance of the MOV

OBJECTIVES

. . . .

1 To determine experimentally the volt-ampere characteristic of a varistor

2 To determine experimentally the relationship between voltage and resistance of a varistor

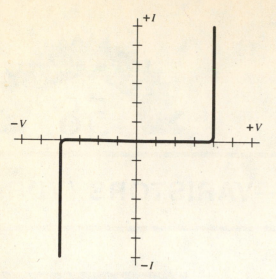

Figure 58–1. Symmetrical volt-ampere characteristic of an MOV.

decreases as its temperature increases. During its conduction stage the voltage across the MOV remains relatively constant despite tremendous increases in current through the MOV. Although the MOV does not behave according to Ohm's law, at conduction the static resistance of the MOV can be defined as V/I. The dynamic, or changing resistance, of the MOV is defined as the incremental change of the instantaneous voltage across the MOV divided by the incremental change of the instantaneous current through the MOV. This can be written as

$$\text{Dynamic } R = \frac{dv}{di}$$

Electrical Specifications and Ratings · · · ·

Specifications for MOV devices are often given in terms of maximum ratings. The following ratings are typical of those given by MOV manufacturers.

Rated (NC applied) Voltage. This is the maximum continuous voltage that can be applied across the varistor. For ac circuits this is usually given as an rms value.

Rated Peak Pulse Current—One Time. This is the maximum peak current that can be applied for one 8 × 20-μs pulse with rated line voltage also applied. The 8 × 20-μs pulse unit is standard that specifies that the pulse wave will reach its peak in 8 μs and fall to one-half its peak value in 20 μs.

Clamping Voltage. Peak voltage across the MOV with a specified waveform (usually 8 × 20-μs pulse) and a peak pulse current applied.

Energy. The maximum energy in joules capable of being handled by the MOV under transient conditions.

Power. Maximum power dissipation in watts.

Operating Ambient Temperature. A range of minimum and maximum temperatures under which the MOV can operate under specified ratings.

SUMMARY

1. Varistors are voltage-depending resistors (VDRs). They are nonlinear devices whose resistance is dependent on the voltage across them.

2. Modern varistors are fabricated from metal oxides, usually zinc oxides. They are referred to as MOVs or, more specifically, as ZNR varistors.

3. Varistors are rated for their continuous power dissipation at a specified temperature and for the maximum voltage they can tolerate.

4. Other electrical specifications include the following:
   (a) Rated peak pulse current—1 time
   (b) Clamping voltage
   (c) Energy
   (d) Operating ambient temperature

5. Varistors have negative temperature coefficients.

6. During conduction the voltage across a varistor remains relatively constant despite tremendous changes in current through the varistor.

7. Varistors are nonpolarized devices that are unaffected by pressure or vibration. They are bilateral and have a symmetrical volt-ampere characteristic curve.

8. Varistors are used as protective devices in lightning arrestors and in ac and dc circuits to protect components against voltage surges.

S E L F   T E S T

Check your understanding by answering the following questions:

1. The resistance of a varistor _____ (increases/decreases) as the voltage across it increases.

2. (True/False) Most modern varistors used are fabricated from silicon carbide. _____

3. Varistors operate independent of current direction because they are _____ (polarized, nonpolarized) devices.

4. If the temperature of a varistor increases, its resistance _____.

5. (True/False) A standard pulse waveshape is often specified as 8 × 20-μs. _____

6. The clamping voltage specified for an MOV refers to the _____ voltage across the MOV with a specified waveform and a peak pulse current applied.

MATERIALS REQUIRED

Power Supply:
- Variable 0–15 V dc, regulated

Instruments:
- DMM
- VOM

Resistors:
- 1   100-Ω, 5-W

Miscellaneous:
- 6-V metal-oxide varistor (MOV) (GE 1V12ZA1 or equivalent)
- SPST switch

PROCEDURE

1. With power **off** and switch S_1 **open,** connect the circuit of Figure 58–2(*a*). Set the power supply for a 0-V output.

2. Turn **on** power; **close** S_1. Slowly increase the power supply voltage until the voltmeter measures 2 V. Watch the ammeter carefully as the voltage is increased; be prepared quickly to increase the current range.

3. With 2 V across the MOV, measure the current and record the value in Table 58–1 (p. 445).

4. Increase the voltage across the MOV in steps of 2 V, 4 V, 6 V, 8 V, 10 V, 12 V. At each step measure the current and record the value in Table 58–1. After completing the measurement, **open** S_1 and turn **off** the power supply.

5. Reverse the leads from the power supply to the circuit so that the polarity to the MOV is reversed, as in Figure 58–2(*b*). Reverse the ammeter connection also, if necessary.

6. Repeat steps 3 and 4, but record the current measurements with the negative voltages across the varistor. Also show the current measurements as negative values.

7. For each voltage (negative and positive values) calculate the static resistance of the MOV using Ohm's law. Record your answers in Table 58–1.

ANSWERS TO SELF TEST

1. decreases
2. false
3. nonpolarized
4. decreases
5. true
6. peak

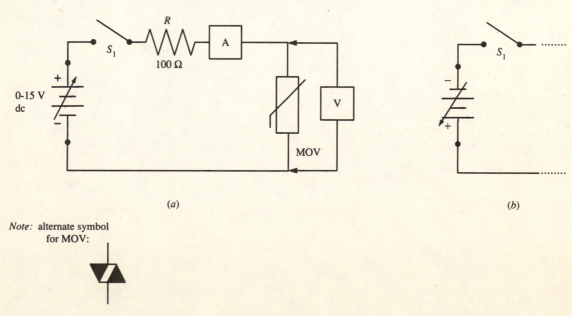

(a)

(b)

Note: alternate symbol for MOV:

Figure 58–2. *(a)* Circuit for procedure step 1. *(b)* Circuit for procedure step 5.

NONLINEAR RESISTORS—VARISTORS (VDRs)   **443**

Name _____ Date _____

TABLE 58–1. Volt-Ampere Characteristic of a Varistor (MOV)

Voltage across MOV V_V, V	Current I_V, A	Calculated Static Resistance of MOV R_V, Ω
+2		
+4		
+6		
+8		
+10		
+12		
−2		
−4		
−6		
−8		
−10		
−12		

QUESTIONS

1. Explain, in your own words, the relationship between voltage and resistance of a varistor.

_____

_____

_____

_____

_____

_____

_____

_____

_____

2. On a separate sheet of 8½ × 11 graph paper, plot a graph of voltage across the varistor versus current through the varistor using your data in Table 58–1. Draw the axes similar to those shown in Figure 58–1. Label all axes.

3. Refer to your data in Table 58–1 and your graph of Question 2. Discuss the voltage-ampere characteristics of the MOV used in this experiment.

_____

_____

_____

_____

_____

_____

_____

_____

_____

_____

_____

_____

_____

_____

_____

_____

4. Does the varistor in this experiment have a positive, negative, or zero temperature coefficient? Refer to specific data to support your answer.

_____

_____

_____

_____

_____

_____

_____

_____

_____

_____

_____

_____

_____

_____

_____

_____

CAPACITOR/INDUCTOR CODING*

---

*This material is provided courtesy of Sencore.

Dipped Tantalum Capacitors

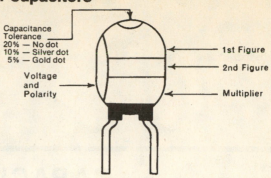

Color	Rated Voltage	Capacitance in Picofarads 1st Figure	Capacitance in Picofarads 2nd Figure	Multiplier
Black	4	0	0	—
Brown	6	1	1	—
Red	10	2	2	—
Orange	15	3	3	—
Yellow	20	4	4	10,000
Green	25	5	5	100,000
Blue	35	6	6	1,000,000
Violet	50	7	7	10,000,000
Gray	—	8	8	—
White	3	9	9	—

Capacitance Tolerance
20% — No dot
10% — Silver dot
5% — Gold dot

Voltage and Polarity

1st Figure
2nd Figure
Multiplier

Ceramic Disc Capacitors

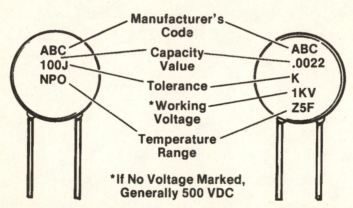

Manufacturer's Code
Capacity Value
Tolerance
*Working Voltage
Temperature Range

ABC
100J
NPO

ABC
.0022
K
1KV
Z5F

*If No Voltage Marked, Generally 500 VDC

Typical Ceramic Disc Capacitor Markings

Z 5 F 1 0 0 J

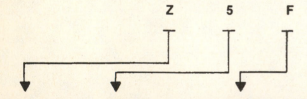

Low Temp.	Letter Symbol	High Temp.	Numerical Symbol	Max. Capac. Change Over Temp. Range	Letter Symbol
+10°C	Z	+45°C	2	+1.0%	A
-30°C	Y	+65°C	4	±1.5%	B
-55°C	X	+85°C	5	±1.1%	C
		+105°C	6	±3.3%	D
		+125°C	7	±4.7%	E
				±7.5%	F
				±10.0%	P
				±15.0%	R
				±22.0%	S
				+22%, -33%	T
				+22%, -56%	U
				+22%, -82%	V

Temperature Range Identification of Ceramic Disc Capacitors

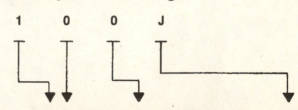

1st & 2nd Fig. of Capacitance	Multiplier	Numerical Symbol	Tolerance on Capacitance	Letter Symbol
	1	0		
	10	1		
	100	2	±5%	J
	1,000	3	±10%	K
	10,000	4	±20%	M
	100,000	5	+100%, -0%	P
	—		+80%, -20%	Z
	.01	8		
	.1	9		

Capacity Value and Tolerance of Ceramic Disc Capacitors

Film Type Capacitors

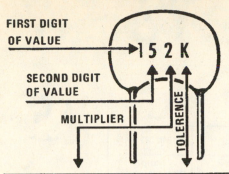

FIRST DIGIT OF VALUE

SECOND DIGIT OF VALUE

MULTIPLIER

TOLERENCE

MULTIPLIER		TOLERANCE OF CAPACITOR		
For the Number	Multiplier	Letter	10 pF or Less	Over 10 pF
0	1	B	± 0.1 pF	
1	10	C	± .25 pF	
2	100	D	± 0.5 pF	
3	1,000	F	± 1.0 pF	± 1%
4	10,000	G	± 2.0 pF	± 2%
5	100,000	H		± 3%
		J		± 5%
8	0.01	K		± 10%
9	0.1	M		± 20%

EXAMPLES:
152K = 15 x 100 = 1500 pF or .0015 uF, ± 10%
759J = 75 x 0.1 = 7.5 pF, ± 5%

NOTE: The letter "R" may be used at times to signify a decimal point; as in: 2R2 = 2.2 (pF or uF).

Ceramic Feed Through Capacitors

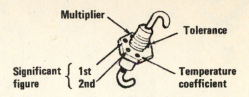

Multiplier

Tolerance

Significant figure { 1st 2nd }

Temperature coefficient

Color	Significant Figure	Multiplier	Tolerance 10 pF or Less	Tolerance Over 10 pF	Temperature Coefficient
Black	0	1	2 pF	20%	0
Brown	1	10	0.1 pF	1%	N30
Red	2	100	—	2%	N60
Orange	3	1,000	—	2.5%	N150
Yellow	4	10,000	—	—	N220
Green	5	—	5 pF	5%	N330
Blue	6	—	—	—	N470
Violet	7	—	—	—	N750
Gray	8	0.001	0.025 pF	—	P30
White	9	0.1	1 pF	10%	+ 120 to -750 (RETMA) + 500 to -330 (JAN)
Gold	—	—	—	—	P100
Silver	—	—	—	—	Bypass or coupling

Postage Stamp Mica Capacitors

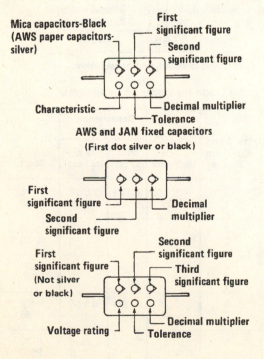

Mica capacitors-Black (AWS paper capacitors-silver)

First significant figure

Second significant figure

Characteristic

Decimal multiplier

Tolerance

AWS and JAN fixed capacitors
(First dot silver or black)

First significant figure

Second significant figure

Decimal multiplier

First significant figure (Not silver or black)

Second significant figure

Third significant figure

Voltage rating

Decimal multiplier

Tolerance

Color	Significant Figure	Multiplier	Tolerance (%)	Voltage Rating
Black	0	1	—	—
Brown	1	10	1	100
Red	2	100	2	200
Orange	3	1,000	3	300
Yellow	4	10,000	4	400
Green	5	100,000	5	500
Blue	6	1,000,000	6	600
Violet	7	10,000,000	7	700
Gray	8	100,000,000	8	800
White	9	1,000,000,000	9	900
Gold	—	0.1	5	1000
Silver	—	0.01	10	2000
No color	—	—	20	500

Standard Button Mica

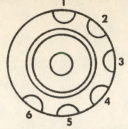

1st DOT	2nd and 3rd DOTS		4th DOT	5th DOT		6th DOT
Identifier	Capacitance in pF		Multiplier	Capacitance Tolerance		Temp. Characteristic
	Color	1st & 2nd Sig. Figs.		Percent	Letter Symbol	
Black	Black	0	1	± 20%	F	
	Brown	1	10	± 1%	F	
	Red	2	100	± 2% or ± 1 pF	G or B	
	Orange	3	1000	± 3%	H	
NOTE: Identifier is omitted if capacitance must be specified to three significant figures.	Yellow	4				+ 100
	Green	5				
	Blue	6				-20 PPM/°C above 50 pF
	Violet	7				
	Gray	8				
	White	9	0.1			± 100 PPM/°C below 50 pF
	Gold			± 5%	J	
	Silver			± 10%	K	

Radial or Axial Lead Ceramic Capacitors
(6 Dot or Band System)

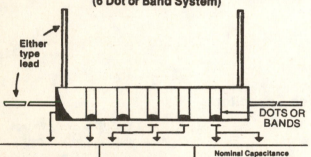

T.C.	1st Color	2nd Color	1st and 2nd Sig. Fig.	Multiplier	Color	10 pF or Less	Over 10 pF	Color
P100	Red	Violet	0	1	Black	± 2.0 pF	± 20%	Black
P030	Green	Blue	1	10	Brown	± 0.1 pF	± 1%	Brown
NP0	Black		2	100	Red		± 2%	Red
N030	Brown		3	1,000	Orange		± 3%	Orange
N080	Red		4	10,000	Yellow		+ 100% -0%	Yellow
N150	Orange		5		Green	± 0.5 pF	± 5%	Green
N220	Yellow		6		Blue			Blue
N330	Green		7		Violet			Violet
N470	Blue		8	.01	Gray	± 0.25 pF	+ 80% -20%	Gray
N750	Violet		9	.1	White	± 1.0 pF	± 10%	White
N1500	Orange	Orange						
N2200	Yellow	Orange						
N3300	Green	Orange						
N4200	Green	Green						
N4700	Blue	Orange						
N5600	Green	Black						
N330 ±500	White							
N750 ±1000	Gray							
N3300 ±2500	Gray	Black						

5 Dot or Band Ceramic Capacitors
(one wide band)

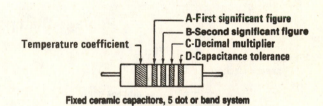

- A-First significant figure
- B-Second significant figure
- C-Decimal multiplier
- D-Capacitance tolerance
- Temperature coefficient

Fixed ceramic capacitors, 5 dot or band system

Color Code for Ceramic Capacitors

Color	1st & 2nd Significant Figure	Multiplier	Capacitance Tolerance Over 10 pF	Capacitance Tolerance 10 pF or Less	Temp. Coeff.
Black	0	1	± 20%	2.0 pF	0
Brown	1	10	± 1%		N30
Red	2	100	± 2%		N80
Orange	3	1000			N150
Yellow	4				N220
Green	5				N330
Blue	6		± 5%	0.5 pF	N470
Violet	7				N750
Gray	8	0.01		0.25 pF	P 30
White	9	0.1	± 10%	1.0 pF	P500

5 Band Ceramic Capacitors
(all bands equal size)

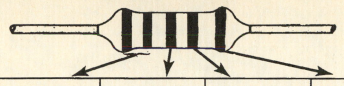

color	1st, 2nd Band	Multiplier	Tolerance	Characteristic
Black Brown	0 1	1 10	±20% (M)	NPO Y5S
Red Orange	2 3	100 1K		Y5T N150
Yellow Green	4 5	10K	N330	N220
Blue Violet	6 7			N470 N750
Grey White	8 9		±30% (N) SL (GP)	Y5R
Gold Silver	- -	0.1 0.01	±5% (J) ±10% (K)	Y5F Y5P

Tubular Encapsulated RF Chokes

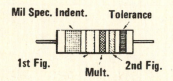

Mil Spec. Indent.　Tolerance

1st Fig.　Mult.　2nd Fig.

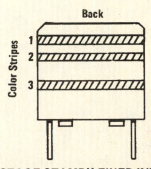

Back

Color Stripes 1 2 3

Color	Figure	Multiplier	Tolerance
Black	0	1	
Brown	1	10	
Red	2	100	
Orange	3	1,000	
Yellow	4		
Green	5		
Blue	6		
Violet	7		
Gray	8		
White	9		
None			20%
Silver			10%
Gold			5%

Multiplier is the factor by which the two color figures are multiplied to obtain the inductance value of the choke coil in uH.
Values will be in uH.

"POSTAGE STAMP" FIXED INDUCTORS

Color	1st Digit 1st Strip	2nd Digit 2nd Strip	Multiplier 3rd Strip
Black or (Blank)	0	0	1
Brown	1	1	10
Red	2	2	100
Orange	3	3	1,000
Yellow	4	4	10,000
Green	5	5	100,000
Blue	6	6	
Violet	7	7	
Gray	8	8	
White	9	9	
Gold			X.1
Silver			X.01

APPENDIX

B

SOLDER AND THE SOLDERING PROCESS*

From Simple Task to Fine Art

Soldering is the process of joining two metals together by the use of a low-temperature melting alloy. Soldering is one of the oldest known joining techniques, first developed by the Egyptians in making weapons such as spears and swords. Since then, it has evolved into what is now used in the manufacturing of electronic assemblies. Soldering is far from the simple task it once was; it is now a fine art, one that requires care, experience, and a thorough knowledge of the fundamentals.

The importance of having high standards of workmanship cannot be overemphasized. Faulty solder joints remain a cause of equipment failure, and because of that, soldering has become a *critical skill.*

The material contained in this appendix is designed to provide the student with both the fundamental knowledge and the practical skills needed to perform many of the high-reliability soldering operations encountered in today's electronics.

Covered here are the fundamentals of the soldering process, the proper selection, and the use of the soldering station.

The key concept in this appendix is *high-reliability soldering.* Much of our present technology is vitally dependent on the reliability of countless, individual soldered connections. High-reliability soldering was developed in response to early failures with space equipment. Since then the concept and practice have spread into military and medical equipment. We have now come to expect it in everyday electronics as well.

The Advantage of Soldering

Soldering is the process of connecting two pieces of metal together to form a reliable electrical path. Why solder them in the first place? The two pieces of metal could be put together with nuts and bolts, or some other kind of mechanical fastening. The disadvantages of these methods are twofold. First, the reliability of the connection cannot be assured because of vibration and shock. Second, because oxidation and corrosion are continually occurring on the metal surfaces, electrical conductivity between the two surfaces would progressively decrease.

A soldered connection does away with both of these problems. There is no movement in the joint and no interfacing surfaces to oxidize. A continuous conductive path is formed, made possible by the characteristics of the solder itself.

The Nature of Solder

Solder used in electronics is a low-temperature melting alloy made by combining various metals in different proportions. The most common types of solder are made from tin and lead. When the proportions are equal, it is known as 50/50 solder—50 percent tin and 50 percent lead. Similarly, 60/40 solder consists of 60 percent tin and 40 percent lead. The percentages are usually marked on the various types of solder available; sometimes only the tin percentage is shown. The chemical symbol for tin is Sn; thus Sn 63 indicates a solder which contains 63 percent tin.

Pure lead (Pb) has a melting point of 327°C (621°F); pure tin, a melting point of 232°C (450°F). But when they are combined into a 60/40 solder, the melting point drops to 190°C (374°F)—lower than either of the two metals alone.

Melting generally does not take place all at once. As illustrated in Fig. B-1, 60/40 solder begins to melt at 183°C (361°F), but it has not fully melted until the temperature reaches 190°C (374°F). Between these two temperatures, the solder exists in a plastic (semiliquid) state—some, but not all, of the solder has melted.

The plastic range of solder will vary, depending on the ratio of tin to lead, as shown in Fig. B-2. Various ratios of tin to lead are shown across the top of this figure. With

---

*This material is provided courtesy of PACE, Inc., Laurel, Maryland.

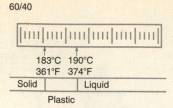

60/40

183°C 190°C
361°F 374°F

Solid	Liquid

Plastic

FIG. B-1 Plastic range of 60/40 solder. Melt begins at 183°C (361°F) and is complete at 190°C (374°F).

most ratios, melting begins at 183°C (361°F), but the full melting temperatures vary dramatically. There is one ratio of tin to lead that has no plastic state and is known as *eutectic solder*. This ratio is 63/37 (Sn 63) and it fully melts and solidifies at 183°C (361°F).

The solder most commonly used for hand soldering in electronics is the 60/40 type, but because of its plastic range, care must be taken not to move any elements of the joint during the cool-down period. Movement may cause a disturbed joint. Characteristically, this type of joint has a rough, irregular appearance and looks dull instead of bright and shiny. It is unreliable and therefore one of the types of joints that is unacceptable in high-reliability soldering.

In some situations, it is difficult to maintain a stable joint during cooling, for example, when wave soldering is used with a moving conveyor line of circuit boards during the manufacturing process. In other cases it may be necessary to use minimal heat to avoid damage to heat-sensitive components. In both of these situations, eutectic solder is the preferred choice, since it changes from a liquid to a solid during cooling with no plastic range.

The Wetting Action

To someone watching the soldering process for the first time, it looks as though the solder simply sticks the metals together like a hot-melt glue, but what actually happens is far different.

A chemical reaction takes place when the hot solder comes into contact with the copper surface. The solder dissolves and penetrates the surface. The molecules of solder and copper blend together to form a new metal alloy, one that is part copper and part solder and that has characteristics all its own. This reaction is called *wetting* and forms the intermetallic bond between the solder and copper (Fig. B-3).

Proper wetting can occur only if the surface of the copper is free of contamination and from oxide films that form when the metal is exposed to air. Also, the solder and copper surfaces need to have reached the proper temperature.

Even though the surface may look clean before soldering, there may still be a thin film of oxide covering it. When solder is applied, it acts like a drop of water on an oily surface because the oxide coating prevents the solder from coming into contact with the copper. No reaction takes place, and the solder can be easily scraped off. For a good solder bond, surface oxides must be removed during the soldering process.

The Role of Flux

Reliable solder connections can be accomplished only on clean surfaces. Some sort of cleaning process is essential in achieving successful soldered connections, but in most cases it is insufficient. This is due to the extremely rapid rate at which oxides form on the surfaces of heated metals, thus creating oxide films which prevent proper soldering. To overcome these oxide films, it is necessary to utilize materials, called *fluxes*, which consist of natural or synthetic rosins and sometimes additives called activators.

It is the function of flux to remove surface oxides and keep them removed during the soldering operation. This is accomplished because the flux action is very corrosive at or near solder melt temperatures and accounts for the flux's ability to rapidly remove metal oxides. It is the fluxing action of removing oxides and carrying them away, as well as preventing the formation of new oxides, that allows the solder to form the desired intermetallic bond.

Flux must activate at a temperature lower than solder so that it can do its job prior to the solder flowing. It volatilizes very rapidly; thus it is mandatory that the flux be activated to flow onto the work surface and not simply be volatilized by the hot iron tip if it is to provide the full benefit of the fluxing action.

There are varieties of fluxes available for many applications. For example, in soldering sheet metal, acid fluxes are used; silver brazing (which requires a much higher

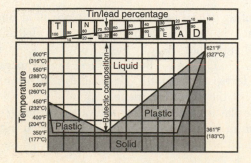

FIG. B-2 Fusion characteristics of tin/lead solders.

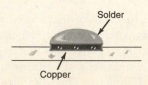

FIG. B-3 The wetting action. Molten solder dissolves and penetrates a clean copper surface, forming an intermetallic bond.

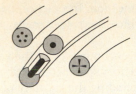

FIG. B-4 Types of cored solder, with varying solder-flux percentages.

temperature for melting than that required by tin/lead alloys) uses a borax paste. Each of these fluxes removes oxides and, in many cases, serves additional purposes. The fluxes used in electronic hand soldering are the pure rosins, rosins combined with mild activators to accelerate the rosin's fluxing capability, low-residue/no-clean fluxes, or water-soluble fluxes. Acid fluxes or highly activated fluxes should never be used in electronic work. Various types of flux-cored solder are now in common use. They provide a convenient way to apply and control the amount of flux used at the joint (Fig. B-4).

Soldering Irons

In any kind of soldering, the primary requirement, beyond the solder itself, is heat. Heat can be applied in a number of ways—conductive (e.g., soldering iron, wave, vapor phase), convective (hot air), or radiant (IR). We are mainly concerned with the conductive method, which uses a soldering iron.

Soldering stations come in a variety of sizes and shapes, but consist basically of three main elements: a resistance heating unit; a heater block, which acts as a heat reservoir; and the tip, or bit, for transferring heat to the work. The standard production station is a variable-temperature, closed-loop system with interchangeable tips and is made with ESD-safe plastics.

Controlling Heat at the Joint

Controlling tip temperature is not the real challenge in soldering; the real challenge is to control the *heat cycle* of the work—how fast the work gets hot, how hot it gets, and how long it stays that way. This is affected by so many factors that, in reality, tip temperature is not that critical.

The first factor that needs to be considered is the *relative thermal mass* of the area to be soldered. This mass may vary over a wide range.

Consider a single land on a single-sided circuit board. There is relatively little mass, so the land heats up quickly. But on a double-sided board with plated-through holes, the mass is more than doubled. Multilayered boards may have an even greater mass, and that's before the mass of the component lead is taken into consideration. Lead mass may vary greatly, since some leads are much larger than others.

Further, there may be terminals (e.g., turret or bifurcated) mounted on the board. Again, the thermal mass is

increased, and will further increase as connecting wires are added.

Each connection, then, has its particular thermal mass. How this combined mass compares with the mass of the iron tip, the "relative" thermal mass, determines the time and temperature rise of the work.

With a large work mass and a small iron tip, the temperature rise will be slow. With the situation reversed, using a large iron tip on a small work mass, the temperature rise of the work will be much more rapid—even though the *temperature of the tip is the same.*

Now consider the capacity of the iron itself and its ability to sustain a given flow of heat. Essentially, irons are instruments for generating and storing heat, and the reservoir is made up of both the heater block and the tip. The tip comes in various sizes and shapes; it's the *pipeline* for heat flowing into the work. For small work, a conical (pointed) tip is used, so that only a small flow of heat occurs. For large work, a large chisel tip is used, providing greater flow.

The reservoir is replenished by the heating element, but when an iron with a large tip is used to heat massive work, the reservoir may lose heat faster than it can be replenished. Thus the *size* of the reservoir becomes important: a large heating block can sustain a larger outflow longer than a small one.

An iron's capacity can be increased by using a larger heating element, thereby increasing the wattage of the iron. These two factors, block size and wattage, are what determine the iron's recovery rate.

If a great deal of heat is needed at a particular connection, the correct temperature with the right size tip is required, as is an iron with a large enough capacity and an ability to recover fast enough. *Relative thermal mass,* then, is a major consideration for controlling the heat cycle of the work.

A second factor of importance is the *surface condition* of the area to be soldered. If there are any oxides or other contaminants covering the lands or leads, there will be a barrier to the flow of heat. Then, even though the iron tip is the right size and has the correct temperature, it may not supply enough heat to the connection to melt the solder. In soldering, a cardinal rule is that a good solder connection cannot be created on a dirty surface. Before attempting to solder, the work should always be cleaned with an approved solvent to remove any grease or oil film from the surface. In some cases pretinning may be required to enhance solderability and remove heavy oxidation of the surfaces prior to soldering.

A third factor to consider is *thermal linkage*—the area of contact between the iron tip and the work.

Figure B-5 shows a cross-sectional view of an iron tip touching a round lead. The contact occurs only at the point indicated by the "X," so the linkage area is very small, not much more than a straight line along the lead.

The contact area can be greatly increased by applying a small amount of solder to the point of contact between

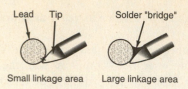

Lead Tip   Solder "bridge"

Small linkage area   Large linkage area

FIG. B-5 Cross-sectional view (left) of iron tip on a round lead. The "X" shows point of contact. Use of a solder bridge (right) increases the linkage area and speeds the transfer of heat.

the tip and workpiece. This solder heat bridge provides the thermal linkage and assures rapid heat transfer into the work.

From the aforementioned, it should now be apparent that there are many more factors than just the temperature of the iron tip that affect how quickly any particular connection is going to heat up. In reality, soldering is a very complex control problem, with a number of variables to it, each influencing the other. And what makes it so critical is *time*. The general rule for high-reliability soldering on printed circuit boards is to apply heat for no more than 2 s from the time solder starts to melt (wetting). Applying heat for longer than 2 s after wetting may cause damage to the component or board.

With all these factors to consider, the soldering process would appear to be too complex to accurately control in so short a time, but there is a simple solution—the *workpiece indicator* (WPI). This is defined as the reaction of the workpiece to the work being performed on it—a reaction that is discernible to the human senses of sight, touch, smell, sound, and taste.

Put simply, workpiece indicators are the way the work talks back to you—the way it tells you what effect you are having and how to control it so that you accomplish what you want.

In any kind of work, you become part of a closed-loop system. It begins when you take some action on the workpiece; then the workpiece reacts to what you did; you sense the change, and then modify your action to accomplish the result. It is in the sensing of the change, by sight, sound, smell, taste, or touch, that the workpiece indicators come in (Fig. B-6).

For soldering and desoldering, a primary workpiece indicator is *heat rate recognition*—observing how fast heat

flows into the connection. In practice, this means observing the rate at which the solder melts, which should be within 1 to 2 s.

This indicator encompasses all the variables involved in making a satisfactory solder connection with minimum heating effects, including the capacity of the iron and its tip temperature, the surface conditions, the thermal linkage between tip and workpiece, and the relative thermal masses involved.

If the iron tip is too large for the work, the heating rate may be too fast to be controlled. If the tip is too small, it may produce a "mush" kind of melt; the heating rate will be too slow, even though the temperature at the tip is the same.

A general rule for preventing overheating is "Get in and get out as fast as you can." That means using a heated iron you can react to—one giving a 1- to 2-s dwell time on the particular connection being soldered.

Selecting the Soldering Iron and Tip · · · ·

A good all-around soldering station for electronic soldering is a variable-temperature, ESD-safe station with a pencil-type iron and tips that are easily interchangeable, even when hot (Fig. B-7).

The soldering iron tip should always be fully inserted into the heating element and tightened. This will allow for maximum heat transfer from the heater to the tip.

The tip should be removed daily to prevent an oxidation scale from accumulating between the heating element and the tip. A bright, thin tinned surface must be maintained on the tip's working surface to ensure proper heat transfer and to avoid contaminating the solder connection.

The plated tip is initially prepared by holding a piece of flux-cored solder to the face so that it will tin the surface when it reaches the lowest temperature at which solder will melt. Once the tip is up to operating temperature, it will usually be too hot for good tinning, because of the rapidity of oxidation at elevated temperatures. The hot tinned tip is maintained by wiping it lightly on a damp sponge to shock off the oxides. When the iron is not being used, the tip should be coated with a layer of solder.

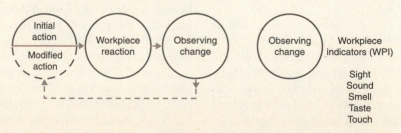

FIG. B-6 Work can be viewed as a closed-loop system (left). Feedback comes from the reaction of the workpiece and is used to modify the action. Workpiece indicators (right)—changes discernible to the human senses—are the way the "work talks back to you."

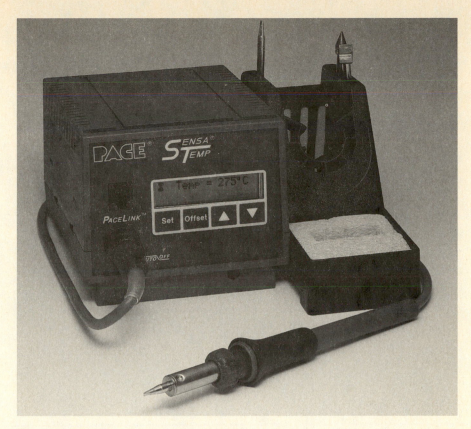

FIG. B-7 Pencil-type iron with changeable tips.

Making the Solder Connection · · · · · · · ·

The soldering iron tip should be applied to the area of maximum thermal mass of the connection being made. This will permit the rapid thermal elevation of the parts being soldered. Molten solder always flows toward the heat of a properly prepared connection.

When the solder connection is heated, a small amount of solder is applied to the tip to increase the thermal linkage to the area being heated. The solder is then applied to the opposite side of the connection so that the work surfaces, not the iron, melt the solder. Never melt the solder against the iron tip and allow it to flow onto a surface cooler than the solder melting temperature.

Solder, with flux, applied to a cleaned and properly heated surface will melt and flow without direct contact with the heat source and provide a smooth, even surface, feathering out to a thin edge (Fig. B-8). Improper sol-

dering will exhibit a built-up, irregular appearance and poor filleting. The parts being soldered must be held rigidly in place until the temperature decreases to solidify the solder. This will prevent a disturbed or fractured solder joint.

Selecting cored solder of the proper diameter will aid in controlling the amount of solder being applied to the connection (e.g., a small-gauge solder for a small connection; a large-gauge solder for a large connection).

Removal of Flux ·

Cleaning may be required to remove certain types of fluxes after soldering. If cleaning is required, the flux residue should be removed as soon as possible, preferably within 1 hour after soldering.

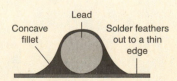

FIG. B-8 Cross-sectional view of a round lead on a flat surface.

ELECTRONICS CAREERS

Following are additional sources which can provide dependable and up-to-date information about careers in electronics and career planning. Catalogs and other information about the organizations can be obtained by writing to them. An abundance of online information is immediately available from the Internet addresses.

Sources of Additional Career Information

American Electronics Association
5201 Great America Parkway, Suite 520
Santa Clara, CA 95054
http://www.aeanet.org

Consumer Electronics Manufacturers Association
2500 Wilson Boulevard
Arlington, VA 22201-3834
http://www.cemacity.org

Electronics Industries Alliance
2500 Wilson Boulevard
Arlington, VA 22201-3834
http://www.eia.org

Institute of Electrical and Electronics Engineers
171 State St.
Framingham, MA 01701
http://www.iee.org

International Society of Certified Electronics Technicians
2708 West Berry Street
Fort Worth, TX 76109
http://www.iscet.org

Standard Occupational Publications · · · ·

Occupational Outlook Handbook (OOH) and *Dictionary of Occupational Titles* (DOT) provide a wealth of data and commentary about all occupational categories. Projections on job opportunities are given through the year 2005. Both publications are maintained by the U.S. Department of Labor and made available through the Government Printing Office. The current edition of OOH is also available online at http://stats.bls.gov/ocohome.htm.

The following is a listing of DOT reference numbers and occupational titles of particular interest to electronics students. The purpose of the list is to demonstrate the diversity of existing working environments. Although most entries are engineering titles, technicians should be aware that their support is required in almost all positions.

002167014 Field-service technician
003061010 Electrical engineer
003061014 Electrical test engineer
003061018 Electrical-design engineer
003061022 Electrical-prospecting engineer
003061026 Electrical-research engineer
003061030 Electronics engineer
003061034 Electronics-design engineer
003061038 Electronics-research engineer
003061042 Electronics-test engineer
003061046 Illuminating engineer
003131010 Supervisor, drafting and printed circuit design
003161010 Electrical technician
003161014 Electronics technician
003161018 Technician, semiconductor development
003167018 Electrical engineer, power system
003167026 Engineer of system development
003167030 Engineer-in-charge, studio operations
003167034 Engineer-in-charge, transmitter
003167038 Induction-coordination power engineer
003167046 Power-distribution engineer
003167050 Power-transmission engineer
003167054 Protection engineer
003167066 Transmission-and-protection engineer
003167070 Engineering manager, electronics
003187010 Central-office equipment enqineer
003187018 Customer-equipment engineer
003261010 Instrumentation technician
003261014 Controls designer

003261018 Integrated circuit layout designer
003261022 Printed circuit designer
003281010 Drafter, electrical
003281014 Drafter, electronic
003362010 Design technician, computer-aided